Saurabh Chandra Maury

Linear Algebra

Also of Interest

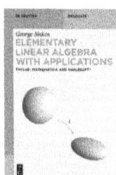

Elementary Linear Algebra with Applications. MATLAB®, Mathematica® and Maplesoft™
George Nakos, 2024
ISBN 978-3-11-133179-9, e-ISBN (PDF) 978-3-11-133185-0,
e-ISBN (EPUB) 978-3-11-133195-9

Lectures on Linear Algebra and its Applications
Philip Korman, 2023
ISBN 978-3-11-108540-1, e-ISBN (PDF) 978-3-11-108650-7,
e-ISBN (EPUB) 978-3-11-108662-0

Optimal Control of ODEs and DAEs
Matthias Gerdts, 2023
ISBN 978-3-11-079769-5, e-ISBN (PDF) 978-3-11-079789-3,
e-ISBN (EPUB) 978-3-11-079793-0

Linear Algebra. A Minimal Polynomial Approach to Eigen Theory
Fernando Barrera-Mora, 2023
ISBN 978-3-11-113589-2, e-ISBN (PDF) 978-3-11-113591-5,
e-ISBN (EPUB) 978-3-11-113614-1

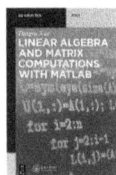

Linear Algebra and Matrix Computations with MATLAB®
Dingyü Xue, 2020
ISBN 978-3-11-066363-1, e-ISBN (PDF) 978-3-11-066699-1,
e-ISBN (EPUB) 978-3-11-066371-6

Saurabh Chandra Maury

Linear Algebra

Vector and Inner Product Spaces

DE GRUYTER

Mathematics Subject Classification 2020
Primary: 15A03, 15A18, 15A04; Secondary: 15A15, 15A63

Author
Dr. Saurabh Chandra Maury
Vill-Jagdishpur, Post-Kanta
Dist-Chandauli 232104
India
saurabhchandra.maury@vit.ac.in

ISBN 978-3-11-151570-0
e-ISBN (PDF) 978-3-11-151603-5
e-ISBN (EPUB) 978-3-11-151688-2

Library of Congress Control Number: 2024942958

Bibliographic information published by the Deutsche Nationalbibliothek
The Deutsche Nationalbibliothek lists this publication in the Deutsche Nationalbibliografie;
detailed bibliographic data are available on the Internet at http://dnb.dnb.de.

© 2025 Walter de Gruyter GmbH, Berlin/Boston
Cover image: Thawatchai Pimairam / iStock / Getty Images Plus
Typesetting: VTeX UAB, Lithuania

www.degruyter.com

Preface

Arguably the most potent mathematical tool ever created, linear algebra is a fundamental branch of mathematics. This book represents a balanced combination of step-by-step explained theory content, solved examples and exercises. The abundant examples and detailed explanations of all topics, such as vector spaces, linear independence, basis, linear transformations, matrices, determinants, inner products, eigenvectors, bilinear forms and canonical forms are what make the text so valuable.

The early introduction of concepts such as fields, rings, group homomorphism and binary operations gives students a proper foundation on which they can understand the remaining part of the text.

The book includes numerous examples at varying levels of difficulty, each carefully selected. The fundamentals are presented in a clear, methodical and straightforward manner, with many worked-out examples illustrating the concepts and problem-solving techniques. This book can serve as a textbook for undergraduate and graduate linear algebra courses in mathematics and engineering at any university. It also includes some of the latest and most challenging questions suitable for competitive exams in higher education mathematics. The text is structured to support students in their degree programs and to assist them in preparing for future higher education competitive exams in linear algebra.

I acknowledge the influence of numerous authors whose works have guided my writing. The results in this book belong to the collective heritage of mathematics, and crediting all contributors would be a difficult task. No theorem presented here should be assumed to be my original contribution. However, I have endeavored to present linear algebra and prove its theorems in the most effective way, without strictly adhering to traditional methods and proofs found in most textbooks.

I express my deep appreciation and thanks to VIT University, Chennai, for providing the infrastructural facilities to write this book. My gratitude extends to my younger brother, Birendra Kumar Maurya, for his assistance in typing, and to my wife, Mahasweta, for her encouragement and support. I am also grateful to my parents for their blessings and hard work, which have brought me to this level.

Finally, I thank De Gruyter for offering the platform to publish this book. Special thanks to Ranis N. Ibragimov for his excellent work as an editor and Nadja Schedensack for her skillful copyediting.

Despite careful efforts, some errors and misprints may have escaped my notice. I welcome any feedback and suggestions for improving the book.

Saurabh Chandra Maury

https://doi.org/10.1515/9783111516035-201

Contents

Symbols used

$\mathbb{N}$	the set of natural numbers
$\mathbb{Z}$	the set of integers
$\mathbb{Q}$	the set of rational numbers
$\mathbb{C}$	the set of complex numbers
$M_{m,n}(\mathbb{F})$	the set of all $m \times n$ matrices with entries in F
$P_n(x)$	the set of all polynomials of degree at most n
$P(x)$	the set of all polynomials
$\mathbb{R}^{[0,1]}$	the set of real-valued functions defined on $[0,1]$
$C(\mathbb{R}^{[0,1]})$	the set of continuous real valued functions on $[0,1]$
$D(\mathbb{R}^{[0,1]})$	the set of differentiable real valued functions on $[0,1]$ and
$\mathcal{R}(\mathbb{R}^{[0,1]})$	the set of Riemann integrable functions on $[0,1]$
$\langle S \rangle$	the subspace generated by S
$UT(M_{n,n}(F))$	the set of all upper triangular matrices of order $n \times n$
$LT(M_{n,n}(F))$	the set of all lower triangular matrices of order $n \times n$
$S(M_{n,n}(F))$	the set of all n-square symmetric matrices
$[T]_{B_1,B_2}$	the matrix of the linear transformation T related to the ordered bases B_1 and B_2
$L_A : F^n \rightarrow F^m$	the left multiplication linear transformation

https://doi.org/10.1515/9783111516035-202

1 Preliminaries

In this chapter, we briefly discuss some necessary topics and related results that are needed for the rest of the book.

1.1 Binary operations

Definition 1.1.1. Let A, B be sets. A binary operation is a function $* : A \times B \to B$. Usually, we use the notation $a * b$ in place of the image of (a, b) under the function $*$. If $A = B$, the binary operation is called internal or closed on A. If $A \neq B$, it is called external. The pair $(A, *)$ is used to denote an internal binary operation $*$ on the set A.

Example 1.1.2. Ordinary addition $+$ is an internal binary operation on $\mathbb{N}$, whereas ordinary subtraction $-$ is not a binary operation on $\mathbb{N}$. In the form of more examples, the pairs $(\mathbb{N}, +)$, $(\mathbb{N}, \cdot)$, $(\mathbb{Z}, +)$, $(\mathbb{Z}, -)$, $(\mathbb{Z}, \cdot)$, $(\mathbb{Q}, +)$, $(\mathbb{Q}, -)$, $(\mathbb{Q}, \cdot)$, $(\mathbb{C}, +)$, $(\mathbb{C}, -)$, $(\mathbb{C}, \cdot)$ are representing internal binary operations.

Example 1.1.3. Let $M_{n,n}(\mathbb{Z})$ denote the set of all $n \times n$ matrices with entries in $\mathbb{Z}$. Then the matrix addition and the matrix multiplication both are binary operations on $M_{n,n}(\mathbb{Z})$.

Definition 1.1.4. Let $*$ be a binary operation on a set A. Then:
(i)   The operation $*$ is called associative if $(a * b) * c = a * (b * c)$ for all $a, b, c \in A$.
(ii) The operation $*$ is called commutative if $a * b = b * a$ for all $a, b \in A$.
(iii) An element $e \in A$ is called the identity of A if for all $a \in A$, $a * e = e * a = a$.
(iv) For a given identity $e \in A$ and $a \in A$, if there exists an element $b \in A$ such that $a * b = b * a = e$, we say that b is an inverse of a and write $b = a^{-1}$.

Example 1.1.5. The internal binary operation '$+$' on $\mathbb{Z}, \mathbb{Q}, \mathbb{R}$ and $\mathbb{C}$ is both associative and commutative. The element zero serves as the identity for all these sets with respect to the operation '$+$'. Additionally, for each $a \in \mathbb{Z}$ (or $a \in \mathbb{Q}$ or $a \in \mathbb{R}$ or $a \in \mathbb{C}$) there exists inverse $-a \in \mathbb{Z}$ (or $-a \in \mathbb{Q}$ or $-a \in \mathbb{R}$ or $-a \in \mathbb{C}$) such that $a + (-a) = (-a) + a = 0$.

1.2 Congruence and residue classes

Definition 1.2.1. Let $a, b, m \in \mathbb{Z}$ with $m > 0$. We say that a is congruent to b modulo m, denoted by $a \equiv b \pmod{m}$, if m divides $a - b$. This means there exists an integer $r \in \mathbb{Z}$ such that $a - b = mr$. Therefore, $a \equiv b \pmod{m}$ is equivalent to $a = b + mr$ for some integer r.

The following result shows that the congruence defined as above is an equivalence relation on $\mathbb{Z}$.

Theorem 1.2.2. *Let m be a positive integer. Then for all $a, b, c \in \mathbb{Z}$:*

https://doi.org/10.1515/9783111516035-001

(i) $a \equiv a \pmod{m}$ *(reflexive)*;

(ii)   *if* $a \equiv b \pmod{m}$, *then* $b \equiv a \pmod{m}$ *(symmetric)*;

(iii) *if* $a \equiv b \pmod{m}$ *and* $b \equiv c \pmod{m}$, *then* $a \equiv c \pmod{m}$ *(transitive)*.

Proof. (i)   Since $a - a = 0$ is divisible by m, it follows that $a \equiv a \pmod{m}$.

(ii)   If $a \equiv b \pmod{m}$, then there exists an integer $q \in \mathbb{Z}$ such that $a - b = mq$.

Now, $b - a = -(a - b) = -mq = m(-q)$, which implies that $b \equiv a \pmod{m}$.

(iii) Given $a \equiv b \pmod{m}$ and $b \equiv c \pmod{m}$, there exist integers q and r such that $a - b = mq$ and $b - c = mr$. Adding these two equations give $a - c = m(q + r) = mk$, where $k = q + r \in \mathbb{Z}$.

Hence, $a \equiv c \pmod{m}$. □

Thus, the relation $R = \{(a, b) \in \mathbb{Z} \times \mathbb{Z} : a \equiv b \pmod{m}\}$ is an equivalence relation on $\mathbb{Z}$.

For any $a \in \mathbb{Z}$, the set $\bar{a} = \{b \in \mathbb{Z} : a \equiv b \pmod{m}\} = \{b \in \mathbb{Z} : m \text{ divides } a - b\}$ is called an equivalence class or residue class of a modulo m.

Let Z_m denote the quotient set $\mathbb{Z}/R$. Then $Z_m = \{\bar{a} : a \in \mathbb{Z}\}$.

In view of the properties of equivalence classes, we have the following result.

Theorem 1.2.3. *Let* Z_m *denotes the set of residue classes modulo m. Then for* $\bar{a}, \bar{b} \in Z_m$ *we have the following:*

(i) $a \in \bar{a} \; \forall a \in \mathbb{Z}$.

(ii) $\bar{a} = \bar{b}$ *if and only if* $a \equiv b \pmod{m}$.

(iii) $\bar{a} \neq \bar{b}$ *if and only if* $\bar{a} \cap \bar{b} = \phi$.

(iv) $\bigcup_{\bar{a} \in Z_m} \bar{a} = \mathbb{Z}$.

Theorem 1.2.4. *For any positive integer m,* $Z_m = \{\bar{0}, \bar{1}, \dots, \overline{m-1}\}$.

Proof. Let $a \in \mathbb{Z}$. Then by the division algorithm, there exist $q, r \in \mathbb{Z}$ such that $a = mq + r$, with $0 \leq r < m$. Or equivalently, $a - r = mq$ is divisible by m. This shows that $a \equiv r \pmod{m}$, and hence from the above theorem, we have $\bar{a} = \bar{r}$. Thus, each residue class $\bar{a}$ is identical with one of the residue classes $\bar{0}, \bar{1}, \dots, \overline{m-1}$. Also, if $0 \leq r < s \leq m - 1$, then $r - s$ is not divisible by m. This shows that $\bar{r} \neq \bar{s}$. That is, all the elements of Z_m are distinct. □

Theorem 1.2.5. *Let* Z_m *be the set of residue classes modulo m. Then the operations* $\oplus :$ $Z_m \times Z_m \to Z_m$ *and* $* : Z_m \times Z_m \to Z_m$ *defined by* $\bar{a} \oplus \bar{b} = \overline{a + b}$ *and* $\bar{a} * \bar{b} = \overline{ab}$, *respectively, are binary operations.*

Proof. To prove that the operation, $\oplus$ is a binary operation, we need to show that $\oplus$ is a function.

Let $(\bar{a}, \bar{b}) = (\bar{c}, \bar{d}) \in Z_m \times Z_m$. Then we need to show that the images of $(\bar{a}, \bar{b})$ and $(\bar{c}, \bar{d})$ under operation $\oplus$ are equal, i. e., $\bar{a} \oplus \bar{b} = \bar{c} \oplus \bar{d}$.

Since $(\bar{a}, \bar{b}) = (\bar{c}, \bar{d})$, we have $\bar{a} = \bar{c}$ and $\bar{b} = \bar{d}$. Hence, $a \equiv c \pmod{m}$ and $b \equiv d$ (mod m) or $a - c = mr$ and $b - d = ms$ for some $r, s \in \mathbb{Z}$.

Adding the above equations, we get

$$(a + b) - (c + d) = m(r + s) \Rightarrow (a + b) \equiv (c + d) \pmod{m} \Rightarrow \overline{a + b} = \overline{c + d} \Rightarrow \bar{a} \oplus \bar{b} = \bar{c} \oplus \bar{d}.$$

Similarly, for the operation $*$ we need to show that $\bar{a} * \bar{b} = \bar{c} * \bar{d}$.

From

$$ab - cd = ab - cd + 0 = ab - cd + (bc - bc) = ab - bc + bc - cd = (a - c)b + (b - d)c$$
$$= mrb + msc = m(rb + sc),$$

we see that m divides $ab - cd$. Hence, $ab \equiv cd \pmod{m}$ and so $\overline{ab} = \overline{cd}$. This proves that $\bar{a} * \bar{b} = \bar{c} * \bar{d}$. ☐

Note 1. The binary operation $\oplus$ on Z_m is associative and commutative. The element $\bar{0}$ is the identity of $(Z_m, \oplus)$ and $\overline{(-a)} = \overline{m - a}$ is the inverse of $\bar{a} \in Z_m$.

Note 2. The binary operation $*$ on Z_m is also associative and commutative. The element $\bar{1}$ is the identity of $(Z_m, *)$.

Example 1.2.6. For $m = 3$, $Z_3 = \{\bar{0}, \bar{1}, \bar{2}\}$, where

$$\bar{0} = \{\ldots, -9, -6, -3, 0, 3, 6, 9, \ldots\} = \{3k : k \in \mathbb{Z}\},$$
$$\bar{1} = \{\ldots, -8, -5, -2, 1, 4, 7, 10, \ldots\} = \{3k + 1 : k \in \mathbb{Z}\},$$
$$\bar{2} = \{\ldots, -7, -4, -1, 2, 5, 8, 11, \ldots\} = \{3k + 2 : k \in \mathbb{Z}\}.$$

It is clear that $\bar{0} \cup \bar{1} \cup \bar{2} = \mathbb{Z}$. Also, we have $\bar{0}, \bar{1}$ and $\bar{2}$ are pairwise disjoint. That is, $\bar{0} \cap \bar{1} = \bar{1} \cap \bar{2} = \bar{2} \cap \bar{0} = \phi$.

Example 1.2.7. In $Z_5 = \{\bar{0}, \bar{1}, \bar{2}, \bar{3}, \bar{4}\}$, the binary operations $\oplus$ and $*$ are defined as follows:

$\oplus$	$\bar{0}$	$\bar{1}$	$\bar{2}$	$\bar{3}$	$\bar{4}$
$\bar{0}$	$\bar{0}$	$\bar{1}$	$\bar{2}$	$\bar{3}$	$\bar{4}$
$\bar{1}$	$\bar{1}$	$\bar{2}$	$\bar{3}$	$\bar{4}$	$\bar{0}$
$\bar{2}$	$\bar{2}$	$\bar{3}$	$\bar{4}$	$\bar{0}$	$\bar{1}$
$\bar{3}$	$\bar{3}$	$\bar{4}$	$\bar{0}$	$\bar{1}$	$\bar{2}$
$\bar{4}$	$\bar{4}$	$\bar{0}$	$\bar{1}$	$\bar{2}$	$\bar{3}$

$*$	$\bar{0}$	$\bar{1}$	$\bar{2}$	$\bar{3}$	$\bar{4}$
$\bar{0}$	$\bar{0}$	$\bar{0}$	$\bar{0}$	$\bar{0}$	$\bar{0}$
$\bar{1}$	$\bar{0}$	$\bar{1}$	$\bar{2}$	$\bar{3}$	$\bar{4}$
$\bar{2}$	$\bar{0}$	$\bar{2}$	$\bar{4}$	$\bar{1}$	$\bar{3}$
$\bar{3}$	$\bar{0}$	$\bar{3}$	$\bar{1}$	$\bar{4}$	$\bar{2}$
$\bar{4}$	$\bar{0}$	$\bar{4}$	$\bar{3}$	$\bar{2}$	$\bar{1}$

Remark. In $Z_m, \bar{0} = \{\ldots, -3m, -2m, -m, 0, m, 2m, 3m, \ldots\}$ and in $Z_n, \bar{0} = \{\ldots, -3n, -2n, -n, 0, n, 2n, 3n, \ldots\}$. This shows that both the zeros are distinct if $m \neq n$. That is, the identity element $\bar{0}$ in $(Z_m, \oplus)$ is not equal to the identity element $\bar{0}$ in $(Z_n, \oplus)$ if $m \neq n$.

1.3 Group

Definition 1.3.1. A pair $(G, *)$, where G is a set and $*$ is a binary operation on G, is called a group if (i) the binary operation $*$ is associative, (ii) there exists an identity element in G with respect to operation $*$ and (iii) for every element of G has an inverse with respect to the identity. That is, the pair $(G, *)$ is a group if it satisfies the following properties:
(i) $(a * b) * c = a * (b * c)\ \forall a, b, c \in G.$
(ii)   There exists an element $e \in G$ such that $e * a = a * e = a\ \forall a \in G.$
(iii) For each element $a \in G$, there exists an element $b \in G$ such that $a * b = b * a.$

If the binary operation $*$ is commutative, i. e., $a * b = b * a\ \forall a, b \in G$, then the group $(G, *)$ is called a commutative group or an Abelian group.

Example 1.3.2. From Example 1.1.5, it is clear that the pairs $(\mathbb{Z}, +)$, $(\mathbb{Q}, +)$, $(\mathbb{R}, +)$ and $(\mathbb{C}, +)$ are infinite Abelian groups.

Example 1.3.3. In $(\mathbb{Z}, \cdot)$, for any $a \in \mathbb{Z}$ other than ± 1, there is no inverse. That is, for $a \neq \pm 1$, there is no $b \in \mathbb{Z}$ such that $a \cdot b = 1$, where 1 is the multiplicative identity. In $(\mathbb{Q}, \cdot), (\mathbb{R}, \cdot)$ and $(\mathbb{C}, \cdot)$, the element 0 has no inverse. Therefore, the sets $(\mathbb{Z}, \cdot), (\mathbb{Q}, \cdot), (\mathbb{R}, \cdot)$ and $(\mathbb{C}, \cdot)$ are not groups under the usual multiplication.

After removing the element 0 from the sets $\mathbb{Q}, \mathbb{R}$ and $\mathbb{C}$, the resulting sets become Abelian groups under usual multiplication. In other words, $(\mathbb{Q} - \{0\}, \cdot)$, $(\mathbb{R} - \{0\}, \cdot)$ and $(\mathbb{C} - \{0\}, \cdot)$ are Abelian groups.

Example 1.3.4. The pair $(Z_m, \oplus)$ is an Abelian group.
We will demonstrate that $(Z_m, \oplus)$ satisfies all the properties of a group.
Associativity: Let $\bar{a}, \bar{b}, \bar{c} \in Z_m$. Then

$$(\bar{a} \oplus \bar{b}) \oplus \bar{c} = \overline{(a+b)} \oplus \bar{c} = \overline{(a+b)+c} = \overline{a+(b+c)} = \bar{a} \oplus \overline{(b+c)} = \bar{a} \oplus (\bar{b} \oplus \bar{c}).$$

Here, we have used the associativity of $(\mathbb{Z}, +)$.

Identity: Let $\bar{a} \in Z_m$. Then

$$\bar{0} \oplus \bar{a} = \overline{(0+a)} = \bar{a} \quad \text{and} \quad \bar{a} \oplus \bar{0} = \overline{(a+0)} = \bar{a}.$$

Inverse: Let $\bar{a} \in Z_m$. Then $0 \le a \le m-1$. Now,

$$\bar{a} \oplus \overline{m-a} = \overline{a+m-a} = \overline{a-a+m} = \overline{0+m} = \bar{m} = \bar{0}$$

and similarly,

$$\overline{m-a} \oplus \bar{a} = \bar{0}$$

shows that for any $\bar{a} \in Z_m$ there exists an inverse $\overline{m-a} = \overline{-a}$.

Commutative: For all $\bar{a}, \bar{b} \in Z_m$,

$$\bar{a} \oplus \bar{b} = \overline{a+b} = \overline{b+a} = \bar{b} \oplus \bar{a}.$$

Example 1.3.5. The set Z_m does not form a group under the binary operation $*$ defined as $\bar{a} * \bar{b} = \overline{ab}$, where $\bar{a}, \bar{b} \in Z_m$, because $\bar{0}$ has no inverse in Z_m. However, one can consider the structure $(Z_m - \{\bar{0}\}, *)$. For example, let $Z_6 - \{\bar{0}\} = \{\bar{1}, \bar{2}, \bar{3}, \bar{4}, \bar{5}\}$. Then, from the calculations,

$$\bar{2} * \bar{1} = \overline{2 \cdot 1} = \bar{2}, \quad \bar{2} * \bar{2} = \bar{4}, \quad \bar{2} * \bar{3} = \bar{0}, \quad \bar{2} * \bar{4} = \bar{2}, \quad \bar{2} * \bar{5} = \bar{4},$$

it is clear that $\bar{2}$ has no inverse in $(Z_6 - \{\bar{0}\}, *)$ with respect to the identity $\bar{1}$.

This proves that $(Z_6 - \{\bar{0}\}, *)$ is not a group with respect to the binary operation $*$, and hence $(Z_m - \{\bar{0}\}, *)$ may not be a group.

Let p be a positive prime in $\mathbb{Z}$. Then it can be proven that $(Z_p - \{\bar{0}\}, *)$ is an Abelian group. For example, consider $Z_5 - \{\bar{0}\} = \{\bar{1}, \bar{2}, \bar{3}, \bar{4}\}$. Then

$$\bar{2} * \bar{3} = \bar{3} * \bar{2} = \bar{6} = \bar{1},$$

i. e., $\bar{2}$ and $\bar{3}$ are inverses of each other, and

$$\bar{4} * \bar{4} = \overline{16} = \bar{1}$$

shows that $\bar{4}$ is its own inverse. Hence, $(Z_5 - \{\bar{0}\}, *)$ is an Abelian group.

Example 1.3.6. Let $(G_1, *_1), (G_2, *_2), \ldots, (G_n, *_n)$ be a family of groups. Then their Cartesian product $G_1 \times G_2 \times \cdots \times G_n$ is a group under the binary operation $*$ defined by $(g_1, g_2, \ldots, g_n) * (\acute{g}_1, \acute{g}_2, \ldots, \acute{g}_n) = (g_1 *_1 \acute{g}_1, g_2 *_2 \acute{g}_2, \ldots, g_n *_n \acute{g}_n)$, where $(g_1, g_2, \ldots, g_n)$ and $(\acute{g}_1, \acute{g}_2, \ldots, \acute{g}_n)$ are members of $G_1 \times G_2 \times \cdots \times G_n$.

The identity element of the group $G_1 \times G_2 \times \cdots \times G_n$ is $(e_1, e_2, \ldots, e_n)$ where $e_i \in G_i, 1 \le i \le n$ is the identity of G_i and the inverse of any element $(g_1, g_2, \ldots, g_n)$ of $G_1 \times G_2 \times \cdots \times G_n$ is $(g_1^{-1}, g_2^{-1}, \ldots, g_n^{-1})$.

For example, the Cartesian product $\mathbb{R}^n = \mathbb{R} \times \mathbb{R} \times \cdots \times \mathbb{R}$ is a group under the operation coordinatewise addition.

Below, we mention some important properties of groups without proof.

Theorem 1.3.7. *Let $(G, *)$ be a group. Then*
(a) *identity element in $(G, *)$ is unique;*
(b) *inverse element a^{-1} of an element a in $(G, *)$ is unique;*
(c) *$(a * b)^{-1} = b^{-1} * a^{-1}$, for all $a, b \in G$;*
(d) *both (left and right) the cancellation laws hold in $(G, *)$, i. e., $a * b = a * c \Rightarrow b = c$ and $b * a = c * a \Rightarrow b = c$.*

Definition 1.3.8. Let $(G, *)$ be a group. The number of elements in G is called the order of the group. The order of the group $(G, *)$ is denoted by $|G|$ or $O(G)$.

The groups $(\mathbb{Z}, +)$, $(\mathbb{Q}, +)$, $(\mathbb{R}, +)$ and $(\mathbb{C}, +)$ have infinite order, while $(Z_m, \oplus)$ is a group of finite order m.

Definition 1.3.9. Let $(G, *)$ be a group. The order of an element $g \in G$ is the smallest positive integer n such that $g^n = g * g * \cdots * g$ (n times) $= e$. The order of an element $g \in G$ is denoted by $|g|$ or $O(g)$.

The order of the identity element e in a group is 1. The order of $\bar{1}$ in $(Z_m, \oplus)$ is m, and in $Z_6 = \{\bar{0}, \bar{1}, \bar{2}, \bar{3}, \bar{4}, \bar{5}\}$, the order of $\bar{2}$ is 3 $(\bar{2}^3 = \bar{2} \oplus \bar{2} \oplus \bar{2} = \bar{6} = \bar{0})$. Every nonzero element in $(\mathbb{Z}, +)$ has infinite order.

Definition 1.3.10. Let $(G, *)$ be a group. A subset H of G is called a subgroup of $(G, *)$ if H is itself a group under the operation '$*$' of G.

We use the notation $H \leq G$ to indicate that H is a subgroup of G. Let $(G, *)$ be a group with identity e. Then $\{e\}$ and G are subgroups of G, called the trivial subgroups. $(\mathbb{Z}, +)$ is a subgroup of $(\mathbb{R}, +)$, and if $m \in \mathbb{N} \cup \{0\}$, then $m\mathbb{Z} = \{mk : k \in \mathbb{Z}\}$ is a subgroup of $(\mathbb{Z}, +)$. The following results are useful for testing whether a subset of a group is a subgroup.

Test 1. Let $(G, *)$ be a group. A nonempty subset H of G is a subgroup of G if and only if (i) $e \in H$, (ii) $a * b \in H$ and (iii) $a \in H \Rightarrow a^{-1} \in H$ for all $a, b \in H$.

Test 2. Let $(G, *)$ be a group. A nonempty subset H of G is a subgroup of G if and only if $a * b^{-1} \in H$ for all $a, b \in H$.

Test 3. Let $(G, *)$ be a group. A nonempty finite subset H of G is a subgroup of G if and only if $a * b \in H$ for all $a, b \in H$.

Definition 1.3.11. The set $Z(G) = \{g \in G : g * x = x * g \; \forall x \in G\}$ is called the center of the group $(G, *)$.

The center $Z(G)$ of G is a subgroup of $(G, *)$.

Definition 1.3.12. Let $(G, *)$ be a group and H be its subgroup. Then for any $g \in G$, the set $g * H = \{g * h : h \in H\}$ is called the left coset of H in G. Similarly, $H * g = \{h * g : h \in H\}$ is called the right coset of H in G.

Here, we list some properties of left cosets in the form of the following theorem, which similarly apply to right cosets.

Theorem 1.3.13. *Let $(G, *)$ be a group and H be its subgroup. Then we have the following properties:*

(i) $g \in g * H$;

(ii) $g * H = H$ *if and only if* $g \in H$;

(iii) $g_1 * H = g_2 * H$ *if and only if* $g_1^{-1} g_2 \in H$;

(iv) $g_1 * H \neq g_2 * H$ *if and only if* $g_1 * H \cap g_2 * H = \phi$;

(v) $\bigcup_{g \in G} g * H = G$.

Definition 1.3.14. Let $(G, *)$ be a group. Then a subgroup H of G is called a normal subgroup of G if $g * H = H * g \ \forall g \in G$. The normal subgroup H of G is denoted by $H \trianglelefteq G$.

It is clear that every subgroup of an Abelian group is a normal subgroup. The center $Z(G)$ is a normal subgroup of G.

Definition 1.3.15. Let $(G, *)$ be a group and H be its normal subgroup. Then the quotient set $G/H = \{g * H : g \in G\}$ is a group under the operation $(g_1 * H)(g_2 * H) = (g_1 * g_2) * H$ is called the quotient group or the factor group of G modulo H. The identity of the quotient group G/H is H.

1.4 Homomorphism

Definition 1.4.1. A homomorphism f from a group $(G_1, *_1)$ to a group $(G_2, *_2)$ is a mapping $f : G_1 \to G_2$ that satisfies $f(x *_1 y) = f(x) *_2 f(y)$ for all $x, y \in G_1$. A homomorphism f is called an isomorphism if f is bijective (one-one onto), and in this case G_1 is said to be isomorphic to G_2 (denoted $G_1 \cong G_2$). If $G_1 = G_2$, the homomorphism f is called an endomorphism. A bijective endomorphism is known as an automorphism.

Below, we list some properties of homomorphism without proof.

Theorem 1.4.2. *Let $(G_1, *_1)$ and $(G_2, *_2)$ be groups with identities e_1 and e_2, respectively, and let $f : G_1 \to G_2$ be a homomorphism. Then*

(a) $f(e_1) = e_2$;

(b) $f(g_1^{-1}) = (f(g_1))^{-1}$ *for all* $g_1 \in G_1$;

(c) $f(g_1^n) = (f(g_1))^n$ *for all* $g_1 \in G_1$ *and* $n \in \mathbb{Z}$;

(d) *for a homomorphism* $g : G_2 \to G_3$, *the composition* $g \circ f : G_1 \to G_3$ *is a homomorphism;*

(e) *inverse of an isomorphism is an isomorphism, i. e., if $f : G_1 \to G_2$ is an isomorphism, then $f^{-1} : G_2 \to G_1$ is an isomorphism;*

(f) *if H_1 is a subgroup of $(G_1, *_1)$, then $f(H_1)$ is a subgroup of $(G_2, *_2)$ and for the subgroup H_2 of $(G_2, *_2)$, $f^{-1}(H_2)$ is a subgroup of $(G_1, *_1)$.*

Example 1.4.3.

(a) Let $(G_1, *_1)$ and $(G_2, *_2)$ be groups. Then the map $f : G_1 \to G_2$, defined by $f(g_1) = e_2$ for all $g_1 \in G_1$ (e_2 is the identity of $(G_2, *_2)$) is called zero homomorphism or trivial homomorphism.

(b) Define a map $f : \mathbb{Z} \to \mathbb{Z}_m$ by $f(n) = \overline{nr}$, where $(\mathbb{Z}_m, \oplus)$ and $(\mathbb{Z}, +)$ are groups and $\overline{r} \in \mathbb{Z}_m$. Then $f(n_1 + n_2) = \overline{(n_1 + n_2)r} = \overline{n_1 r + n_2 r} = \overline{n_1 r} \oplus \overline{n_2 r} = f(n_1) \oplus f(n_2)$ shows that f is a homomorphism.

Definition 1.4.4. Let f be a homomorphism from a group $(G_1, *_1)$ to a group $(G_2, *_2)$. Then the set $\mathrm{Ker}(f) = \{g \in G_1 : f(g) = e_2\}$ is called the kernel of f.

It is easy to prove that $\mathrm{Ker}(f) = f^{-1}(e_2)$ is a subgroup of $(G_1, *_1)$.

Theorem 1.4.5. *Let $(G_1, *_1)$ and $(G_2, *_2)$ be groups. Then the homomorphism $f : G_1 \to G_2$, is injective (one-one) if and only if $\mathrm{Ker}(f) = \{e_1\}$, where e_1 is the identity of $(G_1, *_1)$.*

Proof. Suppose f is injective.

Let $g \in \mathrm{Ker}(f)$. Then $f(g) = e_2 = f(e_1)$.

Since f is injective, $g = e_1$. Hence, $\mathrm{Ker}(f) = \{e_1\}$.

Conversely, suppose that $\mathrm{Ker}(f) = \{e_1\}$.

Let $f(g) = f(h)$, where $g, h \in G_1$. Then

$$e_2 = f(g)(f(g))^{-1} = f(g)(f(h))^{-1} = f(g)f(h^{-1}) = f(gh^{-1})$$
$$\Rightarrow gh^{-1} \in \mathrm{Ker}(f) = \{e_1\}$$
$$\Rightarrow gh^{-1} = e_1 \Rightarrow g = h.$$

Hence, f is injective. ☐

Here are three fundamental theorems of isomorphism.

Theorem 1.4.6 (First isomorphism theorem). *Let $(G_1, *_1)$ and $(G_2, *_2)$ be groups and let $f : G_1 \to G_2$ be a surjective homomorphism. Then the quotient group $G_1/\mathrm{Ker}(f)$ is isomorphic to G_2.*

Theorem 1.4.7 (Second isomorphism theorem). *Let $(G, *)$ be a group, H be a subgroup of G and K be a normal subgroup of G. Then $H/H \cap K$ is isomorphic to HK/K.*

Theorem 1.4.8 (Third isomorphism theorem). *Let H and K be normal subgroups of a group $(G, *)$ and $H \leq K$. Then $(G/H)/(K/H)$ is isomorphic to G/K.*

Example 1.4.9.

(a) Let $(\mathbb{Z}, +)$ and $(\mathbb{Z}_m, \oplus)$ be groups. Then the map $f : \mathbb{Z} \to \mathbb{Z}_m$ defined by $f(n) = \bar{n}$ is a homomorphism. This follows from the calculation:

$$f(n_1 + n_2) = \overline{n_1 + n_2} = \bar{n_1} \oplus \bar{n_2} = f(n_1) \oplus f(n_2).$$

Moreover, for any $\bar{k} \in \mathbb{Z}_m$, we have $k \in \mathbb{Z}$ such that $f(k) = \bar{k}$, i. e., f is surjective. Now,

$$\ker f = \{n \in \mathbb{Z} \mid f(n) = \bar{0}\} = \{n \in \mathbb{Z} \mid \bar{n} = \bar{0}\} = \{mk : k \in \mathbb{Z}\} = m\mathbb{Z}.$$

Thus, by the first isomorphism theorem, we have $\mathbb{Z}/\ker f \cong \mathbb{Z}_m$, i. e., $\mathbb{Z}/m\mathbb{Z} \cong \mathbb{Z}_m$.

(b) It is easy to prove that $m\mathbb{Z}, m \in \mathbb{Z}$ is a subgroup of the group $(\mathbb{Z}, +)$. Suppose n is an integer such that m divides n. Then $n\mathbb{Z}$ is a normal subgroup of $m\mathbb{Z}$. Define a map $f : \mathbb{Z} \to m\mathbb{Z}/n\mathbb{Z}$ by $f(k) = mk + n\mathbb{Z}$. Now,

$$f(k_1 + k_2) = m(k_1 + k_2) + n\mathbb{Z} = (mk_1 + mk_2) + n\mathbb{Z} = (mk_1 + n\mathbb{Z}) + (mk_2 + n\mathbb{Z})$$
$$= f(k_1) + f(k_2),$$

showing that f is a homomorphism.

For all $mk + n\mathbb{Z} \in m\mathbb{Z}/n\mathbb{Z}$, there exists $k \in \mathbb{Z}$ such that $f(k) = mk + n\mathbb{Z}$. Hence, f is a surjective homomorphism.

Also, $\text{Ker} f = \{k \in \mathbb{Z} : f(k) = n\mathbb{Z}\}$, where $n\mathbb{Z}$ is the identity of the quotient group $m\mathbb{Z}/n\mathbb{Z}$. Thus,

$$\begin{aligned}
\text{Ker} f &= \{k \in \mathbb{Z} : mk + n\mathbb{Z} = n\mathbb{Z}\} \\
&= \{k \in \mathbb{Z} : mk \in n\mathbb{Z}\} \\
&= \{k \in \mathbb{Z} : mk = nt \text{ for some } t \in \mathbb{Z}\} \\
&= \left\{k \in \mathbb{Z} \mid k = \frac{n}{m}t, t \in \mathbb{Z}\right\} \\
&= \left\{\frac{n}{m}t : t \in \mathbb{Z}\right\} \\
&= \frac{n}{m}\mathbb{Z}.
\end{aligned}$$

Hence, by the first isomorphism theorem,

$$\mathbb{Z}/\text{Ker} f \cong m\mathbb{Z}/n\mathbb{Z} \quad \text{or} \quad m\mathbb{Z}/n\mathbb{Z} \cong \mathbb{Z}/\frac{n}{m}\mathbb{Z} \cong \mathbb{Z}_{\frac{n}{m}}.$$

(c) Let $m, n \in \mathbb{Z}$ such that m divides n. Then $m\mathbb{Z}$ and $n\mathbb{Z}$ are normal subgroups of $(\mathbb{Z}, +)$, and $n\mathbb{Z}$ is also a normal subgroup of $m\mathbb{Z}$. By the third isomorphism theorem, we have

$$\frac{\mathbb{Z}/n\mathbb{Z}}{m\mathbb{Z}/n\mathbb{Z}} \cong \mathbb{Z}/m\mathbb{Z} \cong \mathbb{Z}_m.$$

(d) Let $m, n \in \mathbb{Z}$. Then $m\mathbb{Z}$ and $n\mathbb{Z}$ are normal subgroups of $(\mathbb{Z}, +)$. By the second isomorphism theorem, we have

$$\frac{m\mathbb{Z} + n\mathbb{Z}}{n\mathbb{Z}} \cong \frac{m\mathbb{Z}}{m\mathbb{Z} \cap n\mathbb{Z}} \quad \text{and} \quad \frac{m\mathbb{Z} + n\mathbb{Z}}{m\mathbb{Z}} \cong \frac{n\mathbb{Z}}{m\mathbb{Z} \cap n\mathbb{Z}}.$$

In particular, consider $8\mathbb{Z}$ and $4\mathbb{Z}$. Clearly,

$$8\mathbb{Z} = \{\dots, -24, -16, -8, 0, 8, 16, 24, \dots\},$$

is a normal subgroup of

$$4\mathbb{Z} = \{\dots, -16, -12, -8, -4, 0, 4, 8, 12, 16, 20, 24, \dots\}.$$

Take $m = 4$ and $n = 8$. Then

$$\frac{4\mathbb{Z} + 8\mathbb{Z}}{8\mathbb{Z}} \cong \frac{4\mathbb{Z}}{8\mathbb{Z}} \cong \mathbb{Z}_{\frac{8}{4}} \cong \mathbb{Z}_2,$$

and

$$\frac{4\mathbb{Z}}{4\mathbb{Z} \cap 8\mathbb{Z}} \cong \frac{4\mathbb{Z}}{8\mathbb{Z}} \cong \mathbb{Z}_2,$$

are isomorphic.
Also,

$$\frac{4\mathbb{Z} + 8\mathbb{Z}}{4\mathbb{Z}} \cong \frac{4\mathbb{Z}}{4\mathbb{Z}} = \{\text{identity element}\},$$

and

$$\frac{8\mathbb{Z}}{4\mathbb{Z} \cap 8\mathbb{Z}} \cong \frac{8\mathbb{Z}}{8\mathbb{Z}} = \{\text{identity element}\},$$

are isomorphic.

1.5 Rings and fields

Definition 1.5.1. A set R with two internal binary operations $+$ (addition) and $\cdot$ (multiplication) is called a ring if:
(a) $(R, +)$ is an Abelian group;
(b) binary operation $\cdot$ on R is associative, i. e., $(a \cdot b) \cdot c = a \cdot (b \cdot c) \; \forall a, b, c \in R$;
(c) the binary operation $\cdot$ is distributive over $+$ from both (left and right) sides, i. e., for all $a, b, c \in R$, $a \cdot (b + c) = a \cdot b + a \cdot c$ and $(a + b) \cdot c = a \cdot c + b \cdot c$.

A ring R is called commutative if $a \cdot b = b \cdot a$ $\forall a, b \in R$. An identity element with respect to multiplication $\cdot$ in R is called the identity of the ring. That is, if $1 \in R$ is the identity of R, then $1 \cdot a = a \cdot 1 = a$, $\forall a \in R$. In a ring, the binary operation $\cdot$ need not be commutative and also a ring need not have an identity element.

Note. We use the notation $(R, +, \cdot)$ for a ring with respect to binary operations $+$ and $\cdot$.

Example 1.5.2. With the help of Examples 1.3.2, 1.3.3, 1.3.4 and 1.3.5 it is easy to prove that $(\mathbb{Z}, +, \cdot)$, $(\mathbb{Q}, +, \cdot)$, $(\mathbb{R}, +, \cdot)$, $(\mathbb{C}, +, \cdot)$ and $(Z_m, \oplus, *)$ are commutative rings with identity.

Example 1.5.3. Let $(R_1, +_1, \cdot_1), (R_2, +_2, \cdot_2), \ldots, (R_n, +_n, \cdot_n)$ be rings. Then the product $R_1 \times R_2 \times \ldots \times R_n$, is also a ring with respect to the operations $\oplus$ and $*$ defined as follows:

$$(x_1, x_2, \ldots, x_n) \oplus (y_1, y_2, \ldots, y_n) = (x_1 +_1 y_1, x_2 +_2 y_2, \ldots, x_n +_n y_n)$$

and

$$(x_1, x_2, \ldots, x_n) * (y_1, y_2, \ldots, y_n) = (x_1 \cdot_1 y_1, x_2 \cdot_2 y_2, \ldots, x_n \cdot_n y_n).$$

Theorem 1.5.4 (Elementary properties of rings). *Let $(R, +, \cdot)$ be a ring. Then for $a, b, c \in R$:*
(i) $a \cdot 0 = 0 \cdot a = 0$, *where 0 is the additive identity,*
(ii) $a \cdot (-b) = -(a \cdot b) = (-a) \cdot b$,
(iii) $(-a) \cdot (-b) = a \cdot b$,
(iv) $a \cdot (b - c) = a \cdot b - a \cdot c$.

Proof. (i) $a \cdot 0 = a \cdot (0 + 0) = a \cdot 0 + a \cdot 0$ or $a \cdot 0 + 0 = a \cdot 0 + a \cdot 0$. By the cancellation law in the group $(R, +)$, we have $0 = a \cdot 0$. Similarly, we can show that $0 = 0 \cdot a$.
(ii) We have to prove that $a \cdot (-b) = -(a \cdot b)$, i. e., we have to prove that the additive inverse of $a \cdot b$ in $(R, +)$, denoted by $-(a \cdot b)$, is $a \cdot (-b)$. Then $a \cdot b + a \cdot (-b) = a \cdot (b + (-b)) = a \cdot 0 = 0$, which gives that $-(a \cdot b) = a \cdot (-b)$. Similarly, $-(a \cdot b) = (-a) \cdot b$.
(iii) $(-a) \cdot (-b) = -(a \cdot (-b)) = a \cdot b$, by (ii).
(iv) $a \cdot (b - c) = a \cdot (b + (-c)) = a \cdot b + a \cdot (-c) = a \cdot b - a \cdot c$.   □

Definition 1.5.5. A ring $(R, +, \cdot)$ is called an integral domain if it satisfies the condition $a \cdot b = 0 \Rightarrow a = 0$ or $b = 0$ for all $a, b \in R$.

Example 1.5.6. Rings $(\mathbb{Z}, +, \cdot)$, $(\mathbb{Q}, +, \cdot)$, $(\mathbb{R}, +, \cdot)$ and $(\mathbb{C}, +, \cdot)$ are integral domains but the ring $(Z_4, \oplus, *)$ is not an integral domain because $\bar{2} * \bar{2} = \bar{0}$.

Definition 1.5.7. A ring $(R, +, \cdot)$ is called a field if the set $R - \{0\}$ forms a commutative group under multiplication. In other words, the triple $(R, +, \cdot)$, where $+$ and $\cdot$ are binary operations, is called a field if the following conditions hold:
(a) $(R, +)$ is an Abelian group;
(b) $(R - \{0\}, \cdot)$ is an Abelian group;
(c) $a \cdot (b + c) = a \cdot b + a \cdot c$ and $(a + b) \cdot c = a \cdot c + b \cdot c$ for all $a, b, c \in R$.

Example 1.5.8. The rings $(\mathbb{Q}, +, \cdot)$, $(\mathbb{R}, +, \cdot)$ and $(\mathbb{C}, +, \cdot)$ are fields.

Example 1.5.9. From Example 1.3.5, the ring $(Z_5, \oplus, *)$ is a field. In general, $(Z_m, \oplus, *)$ is a field if and only if m is a prime.

Definition 1.5.10. Let $R, +, \cdot$ be a ring. Then the small positive integer n is called the characteristic of R if $nx = 0$ for all $x \in R$. If no such integer exist, we say that R is of characteristic zero. The characteristic of R is denoted by Char R.

It is easy to prove that the characteristic of an integral domain is zero or a prime and order of a finite integral domain or field is p^n for some prime p and $n > 0$.

ℹ Exercises

(1.1)   Show that ordinary substraction '−' is not a binary operation on the set of natural numbers $\mathbb{N}$.

(1.2)   Let $M_{3,3}(\mathbb{Z})$ denote the set of all 3×3 matrices with entries in $\mathbb{Z}$. Then show that the matrix addition and matrix multiplication both are binary operations on $M_{3,3}(\mathbb{Z})$.

(1.3)   Do all four conditions of Definition 1.1.4 hold for the pairs $(\mathbb{N}, +)$, $(\mathbb{N}, \cdot)$, $(\mathbb{Z}, \cdot)$, $(\mathbb{Q}, \cdot)$, $(\mathbb{R}, \cdot)$ and $(\mathbb{C}, \cdot)$? If not, provide the appropriate reasons.

(1.4)   Construct binary operation tables for the following pairs:
$(Z_7, \oplus)$, $(Z_7, *)$, $(Z_8, \oplus)$, $(Z_8, *)$.

(1.5)   Find the inverses of $\bar{5}$ in the groups $(Z_{12}, \oplus)$, $(Z_{12}, *)$.

(1.6)   Find the order of $\bar{6}$ in the group $(Z_{10}, \oplus)$ and order of $\bar{4}$ in the group $(Z_7, *)$.

(1.7)   Prove that $(Z_p - \{0\}, *)$ is a group, where p is a prime.

(1.8)   Prove that the set $\{\bar{0}, \bar{3}, \bar{6}, \bar{9}\}$ is a subgroup of the group $(Z_{12}, \oplus)$.

(1.9)   Prove that for any divisor d of an integer m, the set $\{\bar{0}, \bar{d}, \overline{2d}, \overline{3d}, \dots, \overline{(r-1)d}\}$, is a subgroup of $(Z_m, \oplus)$, where $r = \frac{m}{d}$.

(1.10) Let H_1 and H_2 be subgroups of a group $(G, *)$. Then prove that $H_1 \cap H_2$ is also, a subgroups of $(G, *)$.

(1.11) The set $S^1 = \{e^{it} : t \in \mathbb{R}\}$ is a group under the multiplication of complex numbers called the unit circle group. Using the isomorphism theorem, prove that the quotient group $\mathbb{R}/\mathbb{Z}$ is isomorphic to S^1.

(1.12) Prove that the ring $(Z_m, \oplus, *)$, is a field if and only if m is a prime.

(1.13) Let p be a prime and $n > 0$. Then prove that the order of a finite integral domain, and hence a finite field is p^n.

(1.14) Let $C(\mathbb{R}^{[0,1]})$ be the set of all continuous real valued functions on $[0, 1]$. Then show that $C(\mathbb{R}^{[0,1]})$ is a commutative ring with identity under the operations pointwise addition and pointwise multiplication. Also, prove that it is not an integral domain.

(1.15) Is the ring $(Z_{12}, \oplus, *)$ an integral domain?

(1.16) Is the ring $(Z_{31}, \oplus, *)$ a field?

---

2 Vector spaces

2.1 Definition and examples

Before delving into the formal definition of a vector space, it is important to grasp the underlying concept. When we refer to a vector space, we are essentially referring to a collection of vectors. This notion becomes clear when considering $\mathbb{R}^n$, where the elements are referred to as vectors, making $\mathbb{R}^n$ itself a vector space. Upon observing the properties of $\mathbb{R}^n$, we notice several key characteristics:

(i)   The structure $(\mathbb{R}^n, +)$ forms an Abelian group, where the operation $+$, defined as $(x_1, x_2, \ldots, x_n) + (y_1, y_2, \ldots, y_n) = (x_1 + y_1, x_2 + y_2, \ldots, x_n + y_n)$ represents vector addition.

(ii) For the real field $(\mathbb{R}, +, \cdot)$, an external binary operation $\cdot : \mathbb{R} \times \mathbb{R}^n \to \mathbb{R}^n$ is defined as $a \cdot (x_1, x_2, \ldots, x_n) = (ax_1, ax_2, \ldots, ax_n)$, representing scalar multiplication.

(iii) The zero vector $(0, 0, \ldots, 0) \in \mathbb{R}^n$ serves as the additive identity, while $0 \in \mathbb{R}$ acts as the identity element for $(\mathbb{R}, +)$.

(iv) Scalar multiplication by 0 yields the zero vector $0 \cdot (x_1, x_2, \ldots, x_n) = (0, 0, \ldots, 0) \in \mathbb{R}^n$, and multiplication by 1 results in the original vector, $1 \cdot (x_1, x_2, \ldots, x_n) = (x_1, x_2, \ldots, x_n) \in \mathbb{R}^n$, among other favorable properties.

Similarly, when considering other sets such as $M_{n,n}(\mathbb{R})$ (the collection of all $n \times n$ matrices with real entries) and $P_n(x)$ (the set of all polynomials of degree at most n with real coefficients), it can be verified that they satisfy the same properties as $\mathbb{R}^n$ over the field $(\mathbb{R}, +, \cdot)$. Instead of individually analyzing these types of spaces, it is more convenient to provide a general definition containing most of the common properties.

Definition 2.1.1. A nonempty set V is called a vector space over a field F if it possesses an internal binary operation $+$, referred to as vector addition and an external binary operation $\cdot : F \times V \to V$ defined as $(a, x) \to a \cdot x$, known as scalar multiplication. These operations must satisfy the following properties:

(i) $(V, +)$ forms an Abelian group;

(ii) $a \cdot (x + y) = a \cdot x + a \cdot y$;

(iii) $(a + \beta) \cdot x = a \cdot x + \beta \cdot x$;

(iv) $(a\beta) \cdot x = a \cdot (\beta \cdot x)$;

(v) $1 \cdot x = x$, where 1 represents the multiplicative identity of F, for all $a, \beta \in F$ and $x \in V$.

The elements within a vector space V are denoted as vectors, while the elements of F are called scalars. Throughout this text, ax is used to denote $a \cdot x$.

Remark. It may be asked, "Why is the above vector space not called left vector space due to left scalar multiplication?" The answer is that $(a\beta)x = (\beta a)x$, and the vector space V remains the same if we define $xa = ax$.

https://doi.org/10.1515/9783111516035-002

Theorem 2.1.2. *Let V represent a vector space over the field F. The following properties hold:*

(a) $0x = \mathbf{0}$, *for all* $x \in V$,

(b) $a\mathbf{0} = \mathbf{0}$, *for all* $a \in F$,

(c) $(-a)x = -(ax) = a(-x)$, *for all* $a \in F$ *and* $x \in V$,

(d) *If* $ax = \mathbf{0}$, *for* $a \in F$ *and* $x \in V$, *then either* $a = 0$ *or* $x = \mathbf{0}$, *where* $\mathbf{0} \in V$ *represents the zero vector.*

Proof. (a) We have $0x = (0 + 0)x = 0x + 0x$, or equivalently, $0x + \mathbf{0} = 0x + 0x$. By the cancellation law in an Abelian group, we conclude $\mathbf{0} = 0x$.

(b) Similarly, $a\mathbf{0} = a(\mathbf{0} + \mathbf{0}) = a\mathbf{0} + a\mathbf{0}$. Applying cancellation law, we get $\mathbf{0} = a\mathbf{0}$.

(c) To prove $(-a)x = -(ax)$, it is sufficient to show that $(-a)x$ serves as the inverse of ax in the group $(V, +)$. Therefore, $(-a)x + ax = (-a + a)x = 0x = \mathbf{0}$, satisfying the requirement. Similarly, $-(ax) = a(-x)$.

(d) Suppose $ax = \mathbf{0}$ and $a \neq 0$. Then a^{-1} exists in F. Consequently, $a^{-1}ax = a^{-1}\mathbf{0} \Rightarrow 1x = \mathbf{0} \Rightarrow x = \mathbf{0}$.

Now, assume $x \neq \mathbf{0}$, and we aim to prove $a = 0$. If $a \neq 0$, then $x = \mathbf{0}$, contradicting the assumption $x \neq \mathbf{0}$. Thus, a must be 0. □

Hereafter, the notation 0 will be used to denote both the zero vector and the zero element of the field.

Example 2.1.3. Let $\mathbb{R}^n = \{(x_1, x_2, \ldots, x_n) : x_i \in \mathbb{R}\}$, the set of all ordered n-tuples of elements of $\mathbb{R}$. Then $\mathbb{R}^n$ is a vector space over the field $\mathbb{R}$, under the operations addition and scalar multiplication defined as follows:

$$(x_1, x_2, \ldots, x_n) + (y_1, y_2, \ldots, y_n) = (x_1 + y_1, x_2 + y_2, \ldots, x_n + y_n), \quad \text{and}$$
$$a(x_1, x_2, \ldots, x_n) = (ax_1, ax_2, \ldots, ax_n), \quad \text{where } a \in \mathbb{R}.$$

Similarly, the set $\mathbb{C}^n = \{(z_1, z_2, \ldots, z_n) : z_i \in \mathbb{C}\}$ constitutes a vector space over the field of complex numbers $\mathbb{C}$, and Z_p^n serves as a vector space over Z_p. In general, for any field F, the set F^n forms a vector space over F, under the operations defined as above. Moreover, if $n = 1$, then F^1 represents a vector space over F. Consequently, every field can be considered a vector space over itself.

Example 2.1.4. Let F be a subfield of the field L. Consequently, $(L, +)$ forms an Abelian group. When scalar multiplication is defined as ax for $a \in F$ and $x \in L$, L emerges as a vector space over the field F. Thus, every field serves as a vector space over its subfield, and particularly, every field is a vector space over itself. In particular, $\mathbb{C}$ over $\mathbb{R}$, $\mathbb{R}$ over $\mathbb{Q}$ and $\mathbb{C}$ over $\mathbb{Q}$ are vector spaces. However, the converse of this result does not hold true. For instance, considering the subfield $\mathbb{Q}$ of the field $\mathbb{R}$, $(\mathbb{Q}, +)$ forms an Abelian group, but for an irrational number $i \in \mathbb{R}$ and $x \in \mathbb{Q}$, the product $ix \notin \mathbb{Q}$. Hence, $\mathbb{Q}$ is not a vector space over $\mathbb{R}$. Similarly, $\mathbb{R}$ is not a vector space over $\mathbb{C}$.

Example 2.1.5. Let $P_n(x)$ denote the set of all polynomials of degree at most $n \in \mathbb{N}$ with coefficients in a field F. Then $P_n(x)$ is a vector space over F under the operations given below:

Let $f(x) = a_0 + a_1x + a_2x^2 + \cdots + a_nx^n$, $g(x) = b_0 + b_1x + b_2x^2 + \cdots + b_nx^n \in P_n(x)$, and $c \in F$. Then

$$f(x) + g(x) = (a_0 + b_0) + (a_1 + b_1)x + (a_2 + b_2)x^2 + \cdots + (a_n + b_n)x^n, \quad \text{and}$$
$$cf(x) = ca_0 + ca_1x + ca_2x^2 + \cdots + ca_nx^n.$$

Particularly, $P_n(x)$ is a vector space over each of the following fields: $\mathbb{Q}, \mathbb{R}, \mathbb{C}$ and Z_p.

Example 2.1.6. Let F be a field. Then the set $M_{m,n}(F)$ of all $m \times n$ matrices with entries in F, constitutes a vector space under matrix addition and scalar multiplication of matrices. In particular, we can replace F by any one of the following fields $\mathbb{Q}, \mathbb{R}, \mathbb{C}, Z_p$.

The example given below plays an important role to obtain various examples of vector spaces.

Example 2.1.7. Consider V as a vector space over the field F, and let S be a nonempty set. Then $V^S = \{f : f : S \to V \text{ be a function}\}$ forms a vector space over F with the operations $f + g$ defined by $(f + g)(x) = f(x) + g(x)$ and af defined by $(af)(x) = af(x)$ for $f, g \in V^S$ and $a \in F$. It can be shown that $(V^S, +)$ constitutes an Abelian group, and the external binary operation $\cdot : F \times V^S \to V^S$ defined by $a \cdot f = af$ represents scalar multiplication.

Remark. Since every field is a vector space over itself, V can be replaced by F in Example 2.1.7. Consequently, F^S serves as a vector space over the field F. This observation, along with the Cartesian product, leads to the generation of numerous examples of vector spaces.

Definition 2.1.8. Consider a family of nonempty sets:

$$\{X_\alpha : \alpha \in I, I \text{ is an index set}\}.$$

The Cartesian product of sets X_α, denoted by $\prod_{\alpha \in I} X_\alpha$, is defined as the collection of all functions $\{x : I \to \bigcup_{\alpha \in I} X_\alpha \text{ such that } x(\alpha) \in X_\alpha\}$, for each $\alpha \in I$. We denote $x(\alpha)$ by x_α and represent x by $(x_\alpha)_{\alpha \in I}$. Therefore, $\prod_{\alpha \in I} X_\alpha = \{x : x : I \to \bigcup_{\alpha \in I} X_\alpha\} = \{(x_\alpha)_{\alpha \in I}\}$. The function x is referred to as an I-tuple of the collection $\{X_\alpha : \alpha \in I\}$.

Example 2.1.9.
(a) Let $S = [0, 1]$ be the unit closed internal and $F = \mathbb{R}$ be the field of real numbers. Then $\mathbb{R}^{[0,1]}$, the set of real-valued functions defined on $[0, 1]$, forms a vector space over $\mathbb{R}$. Utilizing the preserving property of continuity, differentiability and Riemann integrability under addition, we can define the following vector spaces over $\mathbb{R}$:
$C(\mathbb{R}^{[0,1]})$, the set of continuous real valued functions on $[0, 1]$;
$D(\mathbb{R}^{[0,1]})$, the set of differentiable real valued functions on $[0, 1]$ and

$\mathcal{R}(\mathbb{R}^{[0,1]})$, the set of Riemann integrable functions on $[0,1]$.
Here, $[0,1]$ can be replaced by any closed interval $[a,b] \subset \mathbb{R}$.

(b) Let the index set $I = \{1, 2, \ldots, n\}$. Then the Cartesian product

$$\mathbb{R}^n = \{x : x : I \to \mathbb{R} \cup \mathbb{R} \cup \cdots \cup \mathbb{R}(n \text{ times})\}$$
$$= \{x : x : I \to \mathbb{R}\} = \{(x_1, x_2, \ldots, x_n) : x_i \in \mathbb{R}, 1 \le i \le n\},$$

represents all functions from $\{1, 2, \ldots, n\}$ to the field of real numbers $\mathbb{R}$. Hence, $\mathbb{R}^n$ is a vector space over $\mathbb{R}$.

(c) If the index set $I = \mathbb{N}$, then $\mathbb{R}^{\mathbb{N}} = \{x : x : \mathbb{N} \to \mathbb{R}\} = \{(x_n) : x_n \in \mathbb{R}\}$ represents the set of real sequences.

Thus, $\mathbb{R}^{\mathbb{N}}$ is a vector space over $\mathbb{R}$. If $\mathrm{Con}(\mathbb{R}^{\mathbb{N}})$ denotes the set of all convergent real sequences, then $\mathrm{Con}(\mathbb{R}^{\mathbb{N}})$ also forms a vector space over $\mathbb{R}$ due to the property that the sum of convergent sequences is convergent.

(d) Define $F(\mathbb{R}^{\mathbb{N}})$ as the set of all sequences with finite nonzero elements only. Then $F(\mathbb{R}^{\mathbb{N}}) \subset \mathbb{R}^{\mathbb{N}}$ is a vector space over $\mathbb{R}$.

(e) Let $P(x)$ denote the set of all polynomials with real coefficients. It is equivalent to $F(\mathbb{R}^{\mathbb{N}})$ under the mapping $(a_0, a_1, a_2, \ldots, a_n, \ldots) \to a_0 + a_1 x + a_2 x^2 + \cdots + a_n x^n + \cdots$, where only a finite number of a_n's are nonzero. Thus, $P(x)$ is a vector space over $\mathbb{R}$. Similarly, $P_n(x)$ is equivalent to $\mathbb{R}^{n+1}$ under the mapping $(a_0, a_1, a_2, \ldots, a_n) \to a_0 + a_1 x + a_2 x^2 + \cdots + a_n x^n$.

(f) The vector space $\mathbb{R}^{m \times n}$ is equivalent to $M_{m,n}(\mathbb{R})$ under the mapping $(a_{ij})_{m \times n} \to (a_{11}, a_{12}, \ldots, a_{1n}, a_{21}, \ldots, a_{2n}, \ldots, a_{m1}, \ldots, a_{mn})$.

Remark. These examples provide an alternative approach to prove many well-known vector spaces. In these examples, $\mathbb{R}$ can be replaced by any field F.

🛈 Exercises

(2.1.1) Let V be a vector space over the field F. If $ax = bx$, where $x(\ne 0) \in V$ and $a, b \in F$, then show that $a = b$.

(2.1.2) Show that $\mathbb{R}^3$ is not a vector space over the field $\mathbb{R}$ under the operations; $(x_1, y_1, z_1) + (x_2, y_2, z_2) = (x_1 + x_2, y_1 + y_2, z_1 + z_2)$ and $a(x, y, z) = (a^2 x, a^2 y, a^2 z)$, where $a \in \mathbb{R}$.

(2.1.3) Let n be a positive integer. Then prove that $P_n(x)$, the set of polynomials of degree less than or equal to n with coefficients in Z_5 is a vector space over the field Z_5.

(2.1.4) If $V_1 \subseteq V_2 \subseteq V_3 \subseteq \cdots$ is a chain of subspaces of a vector space V, then prove that their union is a subspace of V.

(2.1.5) Using the definition of a vector space show that:
   (i) $\mathbb{R}^{[0,1]}$, the collection of real valued functions on $[0,1]$,
   (ii) $\mathbb{R}^{\mathbb{N}}$, the set of real sequences and
   (iii) $P(x)$, the set of all polynomials with real coefficients are vector spaces over the field of real numbers.

2.2 Vector subspaces

A subspace of a vector space V refers to a subset W of V that also functions as a vector space under the same operations. More precisely, we have the following definition.

Definition 2.2.1. Given a vector space V over a field F, a nonempty subset W of V is called a subspace of V if W constitutes a subgroup of $(V, +)$ and is closed under scalar multiplication, implying that $ax \in W$ for all $a \in F$ and $x \in W$.

For a given set W if we are asked to show that W is a vector space, then we have to show that W satisfies all the axioms of a vector space. But if W is a subset of a vector space V, then some of the axioms automatically hold in W. Hence, to show that a subset W of a vector space V over a field F is a vector space it is sufficient to employ certain straightforward criteria. With consideration to Definition 2.2.1 and the diverse approaches for testing a subgroup, several subspace tests listed below can be effectively utilized interchangeably.

Let W be a nonempty subset of a vector space V over a field F. Then W is considered a subspace of V if it satisfies any of the following criteria:

Test 1. (i) 0 belongs to W, and the sum of any two elements in W is also in W, i. e., $0 \in W$ and $x + y \in W$, $\forall x, y \in W$.
(ii) Scalar multiplication holds, i. e., for any scalar a in F and any vector x in W, $ax \in W$. The condition for including $-x$ is omitted here as $(-x = (-1) \cdot x)$ it is covered in (ii).

Test 2. (i) $x - y \in W$ for all $x, y \in W$.
(ii) Scalar multiplication holds, i. e., $ax \in W$ $\forall a \in F$ and $x \in W$.

Test 3 (If W is a finite subset). (i) $x + y \in W$ $\forall x, y \in W$.
(ii) Scalar multiplication holds, i. e., $ax \in W$ $\forall a \in F$ and $x \in W$.

The following theorem demonstrates that these conditions can be unified into a single condition.

Theorem 2.2.2. *Let V be a vector space over a field F and let W be a nonempty subset of V. Then W is a subspace of V if and only if $ax + by \in W$ for all $a, b \in F$ and $x, y \in W$.*

Proof. Suppose W is a subspace of V. Then, for any a, b in F and x, y in W, both ax and by are in W, leading to $ax + by$ being in W as well.

Conversely, if W is a nonempty subset of V where $ax + by \in W$ for all a, b in F and x, y in W, it can be observed that $1x + (-1)y = x - y$ is in W. Therefore, $(W, +)$ forms a subgroup of $(V, +)$. Furthermore, for any a in F and $b = 0$, $ax + 0y = ax \in W$, establishing that W is a subspace of V. □

Theorem 2.2.3. *Let $\{W_\alpha : \alpha \in I, I$ is an index set$\}$ be a family of subspaces of a vector space V over a field F. Then their intersection $\bigcap_{\alpha \in I} W_\alpha$ is a subspace of V.*

Proof. Since each W_α is a subspace of the vector space $V, 0 \in W_\alpha \, \forall \alpha \in I$, and hence $0 \in \bigcap_{\alpha \in I} W_\alpha$. If $\bigcap_{\alpha \in I} W_\alpha = \{0\}$. Then it is a subspace of V.

Suppose x and y belong to $\bigcap_{\alpha \in I} W_\alpha$, and let a and b be elements of F. Since x and y are in $\bigcap_{\alpha \in I} W_\alpha$, it implies that x and y are in W_α for all $\alpha \in I$. As each W_α is a subspace of V, it follows that $ax + by$ is in W_α for all $\alpha \in I$, and consequently, $ax + by$ belongs to $\bigcap_{\alpha \in I} W_\alpha$. Thus, according to Theorem 2.2.2, $\bigcap_{\alpha \in I} W_\alpha$ is a subspace of the vector space V. ☐

Theorem 2.2.4. *Suppose W_1 and W_2 are two subspaces of a vector space V over a field F. Then $W_1 \cup W_2$ forms a subspace of V if and only if either $W_1 \subseteq W_2$ or $W_2 \subseteq W_1$.*

Proof. Let us first assume $W_1 \subseteq W_2$. In this case, $W_1 \cup W_2 = W_2$, which is indeed a subspace of V. Similarly, if $W_2 \subseteq W_1$, then $W_1 \cup W_2 = W_1$ forms a subspace of V.

Now, let us consider the converse. Suppose $W_1 \cup W_2$ is a subspace of V. Our goal is to show that either $W_1 \subseteq W_2$ or $W_2 \subseteq W_1$. Suppose W_1 is not a subset of W_2. Then there exists $x \in W_1$ such that $x \notin W_2$.

We aim to prove $W_2 \subseteq W_1$. Let y be an arbitrary element of W_2. Since $W_1 \cup W_2$ is a subspace, $x + y$ must belong to $W_1 \cup W_2$, which implies $x + y \in W_1$ or $x + y \in W_2$. If $x + y \in W_2$, then $(x+y) - y = x \in W_2$, contradicting our assumption. Therefore, $x + y \in W_1$, and subsequently $(x + y) - x = y \in W_1$, proving $W_2 \subseteq W_1$.

Similarly, by symmetry, if W_2 is not a subset of W_1, then W_1 must be contained in W_2. ☐

Example 2.2.5. In a vector space V, the set containing only the zero vector, denoted as $\{0\} \subseteq V$, forms a subspace of V.

Example 2.2.6. Let $V = \mathbb{R}^n$. Then $W = \{(0, x_2, \dots, x_n) : x_2, \dots, x_n \in \mathbb{R}\}$ is a subspace of V. Let $a, b \in \mathbb{R}$ and $(0, x_2, x_3, \dots, x_n), (0, y_2, y_3, \dots, y_n) \in \mathbb{R}^n$. Then

$$a(0, x_2, x_3, \dots, x_n) + b(0, y_2, y_3, \dots, y_n)$$
$$= (0, ax_2, ax_3, \dots, ax_n) + (0, by_2, by_3, \dots, by_n)$$
$$= (0, ax_2 + by_2, ax_3 + by_3, \dots, ax_n + by_n).$$

Since $a, b, x_i, y_i \in \mathbb{R}$, for $2 \leq i \leq n$, it follows that $ax_i + by_i = z_i \in \mathbb{R}$ and, therefore, $(0, z_2, \dots, z_n) \in W$. This demonstrates that W is a subspace of V. Similarly,

$$W_1 = \{(x_1, x_2, \dots, x_{n-1}, 0) : x_1, x_2, \dots, x_{n-1} \in \mathbb{R}\},$$
$$W_2 = \{(x_1, x_2, 0, x_4, \dots, x_n) : x_1, x_2, \dots, x_n \in \mathbb{R}\},$$
$$W_3 = \{(x_1, x_2, \dots, x_{n-2}, 0, 0) : x_1, x_2, \dots, x_{n-2} \in \mathbb{R}\},$$
$$W_4 = \{(x_1, 0, 0, \dots, 0) : x_1 \in \mathbb{R}\}, \quad \text{are subspaces of } \mathbb{R}^n.$$

All the above results are also true for any vector space F^n.

Example 2.2.7.
(a) $D(\mathbb{R}^{[0,1]})$ is a subspace of $C(\mathbb{R}^{[0,1]})$, which is a subspace of $\mathbb{R}^{[0,1]}$.
(b) $C(\mathbb{R}^{[0,1]})$ is a subspace of $R(\mathbb{R}^{[0,1]})$, which is a subspace of $\mathbb{R}^{[0,1]}$.
(c) $\mathrm{Con}(\mathbb{R}^{\mathbb{N}})$ is a subspace of $\mathbb{R}^{\mathbb{N}}$ and $F(\mathbb{R}^{\mathbb{N}})$ is a subspace of $\mathbb{R}^{\mathbb{N}}$.

Example 2.2.8. Let $V = \mathbb{R}^n$. Then:
(i) The set $W_1 = \{(x_1, x_2, \ldots, x_n) \in \mathbb{R}^n : x_n \geq 0\}$ is not a subspace of V because for any
$(x_1, x_2, \ldots, x_n) \in W_1, -2(x_1, x_2, \ldots, x_n)$ does not belong to W_1.
(ii) The set $W_2 = \{(x_1, x_2, \ldots, x_n) \in \mathbb{R}^n : x_1^2 + x_2^2 + \cdots + x_n^2 \leq 2\}$ is not a subspace of V. This is
because $(1, 0, \ldots, 0) \in \mathbb{R}^n$, and $1^2 + 0^2 + \cdots + 0^2 = 1 \leq 2$, which means $(1, 0, \ldots, 0)$ is in
W_2 but $2(1, 0, \ldots, 0) = (2, 0, \ldots, 0)$ does not belong to W_2 since $2^2 + 0^2 + \cdots + 0^2 = 4 \geq 2$.

Example 2.2.9. Let $P_{20}(x)$ denote the vector space consisting of all polynomials with
degrees at most 20 and real coefficients over the real field. Now, consider the subset
$W = \{p(x) \in P_{20}(x) : p(1) = 0, p(1/2) = 0, p(5) = 0, p(7) = 0\}$. This subset W is a subspace
of $P_{20}(x)$.

For any polynomials $p(x), q(x) \in W$ and scalars $a, b \in \mathbb{R}$, let $h(x) = ap(x) + bq(x)$.
It can be observed that:

$$h(1) = ap(1) + bq(1) = 0,$$
$$h(1/2) = ap(1/2) + bq(1/2) = 0,$$
$$h(5) = ap(5) + bq(5) = 0, \quad \text{and}$$
$$h(7) = ap(7) + bq(7) = 0.$$

Thus, $h(x) = ap(x) + bq(x) \in W$. This proves that W is a subspace of $P_{20}(x)$.

The following definition outlines a method for constructing a subspace from a given
subset of a vector space.

Definition 2.2.10. Let V be a vector space over a field F and let S be a nonempty subset
of V. Then the subspace generated by S, denoted as $\langle S \rangle$, is defined as the intersection
of all subspaces of V that contain S. Formally, if $W_i : i \in I$ is a family of subspaces of V
containing S, then $\langle S \rangle = \bigcap_{i \in I} W_i$.

This definition relies on the fact that the intersection of subspaces of a vector space
remains a subspace, and the assurance that V itself is a subspace containing S.

Definition 2.2.11. A linear combination of vectors $x_1, x_2, \ldots, x_n$ in a vector space V over
a field F is an expression of the form $a_1 x_1 + a_2 x_2 + \cdots + a_n x_n$, where $a_1, a_2, \ldots, a_n \in F$.
More generally, for a subset S of V, not necessarily finite, an element $x \in V$ is called
a linear combination of elements of S if there exist finitely many $x_1, x_2, \ldots, x_n \in S$ such
that $x = a_1 x_1 + a_2 x_2 + \cdots + a_n x_n$, where $a_1, a_2, \ldots, a_n \in F$.

Theorem 2.2.12. *In a vector space V over a field F, if S is a nonempty subset of V, then* $\langle S \rangle$*, the subspace generated by S, consists precisely of all linear combinations of elements of S.*

Proof. Let us denote the set of all linear combinations of S by W. For any $\alpha, \beta \in F$ and $a_1 x_1 + a_2 x_2 + \cdots + a_n x_n, b_1 y_1 + b_2 y_2 + \cdots + b_m y_m \in W$,

$$\alpha(a_1 x_1 + a_2 x_2 + \cdots + a_n x_n) + \beta(b_1 y_1 + b_2 y_2 + \cdots + b_m y_m)$$
$$= \alpha a_1 x_1 + \alpha a_2 x_2 + \cdots + \alpha a_n x_n + \beta b_1 y_1 + \beta b_2 y_2 + \cdots + \beta b_m y_m$$

is a linear combination of elements of S, and hence belongs to W. This proves that W is a subspace of V.

Now, let U be a subspace of V containing S. It follows that all linear combinations of elements of S are contained in U. Therefore, $W \subseteq U$, indicating that W is the smallest subspace containing S. Thus, $W = \langle S \rangle$. □

Note that if S is a finite subset of V such that $V = \langle S \rangle$, then V is said to be finitely generated.

Example 2.2.13. Let V be a vector space over a field F. Then

$$\text{for } S = \{x\} \subset V, \langle S \rangle = \{\alpha x : \alpha \in F\} \quad \text{and}$$
$$\text{for } S = \{s, y\} \subset V, \langle S \rangle = \{\alpha x + \beta y : \alpha, \beta \in F\}.$$

In particular, let $V = \mathbb{R}^3$ and $x = (1, 0, 0)$. Then $\langle \{x\} \rangle = \langle \{(1, 0, 0)\} \rangle = \{\alpha(1, 0, 0) : \alpha \in \mathbb{R}\} = \{(\alpha, 0, 0) : \alpha \in \mathbb{R}\}$ is a subspace of $\mathbb{R}^3$.

If $V = Z_5^2$, a vector space over Z_5, then for $(\bar{2}, \bar{3}) \in Z_5^2$,

$$\langle \{(\bar{2}, \bar{3})\} \rangle = \{\bar{0}(\bar{2}, \bar{3}), \bar{1}(\bar{2}, \bar{3}), \bar{2}(\bar{2}, \bar{3}), \bar{3}(\bar{2}, \bar{3}), \bar{4}(\bar{2}, \bar{3})\}$$
$$= \{(\bar{0}, \bar{0}), (\bar{2}, \bar{3}), (\bar{4}, \bar{1}), (\bar{1}, \bar{4}), (\bar{3}, \bar{2})\}.$$

Example 2.2.14. Let $P(x)$ be the vector space of polynomials over a field F, and let $S = \{1, x, x^2, x^3, \ldots\}$ be an infinite subset of $P(x)$.

Then $\langle S \rangle = \{a_0 + a_1 x + a_2 x^2 + \cdots + a_n x^n : a_0, a_1, \ldots, a_n \in F\} = P(x)$. Thus, the infinite set $\{1, x, x^2, \ldots\}$, generates or spans the vector space of polynomials.

ⓘ Exercises

(2.2.1) Prove that the subsets W_1, W_2, W_3 and W_4 given in Example 2.2.6 are subspaces of the vector space V.

(2.2.2) Let $P(x)$ be the vector space of polynomials with coefficients in the field F. Then show that the following subsets of $P(x)$:

(a) $W_1 = \{f(x) \in P(x) : f(x) \text{ is an even function}\}$,

(b) $W_2 = \{f(x) \in P(x) : f(0) = f(1)\}$,

are subspaces of $P(x)$.

(2.2.3) Show that $W = \{(x,y,z) \in \mathbb{R}^3 : 2x + 3y - 6z = 2\}$ is not a subspace of $\mathbb{R}^3$.

(2.2.4) Show that $P_5(x)$, the set of all polynomials of degree less than or equal to 5 with real coefficients is a vector space over the field $\mathbb{R}$. If $W = \{f(x) \in P_5(x) : f(1) \in \mathbb{Q}\}$, then show that W is not a subspace of $P_5(x)$.

(2.2.5) Let $M_{3,3}(\mathbb{R})$ be the vector space of all 3×3 real matrices over $\mathbb{R}$. Then prove that $W_1 = \{A \in M_{3,3}(\mathbb{R}) : \det(A) \neq 0\}$ and $W_2 = \{A \in M_{3,3}(\mathbb{R}) : \det(A) = 0\}$ are not subspaces of $M_3(\mathbb{R})$.

(2.2.6) Let $V = \mathbb{R}^{[0,1]}$. Then show that $W = \{f(x) \in V : f(x) = f(1-x), \text{for } x \in [0,1]\}$ is a subspace of V.

(2.2.7) Let $P_{15}(x)$ be the vector space of polynomials of degree less than or equal to 15 over the field $\mathbb{R}$. Then show that $W = \{f(x) \in P_{15}(x) : f(1/2) = f(3) = f(4) = f(7) = 0\}$ is a subspace of $P_{15}(x)$.

(2.2.8) Let Z_5^2 be the vector space over the field Z_5. Then find the subspace generated by the set $\{(\bar{1}, \bar{2}), (\bar{3}, \bar{4})\}$.

(2.2.9) Show that the set $W = \{f : \lim_{x \to 5} f(x) = 0\}$ is a subspace of the vector space $\mathbb{R}^{\mathbb{R}}$.

(2.2.10) Let $M_{n,n}(\mathbb{R})$ denote the vector space of all n-square real matrices. Then show that the following sets are subspaces of $M_{n,n}(\mathbb{R})$: (i) $W_1 = \{A \in M_{n,n}(\mathbb{R}) : \text{trace}(A) = 0\}$ and (ii) $W_2 = \{BA : A \in M_{n,n}(\mathbb{R})\}$, where B is a fixed matrix in $M_{n,n}(\mathbb{R})$.

2.3 Linear dependence and linear independence

The concepts of linear dependence and linear independence play an important role in the theory of linear algebra.

Definition 2.3.1. Let V be a vector space over a field F. Then the set $\{x_1, x_2, \ldots, x_n\}$ of vectors in V is said to be linearly dependent if there exist scalars $a_1, a_2, \ldots, a_n \in F$, not all zero, such that $a_1 x_1 + a_2 x_2 + \cdots + a_n x_n = 0$.

Conversely, the set $\{x_1, x_2, \ldots, x_n\}$ is considered linearly independent if it is not linearly dependent. In other words, if $a_1 x_1 + a_2 x_2 + \cdots + a_n x_n = 0$, then each $a_i = 0$ for $1 \leq i \leq n$.

More generally, a subset S of V, not necessarily finite is said to be linearly dependent (linearly independent) if every finite subset of S is linearly dependent (linearly independent).

Example 2.3.2. Let V be a vector space over a field F. Then we have the following illustrations:

(a) A set of vectors containing zero vector of V is linearly dependent.
    Let $S = \{0, x_1, x_2, \ldots, x_n\} \subseteq V$. Then $k0 + 0x_1 + 0x_2 + \cdots + 0x_n = 0$, for all $k \in F$. This implies that there exists a nonzero scalar k such that $k0 + 0x_1 + 0x_2 + \cdots + 0x_n = 0$.

(b) It is evident from the definition that: (i) every subset of a linearly independent set remains linearly independent, and (ii) a set containing a linearly dependent subset is itself linearly dependent.

(c) Let $S = \{x_1, x_2, \ldots, x_n\}$ be a subset of V such that x_1 is the linear combination of remaining vectors. Then S is linearly dependent.
    Since $x_1 = a_2 x_2 + \cdots + a_n x_n$, we have $x_1 - a_2 x_2 - a_3 x_3 - \cdots - a_n x_n = 0$. This gives that at least one coefficient which is 1, the coefficient of x_1 is nonzero.

Example 2.3.3. Let $\mathbb{R}^3$ be a vector space over $\mathbb{R}$. Then the set $\{e_1 = (1,0,0), e_2 = (0,1,0), e_3 = (0,0,1)\}$ in $\mathbb{R}^3$ is linearly independent.

Suppose $a_1 e_1 + a_2 e_2 + a_3 e_3 = (0,0,0)$ where $a_1, a_2, a_3 \in \mathbb{R}$. This implies

$$a_1(1,0,0) + a_2(0,1,0) + a_3(0,0,1) = (0,0,0)$$
$$\Rightarrow (a_1, a_2, a_3) = (0,0,0)$$
$$\Rightarrow a_1 = a_2 = a_3 = 0.$$

More generally, in the vector space F^n over a field F the set $\{e_1, e_2, \ldots, e_n\}$ is linearly independent, where $e_i = (0,0,\ldots 0,1,0,\ldots,0) \in F^n$, with each entry 0 except ith entry which is 1.

Example 2.3.4. Let V be a vector space over a field F and let $\{x_1, x_2, \ldots, x_n\}$ be a linearly independent set in V. Then the set $x_1 + x_2, x_2 + x_3, \ldots, x_{n-1} + x_n, x_n$ is also linearly independent.

Suppose we have scalars $a_1, a_2, \ldots, a_n \in F$ such that

$$a_1(x_1 + x_2) + a_2(x_2 + x_3) + \cdots + a_{n-1}(x_{n-1} + x_n) + a_n x_n = 0.$$

This simplifies to

$$a_1 x_1 + (a_1 + a_2)x_2 + (a_2 + a_3)x_3 + \cdots + (a_n - 1 + a_n)x_n = 0.$$

Since $x_1, x_2, \ldots, x_n$ are linearly independent, we conclude that

$$a_1 = 0, \quad a_1 + a_2 = 0, \quad a_2 + a_3 = 0, \quad \ldots, \quad a_{n-1} + a_n = 0.$$

Solving these equations yields $a_1 = a_2 = \cdots = a_n = 0$. Thus, the set $x_1 + x_2, x_2 + x_3, \ldots, x_{n-1} + x_n, x_n$ is linearly independent.

In Example 2.3.3, it is shown that the vectors $e_1 = (1,0,0)$, $e_2 = (0,1,0)$ and $e_3 = (0,0,1)$ in $\mathbb{R}^3$ are linearly independent. Consequently, the set $\{e_1 + e_2, e_2 + e_3, e_3\} = \{(1,1,0), (0,1,1), (0,0,1)\}$ is also linearly independent.

Example 2.3.5. The set of vectors $\{(2,1,1), (1,-1,1), (3,0,2)\}$ in R^3 is linearly dependent.

For $a_1, a_2, a_3 \in \mathbb{R}$, let $a_1(2,1,1) + a_2(1,-1,1) + a_3(3,0,2) = (0,0,0)$. Then

$$(2a_1 + a_2 + 3a_3, a_1 - a_2, a_1 + a_2 + 2a_3) = (0,0,0)$$
$$\Rightarrow 2a_1 + a_2 + 3a_3 = 0, a_1 - a_2 = 0 \text{ and } a_1 + a_2 + 2a_3 = 0.$$

Solving these equations, we get infinitely many nonzero solutions. As the solution $a_1 = a_2 = 1$ and $a_3 = -1$ is one of them. Hence, the given set is linearly dependent.

Remark. The task of determining linearly dependent and linearly independent sets can be efficiently addressed using the matrix representation of systems of linear equations. We will explore this approach later on.

Theorem 2.3.6. *Let V be a vector space over a field F. Then the set $\{x,y\} \subseteq V$ is a linearly dependent if and only if one vector is a scalar multiple of the other.*

Proof. Suppose the set $\{x,y\} \subseteq V$ is a linearly dependent.Then there exist a and b in F, not both zero, such that $ax + by = 0$. If $a \neq 0$, then $x = a^{-1}(-by) = -a^{-1}by = ky$, where $k = -a^{-1}b \in F$.

Conversely, suppose x is scalar multiple of y, meaning there exists $k \in F$ such that $x = ky$. Then $x - ky = 0$. Hence, the set x, y is linearly dependent. □

In particular, the set $\{(1,-1,3),(1/3,-1/3,1)\}$ in $\mathbb{R}^3$ is linearly dependent because $(1,-1,3) = 3(1/3,-1/3,1)$.

Example 2.3.7. Let Z_3^2 be the vector space over the field Z_3. Then the number of distinct linearly dependent sets of the form $\{x,y\}$, where $x,y \in Z_3^2 - \{(\bar{0},\bar{0})\}$ and $x \neq y$ is 4.

$$Z_3^2 = Z_3 \times Z_3, \quad \text{where } Z_3 = \{\bar{0},\bar{1},\bar{2}\}.$$
$$Z_3^2 = \{(\bar{0},\bar{0}),(\bar{0},\bar{1}),(\bar{0},\bar{2}),(\bar{1},\bar{0}),(\bar{1},\bar{1}),(\bar{1},\bar{2}),(\bar{2},\bar{0}),(\bar{2},\bar{1}),(\bar{2},\bar{2})\}.$$

Since $\bar{1}(\bar{0},\bar{1}) = (\bar{0},\bar{1})$ and $\bar{2}(\bar{0},\bar{1}) = (\bar{0},\bar{2})$, the set $\{(\bar{0},\bar{1}),(\bar{0},\bar{2})\}$ is linearly dependent and both elements are distinct and nonzero.

Similarly, $\{(\bar{1},\bar{0}),(\bar{2},\bar{0})\}$, $\{(\bar{1},\bar{1}),(\bar{2},\bar{2})\}$ and $\{(\bar{1},\bar{2}),(\bar{2},\bar{1})\}$ are linearly dependent sets.

Exercises

(2.3.1) Let V be a vector space over a field F and let the set of vectors $\{x_1,x_2,\ldots,x_n\}$ in V are linearly independent. Then show that the set $\{x_1+x_2, x_2+x_3, x_3+x_4, \ldots, x_{n-1}+x_n, x_n+x_1\}$ is (i) linearly independent if n is odd and (ii) linearly dependent if n is even.

(2.3.2) Let $C(\mathbb{R}^\mathbb{R})$ be the vector space of all continuous functions over the field $\mathbb{R}$. Then show that the set $\{f_0,f_1,f_2,\ldots,f_n,\ldots\}$ is linearly independent in $C(\mathbb{R}^\mathbb{R})$, where $f_n(x) = x^n$ for $n = 0,1,2,\ldots$.

(2.3.3) Let V be a vector space over a field F. Then prove that the set $S = \{x_1,x_2,\ldots,x_n\}$ in V is linearly dependent if and only if one vector of S is a linear combination of remaining vectors of S.

(2.3.4) In the vector space $C(\mathbb{R}^{[0,1]})$ over $\mathbb{R}$, show that the set $\{x, e^x, e^{-x}\}$ is linearly independent.

(2.3.5) Show that the set $\{1 + 2x + x^2, 1 + x + 2x^2, 2 + x + x^2\}$ in polynomial space $P_3(x)$ over real field is linearly independent.

(2.3.6) Show that the following set of vectors in $\mathbb{R}^3$ are linearly independent:
   (i) $\{(1, 2, -2), (-1, 3, 0), (0, -2, 1)\}$;
   (ii) $\{(1, 6, 4), (0, 2, 3), (0, 1, 2)\}$.

(2.3.7) Show that the following set of vectors in $\mathbb{R}^3$ are linearly dependent:
   (i) $\{(2, 2, -3), (3, 1, -4), (0, -4, 1)\}$;
   (ii) $\{(1, 6, 4), (0, -2, 1), (1/2, 9, -1)\}$.

2.4 Basis and dimension of a vector space

In this section, we introduce the fundamental concepts of basis and dimension within a vector space, both of which are pivotal in the realm of linear algebra. Initially, we present the following definition of a basis for a vector space, and subsequently, we will elucidate additional equivalent definitions.

Definition 2.4.1. Let V be a vector space over a field F. Then a subset B of V is called a basis of V if it satisfies the following conditions:
(a) B is linearly independent.
(b) B spans V, meaning that $\langle B \rangle = V$.

Example 2.4.2. Consider the set $\{e_1, e_2, e_3\}$ within the vector space $\mathbb{R}^3$ over $\mathbb{R}$. This set serves as a basis for the vector space $\mathbb{R}^3$.

In Example 2.3.3, it is shown that the set $\{e_1 = (1, 0, 0), e_2 = (0, 1, 0), e_3 = (0, 0, 1)\}$ is linearly independent in $\mathbb{R}^3$.

Now, let (x, y, z) be an arbitrary element in $\mathbb{R}^3$. Then it can be expressed as

$$(x, y, z) = x(1, 0, 0) + y(0, 1, 0) + z(0, 0, 1),$$

or equivalently, $(x, y, z) = xe_1 + ye_2 + ze_3$.

Thus, $\mathbb{R}^3$ is generated by the set $\{e_1, e_2, e_3\}$.

In general, if F^n denotes a vector space over a field F, then the set $\{e_1, e_2, \ldots, e_n\}$ forms a basis for F^n, where $e_1 = (1, 0, 0, \ldots, 0)$, $e_2 = (0, 1, 0, \ldots, 0)$, $e_3 = (0, 0, 1, 0, \ldots, 0)$, $\ldots, e_n = (0, 0, \ldots, 1)$ are elements of F^n. This basis $\{e_1, e_2, \ldots, e_n\}$ is commonly referred to as the standard basis of F^n.

Example 2.4.3. The collection $B = \{1, x, x^2, x^3, \ldots, x^n\}$ forms a basis for the vector space $P_n(x)$, and similarly, the infinite set $\{1, x, x^2, x^3, \ldots, x^n, \ldots\}$ serves as a basis for the vector space of polynomials $P(x)$.

Example 2.4.4. The set $B = \{(2, 1, 1), (1, -1, 1), (3, 0, 2)\}$ does not constitute a basis for $\mathbb{R}^3$. This conclusion follows from the fact shown in Example 2.3.5 that the set B is linearly dependent.

Definition 2.4.5. A subset B of a vector space V is called a maximal linearly independent set if it is linearly independent and if for any element $x \in V$ not already in B, adding x to B results in a linearly dependent set.

Theorem 2.4.6. *Every maximal linearly independent set in a vector space is a basis, and conversely, every basis is a maximal linearly independent set.*

Proof. Let V be a vector space over a field F, and let B be a maximal linearly independent set in V. Since B is linearly independent, it suffices to prove that $V = \langle B \rangle$.

Consider a nonzero element $x \in V$. If $x \in B$, then there is nothing further to demonstrate. Assume $x \notin B$. By the definition of a maximal linearly independent set, $B \cup \{x\}$ is linearly dependent. Thus, there exist $a_0, a_1, a_2, \ldots, a_n$, not all zero in F and $x_1, x_2, \ldots, x_n$ in B such that

$$a_0 x + a_1 x_1 + a_2 x_2 + \cdots + a_n x_n = 0.$$

Since $a_0 \neq 0$, otherwise, some $a_i \neq 0$ for $1 \leq i \leq n$, implying that $a_1 x_1 + a_2 x_2 + \cdots + a_n x_n = 0$ and so B is linearly dependent, a contradiction. Hence,

$$a_0 x = -a_1 x_1 - a_2 x_2, \ldots, -a_n x_n,$$
$$x = -a_0^{-1} a_1 x_1 - a_0^{-1} a_2 x_2 -, \ldots, -a_0^{-1} a_n x_n.$$

Thus, x is a linear combination of elements of B, and so $V = \langle B \rangle$.

Conversely, suppose B is a basis of V. Then B is linearly independent, and $\langle B \rangle = V$. If B is not maximal, there exists an element $0 \neq x \in V \setminus B$ such that the set $B \cup \{x\}$ is linearly independent. However, $x \in \langle B \rangle = V$ implies that x is a linear combination of finitely many elements of B. Therefore, $B \cup \{x\}$ must be linearly dependent. Thus, B is a maximal linearly independent set. □

Definition 2.4.7. An element x in a vector space V is called uniquely represented as a linear combination of elements of a subset B of V, when expressed as $x = a_1 x_1 + a_2 x_2 + \cdots + a_m x_m = b_1 y_1 + b_2 y_2 + \cdots + b_n y_n$, where $a_i \neq 0$, $b_j \neq 0$ and $x_i : 1 \leq i \leq m, y_j : 1 \leq j \leq n$ are sets of distinct elements in B, the following conditions hold:
(i) $m = n$ and $x_i = y_i$ for all i;
(ii) after some rearrangement, $a_i = b_i$ for all i.

Theorem 2.4.8. *Let V be a vector space over a field F. Then a subset B of V is a basis of V if and only if every element of V is uniquely represented by a linear combination of elements of B.*

Proof. Let B be a basis of V. Then $\langle B \rangle = V$, and B is a linearly independent set in V. Suppose for $x \in V$,

$$x = a_1 x_1 + a_2 x_2 + \cdots + a_m x_m = b_1 y_1 + b_2 y_2 + \cdots + b_n y_n,$$

where $a_i \neq 0$, $b_j \neq 0$ and $\{x_i : 1 \leq i \leq m\}$, $\{y_j : 1 \leq j \leq n\}$ are sets of distinct elements in B. Then

$$a_1 x_1 + a_2 x_2 + \cdots + a_m x_m - b_1 y_1 - b_2 y_2 - \cdots - b_n y_n = 0.$$

If the sets $\{x_i : 1 \leq i \leq m\}$ and $\{y_j : 1 \leq j \leq n\}$ are disjoint, then all x_i's and y_j's in B are distinct. Since all a_i's and b_j's are nonzero in the above equation, it is a contradiction that the set $\{x_1, x_2, \ldots, x_m, y_1, y_2, \ldots, y_n\}$ in B is linearly independent or B is linearly independent. Hence, there must be some overlap between the two sets.

Let $x_1 = y_1, x_2 = y_2, \ldots, x_r = y_r$, for some r. Then $\{x_1, x_2, \ldots, x_m\} \cap \{y_1, y_2, \ldots, y_n\} = \{x_1, x_2, \ldots, x_r\}$.

Now we prove that $m = n = r$. Suppose on contrary, $r < m$ or $r < n$. Then from the above equation we have

$$a_1x_1 + a_2x_2 + \cdots + a_rx_r + a_{r+1}x_{r+1} + \cdots + a_mx_m - b_1x_1 - b_2x_2 - \cdots$$
$$- b_rx_r - b_{r+1}y_{r+1} - \cdots - b_ny_n = 0,$$
$$(a_1 - b_1)x_1 + \cdots + (a_r - b_r)x_r + a_{r+1}x_{r+1} + \cdots + a_mx_m - b_{r+1}y_{r+1} - \cdots - b_ny_n = 0.$$

This is again a contradiction that B is a linearly independent set. Hence, $m = n = r$. This gives that $x_i = y_i$ for each i and

$$(a_1 - b_1)x_1 + (a_2 - b_2)x_2 + \cdots + (a_m - b_m)x_m = 0.$$

Since B is linearly independent,

$$a_1 - b_1 = 0, a_2 - b_2 = 0, \ldots, (a_m - b_m) = 0 \Rightarrow a_i = b_i \quad \forall i.$$

Conversely, suppose that every element of V is uniquely represented by a linear combination of elements of B. Then to prove that B is a basis of V, it suffices to show that B is linearly independent. We prove it by contradiction. Suppose $\{x_1, x_2, \ldots, x_n\}$ is linearly dependent in B. Then there exist $a_1, a_2, \ldots, a_n$ not all zero in F such that $a_1x_1 + a_2x_2 + \cdots + a_nx_n = 0$.

Also, $0 = 0x_1 + 0x_2 + \cdots + 0x_n$. By unique representation property, we have $a_1 = a_2 = \cdots = a_n = 0$, which is a contradiction that $\{x_1, x_2, \ldots, x_n\}$ are linearly dependent. Hence, B is linearly independent. □

Theorem 2.4.9. *Let V be a vector space over a field F. Then a subset B of V is a basis of V if and only if B is a minimal set of generators.*

Proof. Let B be a basis of the vector space V. Then $\langle B \rangle = V$ and B is linearly independent. Now, we prove that B is a minimal set of generators of V.

Suppose there exists a proper subset B' of B such that $\langle B' \rangle = V$. Then for $x \in B \setminus B'$, there exist $x_1, x_2, \ldots, x_n$ in B' and $a_1, a_2, \ldots, a_n$ in F such that $x = a_1x_1 + a_2x_2 + \cdots + a_nx_n$.

This implies that the set $\{x, x_1, x_2, \ldots, x_n\}$ in B is linearly dependent. However, this contradicts the fact that B is a basis. Therefore, there is no proper subset of B generating V. This proves that B is a minimal set of generators of V.

Conversely, suppose that B is a minimal set of generators of V. Then $\langle B \rangle = V$. We still need to prove that B is linearly independent. If B is not linearly independent, then there exist distinct elements $x_1, x_2, \ldots, x_n$ in B and scalars $a_1, a_2, \ldots, a_n$, not all zero in F, such that $a_1x_1 + a_2x_2 + \cdots + a_nx_n = 0$. Let us assume $a_1 \neq 0$. Then $x_1 = a_1^{-1}(-a_2x_2 - \cdots - a_nx_n) = -a_1^{-1}a_2x_2 - \cdots - a_1^{-1}a_nx_n$.

This implies that x_1 belongs to $\langle B-\{x_1\} \rangle$. Hence, $B \subseteq \langle B-\{x_1\} \rangle$. Since $\langle B \rangle = V$, we have $\langle B - \{x_1\} \rangle = V$. This contradicts the fact that B is a minimal set of generators. Therefore, B must be linearly independent. □

Considering the implications of the above three theorems, we can regard the following statements as a definition of a basis for a vector space.

Definition 2.4.10. In a vector space V over a field F, a subset B is considered a basis of V if it satisfies any of the following conditions:

B is a maximal linearly independent set in V

OR

every element of V can be uniquely expressed as a linear combination of elements from B

OR

B is a minimal set of generators for V.

Lemma 2.4.11. *Suppose V is a vector space over a field F, and let $S = \{x_1, x_2, \ldots, x_m\}$ be a linearly dependent set of nonzero vectors in V. Then there exists a vector x_i in S, where $1 < i \le m$ that can be expressed as a linear combination of preceding vectors.*

Proof. We construct the chain $S_1 \subset S_2 \subset S_3 \subset \cdots \subset S_m = S$, where $S_i = \{x_1, x_2, \ldots, x_i\}$ for $1 < i \le m$. Since $x_1 \ne 0$, $S_1 = \{x_1\}$ is a linearly independent set. As $S_m = S$ is linearly dependent, there must exist a smallest index i, where $2 \le i \le m$, such that S_i is linearly dependent. Therefore, there exist scalars $a_1, a_2, \ldots, a_i$ in F, not all zero, such that $a_1 x_1 + a_2 x_2 + \cdots + a_i x_i = 0$.

Now, $a_i \ne 0$, otherwise some $a_j \ne 0$, $1 \le j < i$ and then $a_1 x_1 + a_2 x_2 + \cdots + a_{i-1} x_{i-1} = 0$ imply S_{i-1} linearly dependent, which contradicts our assumption that S_i is the first linearly dependent set in the chain. Hence, $a_i x_i = -a_1 x_1 - a_2 x_2 - \cdots - a_{i-1} x_{i-1}$, and so

$$x_i = -a_i^{-1} a_1 x_1 - a_i^{-1} a_2 x_2 - \cdots - a_i^{-1} a_{i-1} x_{i-1}.$$

Therefore, x_i can be expressed as a linear combination of preceding vectors $x_1, x_2, \ldots, x_{i-1}$. □

Theorem 2.4.12. *Let $A = \{x_1, x_2, \ldots, x_m\}$ be a set of generators of a vector space V over a field F. If $B = \{y_1, y_2, \ldots, y_n\}$ is a linearly independent set in V, then $n \le m$.*

Proof. Without loss of generality, we assume that $x_i \in A$, where $1 \le i \le m$ is nonzero. Since $V = \langle A \rangle$, $y_1 \in B \subset V$ can be expressed as a linear combination of elements of A. Thus, the set $\{y_1, x_1, x_2, \ldots, x_m\}$ is linearly dependent and spans V.

By Lemma 2.4.11, there exists x_i in the above set such that x_i can be expressed as a linear combination of preceding vectors. Since y_1 belongs to a linearly independent set, $x_i \ne y_1$. After removing x_i from the above set, let $A_1 = \{y_1, x_1, x_2, \ldots, x_{i-1}, x_{i+1}, \ldots, x_m\}$. Here, A_1 remains a generating set. By repeating this process on A_1 and B, we obtain

$$A_2 = \{y_1, y_2, x_1, \ldots, x_{i-1}, x_{i+1}, \ldots, x_{j-1}, x_{j+1}, \ldots, x_m\}.$$

In this manner, at each step, we add one member of B and delete one member of A. If $n \le m$, then we eventually obtain a generating set:

$$A_n = \{y_1, y_2, \ldots, y_n, x_{r_1}, x_{r_2}, \ldots, x_{r_{m-n}}\}, \quad \text{where } \{x_{r_1}, x_{r_2}, \ldots, x_{r_{m-n}}\} \subset A,$$

and this completes the proof.

Now, we claim that $n > m$ is not possible. If $n > m$, then after m steps we obtain $A_m = \{y_1, y_2, \ldots, y_m\}$, which is a spanning set. This implies that y_{m+1} can be expressed as a linear combination of vectors $y_1, y_2, \ldots, y_m$. However, this contradicts the assumption that the set B is linearly independent. □

Theorem 2.4.13. *Every finitely generated vector space contains a basis.*

Proof. Consider a vector space V over a field F and let S denote the set of generators of V. If $S = \{0\}$ or $S = \emptyset$, then $V = \{0\}$, and ϕ serves as a basis for V.

Let $S = \{x_1, x_2, \ldots, x_m\}$ be a set of nonzero vectors in V such that $\langle S \rangle = V$. If S is linearly dependent, then according to Lemma 2.4.11, there exists $x_i \in S$ such that x_i can be expressed as a linear combination of preceding vectors. If we remove x_i from S, the set $S - \{x_i\}$ still remains a generating set, i. e., $\langle S - \{x_i\} \rangle = V$.

If $S - \{x_i\}$ is linearly independent, then it constitutes a basis for V. If $S - \{x_i\}$ is linearly dependent, we can remove another element, x_j, from $S - \{x_i\}$, yielding $S - \{x_i, x_j\}$. If $S - \{x_i, x_j\}$ is linearly independent, then it forms a basis for V.

If $S - \{x_i, x_j\}$ is linearly dependent, we continue this process iteratively. After a finite number of attempts, we will eventually obtain a basis for V. Since S is finite, the number of elements in the basis is less than or equal to the number of elements in the generating set. □

Theorem 2.4.14. *In a finitely generated vector space V, any two bases of V have the same number of elements.*

Proof. Let $B_1 = \{x_1, x_2, \ldots, x_m\}$ and $B_2 = \{y_1, y_2, \ldots, y_n\}$ be two bases of a vector space V. Since $\langle B_1 \rangle = V$ and B_2 is a linearly independent, it follows that $n \leq m$. Similarly, $\langle B_2 \rangle = V$ and B_1 is a linearly independent imply $m \leq n$. Consequently, we conclude that $m = n$. □

As we have proved that the number of elements in each basis of a finitely generated vector space V is invariant, we have the following definition.

Definition 2.4.15. The number of elements in a basis of a finitely generated vector space is called the dimension of the vector space. A vector space that is not finitely generated is called infinite-dimensional. We represent the dimension of a vector space V over F as $\dim(V)$.

Theorem 2.4.16 (Extension theorem). *Let $S = \{x_1, x_2, \ldots, x_m\}$ be a linearly independent set in a finite-dimensional vector space V over a field F. Then either S is a basis or can be extended to be a basis of V.*

Proof. Given S is a linearly independent set in V, if $\langle S \rangle = V$, then S already constitutes a basis.

Let the dimension of V be n, and let $B = \{y_1, y_2, \ldots, y_n\}$ be its basis. Consequently, $\langle B \rangle = V$, implying $\langle S \cup B \rangle = V$. According to Lemma 2.4.11, we can eliminate each vector from $S \cup B = \{x_1, x_2, \ldots, x_m, y_1, y_2, \ldots, y_n\}$, which is a linear combination of preceding vectors, to obtain a maximal linearly independent set M in V. Since S is linearly independent, none of its elements will be discarded during this process. Therefore, M contains every vector from S, signifying that S is extended to form a basis M of V. □

Theorem 2.4.17. *If the dimension of a vector space is n, the following assertions hold:*
(a) *If a set A contains more than n vectors in V, then it is linearly dependent.*
(b) *A subset B of V containing n vectors serves as a basis of V if and only if B is linearly independent.*

Proof. (a) Given that V is a vector space with dimension n, a maximal linearly independent set within V contains exactly n elements. Thus, any set A containing more than n vectors must be linearly dependent.
(b) If B constitutes a basis, then it is a linearly independent set. Conversely, if B is a linearly independent set with n elements, then being a maximal linearly independent set, B serves as a basis for V. □

Example 2.4.18. In the vector space F^n over the field F, the set $S = \{e_1, e_2, \ldots, e_n\}$ is the standard basis of F^n. Consequently, F^n is an n-dimensional vector space. In particular, $\mathbb{Q}^n, \mathbb{R}^n, \mathbb{C}^n, Z_p^n$ are n-dimensional vector spaces.

The vector spaces $\mathbb{Q}^n, \mathbb{R}^n$ and $\mathbb{C}^n$ are finite-dimensional, despite having an infinite number of vectors, where as in the vector space Z_p^n the number of elements are p^n.

Example 2.4.19. Let V is an n-dimensional vector space over a finite field F containing p elements. Then there exists a basis $\{x_1, x_2, \ldots, x_n\}$ of V and so every element $x \in V$ can be uniquely expressed as $x = a_1 x_1 + a_2 x_2 + \cdots + a_n x_n$, where $a_1, a_2, \ldots, a_n$ are member of F. Thus, for every n-tuple $(a_1, a_2, \ldots, a_n) \in F^n$, there corresponds a unique $x \in V$ and vice versa. Consequently, the number of elements in V equals the number of elements in Z_p^n, which is p^n.

Therefore, the number of elements in a finite-dimensional vector space over a finite field amounts to p^n, where p represents a prime and n denotes the dimension of the vector space.

In particular, there does not exist a vector space with 51 elements since $p^n = 51$ is not possible for any prime p.

Example 2.4.20.
(a) Every field F is a vector space over itself. The singleton set $\{1\}$, where 1 is the identity of the field is the basis of the vector space F. Hence, $\dim(F) = 1$.
(b) The set $\{1, i\}$ is a basis of the vector space of complex numbers over the field $\mathbb{R}$. Since $x + iy \in \mathbb{C}$, can be written as $x.1 + y.i$, where $x, y \in \mathbb{R}$. Hence, the dimension of complex vector space over $\mathbb{R}$ is 2.

(c) Vector space $\mathbb{R}$ over the field $\mathbb{Q}$ is infinite-dimensional. Since $\mathbb{Q}$ is countable and the vector space generated by a finite subset of $\mathbb{R}$ over $\mathbb{Q}$ must be countable but $\mathbb{R}$ is uncountable.

(d) $\dim(P_n(x)) = n + 1$ and $\dim(P(x))$ is infinite.

Example 2.4.21. Let $W = \{(x_1, x_2, x_3) \in \mathbb{R}^3 : x_1 + x_2 + x_3 = 0\}$ be a subset of the vector space $\mathbb{R}^3$ over $\mathbb{R}$. Then $W = \{(x_1, x_2, -x_1 - x_2) : x_1, x_2 \in \mathbb{R}\}$.

Now, observe that

$$\begin{aligned}
(x_1, x_2, -x_1 - x_2) &= x_1(1, 0, 0) + x_2(0, 1, 0) + (-x_1 - x_2)(0, 0, 1) \\
&= x_1(1, 0, 0) + x_2(0, 1, 0) + x_1(0, 0, -1) + x_2(0, 0, -1) \\
&= x_1(1, 0, -1) + x_2(0, 1, -1).
\end{aligned}$$

This demonstrates that W is generated by $(1, 0, -1)$ and $(0, 1, -1)$.
For $a_1, a_2 \in \mathbb{R}$, consider

$$\begin{aligned}
a_1(1, 0, -1) + a_2(0, 1, -1) &= (0, 0, 0), \\
(a_1, 0, -a_1) + (0, a_2, -a_2) &= (0, 0, 0), \\
(a_1, a_2, -a_1 - a_2) &= (0, 0, 0), \\
a_1 = 0, \quad a_2 &= 0.
\end{aligned}$$

Thus, the set $\{(1, 0, -1), (0, 1, -1)\}$ is linearly independent. Hence,

$$\dim(W) = 2.$$

More generally, the dimension of the subspace,

$$W = \{(x_1, x_2, \ldots, x_n) \in \mathbb{R}^n : x_1 + x_2 + \cdots + x_n = 0\} \text{ is } n - 1,$$

where the set $\{(1, 0, 0, \ldots, 0, -1), (0, 1, 0, \ldots, 0, -1), \ldots, (0, 0, \ldots, 1, -1)\}$ forms a basis of W.

Example 2.4.22. Let $W = \{(x_1, x_2, x_3, x_4, x_5) : 3x_1 - x_2 + x_3 = 0\}$ be a subspace of $\mathbb{R}^5$. Then, given $x_3 = -3x_1 + x_2$, we can express W as $W = \{(x_1, x_2, -3x_1 + x_2, x_4, x_5) : x_1, x_2, x_3, x_4 \in \mathbb{R}\}$.

Now, observe

$$\begin{aligned}
(x_1, x_2, -3x_1 + x_2, x_4, x_5) &= x_1(1, 0, 0, 0, 0) + x_2(0, 1, 0, 0, 0) + (-3x_1 + x_2)(0, 0, 1, 0, 0) \\
&\quad + x_4(0, 0, 0, 1, 0) + x_5(0, 0, 0, 0, 1) \\
&= x_1(1, 0, 0, 0, 0) - 3x_1(0, 0, 1, 0, 0) + x_2(0, 1, 0, 0, 0) \\
&\quad + x_2(0, 0, 1, 0, 0) + x_4(0, 0, 0, 1, 0) + x_5(0, 0, 0, 0, 1) \\
&= x_1(1, 0, -3, 0, 0) + x_2(0, 1, 1, 0, 0) + x_4(0, 0, 0, 1, 0) \\
&\quad + x_5(0, 0, 0, 0, 1).
\end{aligned}$$

This shows that the set $B = \{(1,0,-3,0,0),(0,1,1,0,0),(0,0,0,1,0),(0,0,0,0,1)\}$ is a generating set of W.

Let $a,b,c,d \in \mathbb{R}$, such that

$$a(1,0,-3,0,0) + b(0,1,1,0,0) + c(0,0,0,1,0) + d(0,0,0,0,1) = (0,0,0,0,0).$$

Then we have

$$(a,b,-3a+b,c,d) = (0,0,0,0,0) \Rightarrow a = 0, b = 0, c = 0, d = 0.$$

Thus, B is a linearly independent set, and hence B constitutes a basis of W. Consequently, $\dim(W) = 4$.

Theorem 2.4.23. *Let W be a subspace of an n-dimensional vector space V over a field F. Then:*

(a) $\dim(W) \leq \dim(V)$.

(b) *If $B = \{x_1, x_2, \ldots, x_n\}$ is a basis of V, then every subset $\{x_1, x_2, \ldots, x_m\}$, $m \leq n$, of B generates a subspace of dimension m.*

Proof. (a) Let $B = \{x_1, x_2, \ldots, x_m\}$ be a basis of W. As B is linearly independent in W, it is also linearly independent in V. According to the extension theorem, B is either a basis of V or can be extended to form a basis of V. Thus, $\dim(W) \leq \dim(V)$.

(b) Consider $B_1 = \{x_1, x_2, \ldots, x_m\} \subseteq B$. Then $\langle B_1 \rangle$ is a vector space with a maximal linearly independent set B_1. Consequently, B_1 acts as a basis for $\langle B_1 \rangle$, and $\dim(\langle B_1 \rangle) = m$. □

Theorem 2.4.24. *Let F_1 be a subfield of a field F_2, and let V be a vector space over F_2. Suppose the dimension of the vector space F_2 over F_1 is m_1, and the dimension of the vector space V over F_2 is m_2. Then the dimension of the vector space V over F_1 is $m_1 m_2$.*

Proof. Let $\{x_1, x_2, \ldots, x_{m_1}\}$ be a basis of the vector space F_2 over F_1, and let $\{y_1, y_2, \ldots, y_{m_2}\}$ be a basis of the vector space V over F_2. We claim that the set $B = \{x_r y_s : 1 \leq r \leq m_1, 1 \leq s \leq m_2\}$ forms a basis of the vector space V over F_1.

Let $x \in V$. Then x can be expressed as a linear combination of the basis elements $\{y_1, y_2, \ldots, y_{m_2}\}$, such that

$$x = b_1 y_1 + b_2 y_2 + \cdots + b_{m_2} y_{m_2}, \quad \text{where } b_1, b_2, \ldots, b_{m_2} \in F_2.$$

Further, as $\{x_1, x_2, \ldots, x_{m_1}\}$ forms a basis of the vector space F_2 over F_1, each b_s, $1 \leq s \leq m_2$, can be expressed as $b_s = \sum_{r=1}^{m_1} b_{sr} x_r$, for some $b_{sr} \in F_1$.

Hence,

$$X = \left(\sum_{r=1}^{m_1} b_{1r}x_r \right) y_1 + \left(\sum_{r=1}^{m_1} b_{2r}x_r \right) y_2 + \cdots + \left(\sum_{r=1}^{m_1} b_{m_2 r}x_r \right) y_{m_2}$$

$$= \sum_{r=1}^{m_1} \sum_{s=1}^{m_2} b_{rs}x_r y_s.$$

This shows that B spans vector space V over F_1.
To demonstrate linear independence, suppose

$$\sum_{r=1}^{m_1} \sum_{s=1}^{m_2} b_{rs}x_r y_s = 0.$$

Since $\{y_1, y_2, \ldots, y_{m_2}\}$ is linearly independent, we have

$$\sum_{s=1}^{m_2} \left(\sum_{r=1}^{m_1} b_{rs}x_r \right) y_s = 0 \Rightarrow \sum_{r=1}^{m_1} b_{rs}x_r = 0.$$

Additionally, from the linear independence of $\{x_1, x_2, \ldots, x_{m_1}\}$,

$$\sum_{r=1}^{m_1} b_{rs}x_r = 0 \Rightarrow b_{rs} = 0 \quad \forall r, s.$$

Hence, B is linearly independent. □

Example 2.4.25. $\mathbb{C}^n$ is a vector space over $\mathbb{C}$, and $\mathbb{C}$ is a 2-dimensional vector space over $\mathbb{R}$. By the previous theorem, the dimension of the vector space $\mathbb{C}^n$ over $\mathbb{R}$ is equal to the product of the dimension of $\mathbb{C}^n$ over $\mathbb{C}$ and the dimension of $\mathbb{C}$ over $\mathbb{R}$. Thus, $\dim(\mathbb{C}^n)$ over $\mathbb{R} = \dim(\mathbb{C}^n)$ over $\mathbb{C} \times$ dimension of $\mathbb{C}$ over $\mathbb{R} = n \times 2 = 2n$.

The basis of $\mathbb{C}^n$ over $\mathbb{C}$ is $\{e_1, e_2, \ldots, e_n\}$, where $e_1 = (1, 0, 0, \ldots, 0)$, $e_2 = (0, 1, 0, \ldots, 0)$, $\ldots, e_n = (0, 0, \ldots, 1)$ and the basis of $\mathbb{C}$ over $\mathbb{R}$ is $\{1, i\}$.

Hence, the basis of $\mathbb{C}^n$ over $\mathbb{R}$ is $\{e_1, e_2, \ldots, e_n, ie_1, ie_2, \ldots, ie_n\}$, where $ie_1 = (i, 0, 0, \ldots, 0)$, $ie_2 = (0, i, 0, \ldots, 0), \ldots, ie_n = (0, 0, \ldots, i)$.

In the following example, we obtain the number of bases and the number of subspaces of a finite-dimensional vector space over a finite field $\mathbb{Z}_p$.

Example 2.4.26. Let $V = \mathbb{Z}_p^3$ be a 3-dimensional vector space over the field $\mathbb{Z}_p$. Then the number of elements in V is p^3.

First, we determine the number of bases in V. To count the number of linearly independent sets of the form $\{x_1, x_2, x_3\} \subset V$, we employ the fundamental principle of counting. However, this principle yields the number of ordered 3-tuples (x_1, x_2, x_3) in V.
Hence, the number of bases $\{x_1, x_2, x_3\}$ is calculated as

$$\frac{\text{the number of ordered 3-tuples}(x_1, x_2, x_3)}{3!}.$$

To select the ordered 3-tuples (x_1, x_2, x_3), we need to make a series of choices. The first choice, $x_1 \neq 0$, can be made in $p^3 - 1$ ways. The second choice, x_2, can be made in $p^3 - p$ ways. Since the set $\{x_1, x_2\}$ should not be linearly dependent. It means x_2 should not be of the form $x_2 = ax_1, a \in Z_p$ or $x_2 \notin \langle x_1 \rangle$. Next, the third choice x_3 can be made in $p^3 - p^2$ ways. Since the number of elements in 2-dimensional subspace $\langle \{x_1, x_2\} \rangle$ is p^2 and x_3 should not be written as linear combination of x_1 and x_2, i. e., $x_3 \notin \langle \{x_1, x_2\} \rangle$. Hence, the sequence (x_1, x_2, x_3) of choices can be made in $(p^3 - 1)(p^3 - p)(p^3 - p^2)$ ways. Therefore, the number of bases of V is $\frac{(p^3-1)(p^3-p)(p^3-p^2)}{3!}$.

Now, we determine all the subspaces of V:

(i)   0-dimensional subspace of V is $\{0\}$.

(ii)   1-dimensional subspaces of V:

Let $x_1 \neq 0 \in V$. Then $\langle x_1 \rangle$ is a subspace of V of dimension one. Thus, each of the $p^3 - 1$ nonzero vectors of V, generates a 1-dimensional subspace of V. However, each subspace $\langle x_1 \rangle = \{ax_1 : a \in Z_p\}$ contains $p - 1$ nonzero elements of V. Hence, the number of 1-dimensional subspaces of V is $\frac{p^3-1}{p-1}$.

(iii)   2-dimensional subspaces of V:

The number of linearly independent sets in V containing two elements is $\frac{(p^3-1)(p^3-p)}{2!}$. Each 2-dimensional subspace $\langle \{x_1, x_2\} \rangle$ of V is generated by $\frac{(p^2-1)(p^2-p)}{2!}$ linearly independent sets. Hence, the number of 2-dimensional subspaces of V is $\frac{(p^3-1)(p^3-p)}{(p^2-1)(p^2-p)}$.

(iv)   3-dimensional subspace of V is V itself.

Thus, the total number of subspaces of V is $1 + \frac{p^3-1}{p-1} + \frac{(p^3-1)(p^3-p)}{(p^2-1)(p^2-p)} + 1$.

From above observations, we have the following remarks.

Remark. If V is an n-dimensional vector space over a finite field containing p elements, then:

(1)   the number of bases of $V = \frac{(p^n-1)(p^n-p)\cdots(p^n-p^{n-1})}{n!}$;

(2)   the number d_k of $(k \geq 1)$-dimensional subspace of V is

$$d_k = \frac{(p^n - 1)(p^n - p) \cdots (p^n - p^{k-1})}{(p^k - 1)(p^k - p) \cdots (p^k - p^{k-1})},$$

and the total number of subspace of V is

$$1 + d_1 + d_2 + \cdots + d_n, \quad \text{where } d_n = 1.$$

Example 2.4.27. Consider the vector space Z_5^3 over Z_5. Then from the above formula, the number of bases of V is $\frac{(5^3-1)(5^3-5)(5^3-5^2)}{3!} = 248000$.

The total number of subspaces of Z_5^3 is $1 + d_1 + d_2 + d_3$, where

$$d_k = \frac{(p^n - 1)(p^n - p) \cdots (p^n - p^{k-1})}{(p^k - 1)(p^k - p) \cdots (p^k - p^{k-1})}, \quad 1 \le k \le 3,$$

$$d_1 = \frac{(p^3 - 1)}{(p - 1)} = \frac{5^3 - 1}{5 - 1} = \frac{124}{4} = 31,$$

$$d_2 = \frac{(p^3 - 1)(p^3 - p)}{(p^2 - 1)(p^2 - p)} = \frac{(5^3 - 1)(5^3 - 5)}{(5^2 - 1)(5^2 - 5)} = \frac{124 \times 120}{24 \times 20} = 31,$$

$$d_3 = \frac{(p^3 - 1)(p^3 - p)(p^3 - p^2)}{(p^3 - 1)(p^3 - p)(p^3 - p^2)} = 1.$$

Thus, the total number of subspaces of the vector space Z_5^3 over Z_5 is $1+31+31+1 = 64$.

Exercises

(2.4.1) Prove that it is always possible to choose a maximal linearly independent set from a generating set of a vector space.

(2.4.2) Let V be an n-dimensional vector space. Then prove that a subset of V containing $n + 1$ elements is always linearly dependent.

(2.4.3) Let V be a finitely generated vector space and let S be a subset of V such that $\langle S \rangle = V$. Then prove that $\{x\} \cup S$ is linearly dependent and $\langle S \rangle = V$, for any $x \in V$.

(2.4.4) Let V be an n-dimensional vector space and let S be a set of generators of V. Then show that S contains at least n elements.

(2.4.5) Find the dimension of the subspace $W = \{[a_{ij}] : a_{ij} = 0, \text{if } j \text{ is even}\}$ of the vector space of all 10×10 real matrices.

(2.4.6) Show that $W_1 = \{(x_1, 0, x_3, 0) : x_1, x_3 \in \mathbb{R}\}$ and $W_2 = \{(0, x_2, 0, x_4) : x_2, x_4 \in \mathbb{R}\}$ are subspaces of the vector space $\mathbb{R}^4$ and find their bases and dimensions.

(2.4.7) Find the bases and dimensions of the subspaces:
(i) $W_1 = \{(x_1, x_2, x_3, x_4, x_5) : x_1 + 2x_2 - x_3 = 0\}$ and
(ii) $W_2 = \{(x_1, x_2, x_3, x_4, x_5) : 2x_2 + 3x_3 + 4x_4 = 0\}$ of the vector space $\mathbb{R}^5$ over $\mathbb{R}$.

(2.4.8) Prove that the set $\{1, 1+x, 1+x+x^2, 1+x+x^2+x^3, \ldots, 1+x+x^2+\cdots+x^n\}$ is a basis of the polynomial space $P_n(x)$ over the field F.

(2.4.9) Find the number of bases and the number of subspaces of the following vector spaces:

(i) Z_3^3 over Z_3 (ii) Z_5^4 over Z_5 (iii) Z_7^2 over Z_7 (iv) Z_2^4 over Z_2.

(2.4.10) Show that the following sets of vectors are bases of $\mathbb{R}^3$ over $\mathbb{R}$:
(i) $B_1 = \{(3, 2, 0), (-3, -3, -1), (4, 4, 1)\}$,
(ii) $B_2 = \{(1, 0, -2), (-1, 0, 3), (0, 1, -3)\}$,
(iii) $B_3 = \{(1, 2, 1), (-1, 0, 2), (2, 1, -3)\}$,
(iv) $B_4 = \{(1, 1, 2), (1, 2, 2), (2, 2, 3)\}$.

(2.4.11) Let V be an n-dimensional vector space over a field F. Then show that there exists a k-dimensional subspace of V for all $0 \le k \le n$.

(2.4.12) Find the dimension of the following subspaces of the polynomial space: $P_n(x)$:
(i) $V_1 = \{f(x) \in P_n(x) : f(1) = 0\}$,
(ii) $V_2 = \{f(x) \in P_n(x) : f(1) = f(2) = 0\}$.

(2.4.13) Show that the set $\{1, (1-x), (1-x)^2, (1-x)^3, (1-x)^4, (1-x)^5\}$ is a basis of the polynomial space $P_5(x)$ over the field F.

(2.4.14) Let $\mathbb{Q}[\sqrt{2}] = \{a + b\sqrt{2} : a, b \in \mathbb{Q}\}$. Then $\mathbb{Q}[\sqrt{2}]$ is a field and $\mathbb{Q}$ is its subfield. Find a basis and dimension of the vector space $(\mathbb{Q}[\sqrt{2}])^n$ over $\mathbb{Q}$.

(2.4.15) Let $W = \{(x_1, x_2, \ldots, x_n) \in \mathbb{C}^n : \sum a_i x_i = 0, a_i \in \mathbb{C}\}$ be a vector space over $\mathbb{C}$. Then find the dimension of the vector space W over $\mathbb{R}$.

(2.4.16) Show that the set:

    (i) $\{(1, -i), (1, i)\}$ is a basis of $\mathbb{C}^2$,

    (ii) $\{(1, 0, 0), (0, i, 0), (1, 1, i)\}$ is a basis of $\mathbb{C}^3$,

    (iii) $\{(1 - i, 1 + i, i), (-3 + 3i, 2 + 2i, 2i), (0, i, 1 - i)\}$ is a basis of $\mathbb{C}^3$.

2.5 Sum and direct sum of subspaces

Definition 2.5.1. Let U and W be subspaces of a vector space V over a field F. Then the sum $U + W = \{u + w : u \in U, w \in W\}$ is a subspace of V, called the sum of subspaces U and W.

Theorem 2.5.2. *Let U and W be finite-dimensional subspaces of a vector space V over a field F. Then $U + W$ is also finite-dimensional and its dimension is given by* $\dim(U + W) = \dim(U) + \dim(W) - \dim(U \cap W)$.

Proof. Since $U \cap W$ is a finite-dimensional subspace of U and W both, we can write bases of U and W as an extended form of basis of $U \cap W$. Let $\{x_1, x_2, \ldots, x_r\}$ be a basis of $U \cap W$, and $\{x_1, x_2, \ldots, x_r, u_1, u_2, \ldots, u_{m-r}\}$ and $\{x_1, x_2, \ldots, x_r, w_1, \ldots, w_{n-r}\}$ be the bases of U and W, respectively, where $m = \dim(U)$ and $n = \dim(W)$. Then we claim that the set

$$S = \{x_1, x_2, \ldots, x_r, u_1, u_2, \ldots, u_{m-r}, w_1, w_2, \ldots, w_{n-r}\} \text{ is a basis of } U + W.$$

Since S contains bases of both U and W, every element of $U + W$ can be represented as a linear combination of elements from S, implying that $U + W$ is generated by S. We now demonstrate that S is linearly independent.

Suppose

$$a_1 x_1 + a_2 x_2 + \cdots + a_r x_r + b_1 u_1 + b_2 u_2 + \cdots + b_{m-r} u_{m-r} + c_1 w_1 + c_2 w_2 + \cdots + c_{n-r} w_{n-r} = 0,$$

where $a_i, b_j, c_k \in F$.

Then

$$a_1 x_1 + a_2 x_2 + \cdots + a_r x_r + b_1 u_1 + \cdots + b_{m-r} u_{m-r} = -c_1 w_1 - c_2 w_2 - \cdots - c_{n-r} w_{n-r}.$$

Since the left side belongs to U and the right side belongs to W, we have

$$a_1 x_1 + a_2 x_2 + \cdots + a_r x_r + b_1 u_1 + \cdots + b_{m-r} u_{m-r} = -c_1 w_1 - c_2 w_2 - \cdots - c_{n-r} w_{n-r} \in U \cap W.$$

As $\{x_1, x_2, \ldots, x_r\}$ forms a basis of $U \cap W$, there exist $d_1, d_2, \ldots, d_r$ in F such that $-c_1 w_1 - c_2 w_2 \cdots - c_{n-r} w_{n-r} = d_1 x_1 + d_2 x_2 + \cdots + d_r x_r$.

This leads to

$$d_1 x_1 + d_2 x_2 + \cdots + d_r x_r + c_1 w_1 + \cdots + c_{n-r} w_{n-r} = 0.$$

Since the set $\{x_1, x_2, \ldots, x_r, w_1, \ldots, w_{n-r}\}$ forms a basis of W and is linearly independent, we have $d_1 = d_2 = \cdots = d_r = c_1 = \cdots = c_{n-r} = 0$.

Similarly, $a_1 x_1 + a_2 x_2 + \cdots + a_r + x_r + b_1 u_1 + \cdots + b_{m-r} u_r = 0$ implies $b_1 = b_2 = \cdots = b_{m-r} = a_1 = a_2 = \cdots = a_r = 0$.

Thus, S is a linearly independent set, and hence a basis of $U + W$.

Finally, $\dim(U + W) = r + (m - r) + (n - r) = m + n - r = \dim(U) + \dim(W) - \dim(U \cap W)$. $\qquad\square$

Definition 2.5.3. A vector space V is said to be the direct sum of its subspaces U and W if $V = U + W$ and every $x \in V$ can be uniquely expressed as $x = x_1 + x_2$, where $x_1 \in U$ and $x_2 \in W$. If V is the direct sum of U and W, then we write $V = U \oplus W$.

Theorem 2.5.4. *Let U and W be subspaces of a vector space V. Then $V = U \oplus W$ if and only if $V = U + W$ and $U \cap W = \{0\}$.*

Proof. Let us first assume $V = U \oplus W$. This implies that every vector $x \in V$ can be uniquely expressed as $x = x_1 + x_2$, where $x_1 \in U$ and $x_2 \in W$. Thus, $V = U + W$. Now, let $x \in U \cap W$. Then $x \in U$ and $x \in W$. Also, $0 = x + (-x) = 0 + 0$, where $x \in U$ and $-x \in W$. From the definition of a direct sum of subspaces, we know that both expressions of 0 must be unique, implying $x = 0$. This leads to $U \cap W = \{0\}$.

Conversely, assume $V = U + W$ and $U \cap W = \{0\}$. Every vector $x \in V$ can be expressed as $x = x_1 + x_2$, where $x_1 \in U$ and $x_2 \in W$. To show uniqueness, let us assume $x = y_1 + y_2$, where $y_1 \in W$ and $y_2 \in W$. Then $x_1 + x_2 = y_1 + y_2 \Rightarrow x_1 - y_1 = y_2 - x_2$. Since $x_1 - y_1 \in U$ and $y_2 - x_2 \in W$, $x_1 - y_1 = y_2 - x_2 \in U \cap W = \{0\}$. Hence, $x_1 - y_1 = 0 \Rightarrow x_1 = y_1$, and $y_2 - x_2 = 0 \Rightarrow y_2 = x_2$. Thus, $x_1 + x_2 = y_1 + y_2 = x$. This demonstrates that $x = x_1 + x_2$ is unique. $\qquad\square$

Remark. More generally, we say that V is direct sum of its subspaces $W_1, W_2, \ldots, W_k$, if $V = W_1 + W_2 + \cdots + W_k$ and for all $x \in V$ the expression $x = w_1 + w_2 + \cdots + w_k$ is unique, where $w_i \in W_i, 1 \le i \le k$ and we write $V = W_1 \oplus W_2 \oplus \cdots W_k$.

Theorem 2.5.5. *Let W_1 and W_2 be subspaces of a finite-dimensional vector space V. Then $V = W_1 \oplus W_2$ if and only if $\dim(V) = \dim(W_1) + \dim(W_2)$.*

Proof. Assume $V = W_1 \oplus W_2$. From the previous theorem, we know that $V = W_1 + W_2$ and $W_1 \cap W_2 = \{0\}$. Also, $\dim(V) = \dim(W_1) + \dim(W_2) - \dim(W_1 \cap W_2)$. Since $W_1 \cap W_2 = \{0\}$, we have $\dim(V) = \dim(W_1) + \dim(W_2)$.

Conversely, suppose $\dim(V) = \dim(W_1) + \dim(W_2)$. This implies $\dim(W_1 \cap W_2) = 0$, and hence $W_1 \cap W_2 = \{0\}$. Moreover, $\dim(W_1 + W_2) = \dim(W_1) + \dim(W_2) = \dim(V)$, indicating $W_1 + W_2$ is a subspace of V with the same dimension as V. Hence, $V = W_1 + W_2$. $\square$

Example 2.5.6. Let V_1 and V_2 be two distinct subspaces of dimension $m - 1$ of a vector space V of dimension m. Then we shall show that the dimension of their intersection, $V_1 \cap V_2$, is $m - 2$.

Considering $V_1 = V_1 + \{0\}$ and $V_2 = \{0\} + V_2$, it follows that $V_1 \subseteq V_1 + V_2$ and $V_2 \subseteq V_1 + V_2$. Consequently, $\dim(V_1) = \dim(V_2) = (m - 1) \leq \dim(V_1 + V_2)$.

Moreover, since V_1 and V_2 are distinct subspaces, there exists a nonzero vector $v \in V_2 - V_1$. Thus, $\langle v \rangle$ forms a subspace of V with dimension 1, and $\langle v \rangle \cap V_1 = \{0\}$.

This implies that

$$\dim(\langle v \rangle + V_1) = \dim(\langle v \rangle) + \dim(V_1) = 1 + m - 1 = m.$$

Since $(\langle v \rangle + V_1) \subseteq (V_1 + V_2) \subseteq V$, we have $m = \dim(\langle v \rangle + V_1) \leq \dim(V_1 + V_2) \leq \dim(V) = m$, thus yielding $\dim(V_1 + V_2) = m$.

Utilizing the result $\dim(V_1 + V_2) = \dim(V_1) + \dim(V_2) - \dim(V_1 \cap V_2)$, we derive

$$m = (m - 1) + (m - 1) - \dim(V_1 \cap V_2),$$

which simplifies to

$$\dim(V_1 \cap V_2) = m - 2.$$

Example 2.5.7. Let $U = \{(x, 0, z) : x, z \in \mathbb{R}\}$, $W = \{(0, y, 0) : y \in \mathbb{R}\}$ and $S = \{(0, y, z) : y, z \in \mathbb{R}\}$ be subspaces of the vector space $\mathbb{R}^3$ over $\mathbb{R}$.

Then $U + W = \{(x, y, z) : x, y, z \in \mathbb{R}\} = \mathbb{R}^3$ and $U \cap W = \{(0, 0, 0)\}$. This gives that $\mathbb{R}^3 = U \oplus W$, the direct sum of subspaces U and W.

Next, $U + S = \{(x, y, z) : x, y, z \in \mathbb{R}\} = \mathbb{R}^3$ but $U \cap S = \{(0, 0, z) \in \mathbb{R}^3\} \neq \{(0, 0, 0)\}$. Hence, $\mathbb{R}^3$, is not the direct sum of subspaces U and S.

Example 2.5.8. Let $U = \{p(x) \in P_9(x) : p(x) = a_0 + a_2x^2 + a_4x^4 + a_6x^6 + a_8x^8\}$ and $W = \{p(x) \in P_9(x) : p(x) = a_1x + a_3x^3 + a_5x^5 + a_7x^7 + a_9x^9\}$ be subspaces of the polynomial space $P_9(x)$. Then $P_9(x) = U \oplus W$.

Example 2.5.9. Consider two subspaces of $\mathbb{R}^2$, denoted by W_1 and W_2, where $W_1 = \{(x, y) \in \mathbb{R}^2 : x - y = 0\}$ and $W_2 = \{(x, y) \in \mathbb{R}^2 : x + y = 0\}$. Simplifying the conditions, we find that W_1 represents the set of points where $y = x$, resulting in $W_1 = \{(x, x) : x \in \mathbb{R}\}$. Similarly, W_2 can be expressed as $W_2 = \{(x, -x) : x \in \mathbb{R}\}$.

It is evident that $\{(1, 1)\}$ forms the basis of W_1, and $\{(1, -1)\}$ serves as the basis of W_2.

Considering the intersection, $(x, y) \in W_1 \cap W_2$, we obtain $x + y = 0$ and $x - y = 0$. Solving these equations yield $x = 0$ and $y = 0$, implying $W_1 \cap W_2 = \{(0, 0)\}$.

Therefore, applying the formula $\dim(W_1 + W_2) = \dim(W_1) + \dim(W_2) - \dim(W_1 \cap W_2)$, we find $\dim(W_1 + W_2) = 2$, which equals $\dim(\mathbb{R}^2)$. Hence, we conclude that $\mathbb{R}^2 = W_1 \oplus W_2$.

Example 2.5.10. Let $W_1 = \{(x_1, x_2, x_3) : x_1 + x_2 + x_3 = 0\}$ and $W_2 = \{(x_1, x_2, x_3) : x_1 - x_2 + x_3 = 0\}$ be subspaces of the vector space $\mathbb{R}^3$. Then W_1 and W_2 can be written as $W_1 = \{(x_1, -x_1 - x_3, x_3) : x_1, x_3 \in \mathbb{R}\}$ and $W_2 = \{(x_1, x_1 + x_3, x_3) : x_1, x_3 \in \mathbb{R}\}$.

If we consider the standard basis $\{e_1, e_2, e_3\}$ of $\mathbb{R}^3$, any vector $(x_1, -x_1 - x_3, x_3)$ in W_1 can be represented as a linear combination:

$$(x_1, -x_1 - x_3, x_3) = x_1 e_1 + (-x_1 - x_3)e_2 + x_3 e_3 = x_1(e_1 - e_2) + x_3(e_3 - e_2)$$
$$= x_1(1, -1, 0) + x_3(0, -1, 1).$$

This implies that the set $B_1 = \{(1, -1, 0), (0, -1, 1)\}$ generates W_1. It can be easily shown that B_1 is linearly independent, hence forming a basis of W_1, leading to $\dim(W_1) = 2$. Similarly, using a similar approach, we can demonstrate that the set $B_2 = \{(1, 1, 0), (0, 1, 1)\}$ is a basis of W_2, also resulting in $\dim(W_2) = 2$.

Further, $(x_1, x_2, x_3) \in W_1 \cap W_2 \Rightarrow x_1 + x_2 + x_3 = 0$ and $x_1 - x_2 + x_3 = 0, \Rightarrow (x_1, x_2, x_3) = (x_1, 0, -x_1)$.

Thus, $W_1 \cap W_2 = \{(x_1, 0, -x_1) : x_1 \in \mathbb{R}\}$. Consequently, the set $B_3 = \{(1, 0, -1)\}$ is a basis of $W_1 \cap W_2$, resulting in $\dim(W_1 \cap W_2) = 1$.

Therefore, $\dim(W_1 + W_2) = 2 + 2 - 1 = 3 = \dim(\mathbb{R}^3)$, indicating that $\mathbb{R}^3 = W_1 + W_2$, but $\mathbb{R}^3 \neq W_1 \oplus W_2$.

Example 2.5.11. Consider two subspaces of the vector space $P_3(x)$ over $\mathbb{R}$, denoted by U and W, where

$$U = \{p(x) \in P_3(x) : p(0) = 0, p(1) = 0\},$$
$$W = \{p(x) \in P_3(x) : p(1) = 0, p(2) = 0\}.$$

In this example, we determine the dimensions of U, W and $U \cap W$.

For any polynomial $p(x) = a_0 + a_1 x + a_2 x^2 + a_3 x^3 \in U$, we have $p(0) = 0 = a_0$ and $p(1) = 0 = a_0 + a_1 + a_2 + a_3$. This implies $a_1 = -a_2 - a_3$, leading to the expression:

$$p(x) = (-a_2 - a_3)x + a_2 x^2 + a_3 x^3 = a_2(x^2 - x) + a_3(x^3 - x).$$

Thus, the set $B_1 = \{(x^2 - x), (x^3 - x)\}$ forms a basis of U as it spans U and is linearly independent. Hence, $\dim(U) = 2$.

Similarly, for $p(x) = a_0 + a_1 x + a_2 x^2 + a_3 x^3 \in W$, we find $a_0 + a_1 + a_2 + a_3 = 0$ and $a_0 + 2a_1 + 4a_2 + 8a_3 = 0$. Solving these equations, we obtain $a_0 = 2a_2 + 6a_3$ and $a_1 = -3a_2 - 7a_3$. Thus,

$$p(x) = 2a_2 + 6a_3 + (-3a_2 - 7a_3)x + a_2 x^2 + a_3 x^3 = (x^2 - 3x + 2)a_2 + (x^3 - 7x + 6)a_3.$$

This yields $B_2 = \{(x^2 - 3x + 2), (x^3 - 7x + 6)\}$ as a basis of W. Hence, $\dim(W) = 2$.

Next, $U \cap W = \{p(x) \in P_3(x) : p(0) = 0, p(1) = 0 \text{ and } p(2) = 0\}$.

Let $p(x) = a_0 + a_1x + a_2x^2 + a_3x^3 \in U \cap W$. Then:

$$p(0) = 0 = a_0,$$
$$p(1) = 0 = a_0 + a_1 + a_2 + a_3,$$
$$p(2) = 0 = a_0 + 2a_1 + 4a_2 + 8a_3.$$

Solving these equations, we get $a_0 = 0$, $a_1 = 2a_3$ and $a_2 = -3a_3$. Hence, $p(x) = 2a_3x - 3a_3x^2 + a_3x^3 = a_3(2x - 3x^2 + x^3)$. This gives that $B_3 = \{2x - 3x^2 + x^3\}$ is a basis of $U \cap W$, and hence $\dim(U \cap W) = 1$.

Consequently, $\dim(U + W) = \dim(U) + \dim(W) - \dim(U \cap W) = 2 + 2 - 1 = 3$.

Example 2.5.12. Consider the subset $W = \{p(x) \in P_3(x) : p'(0) = 0 = p''(1)\}$ of the vector space $P_3(x)$.

Let $f(x), g(x) \in W$ and $a, b \in \mathbb{R}$. Then

$$(af(x) + bg(x))'(0) = (af'(x) + bg'(x))(0) = af'(0) + bg'(0) = 0,$$

and

$$(af(x) + bg(x))''(1) = af''(1) + bg''(1) = 0.$$

This implies $af(x) + bg(x) \in W$, and hence W is a subspace of $P_3(x)$.

For any $p(x) = a_0 + a_1x + a_2x^2 + a_3x^3 \in W$, we find $p'(x) = a_1 + 2a_2x + 3a_3x^2$ and $p''(x) = 2a_2 + 6a_3x$. From $p'(0) = 0 = a_1$ and $p''(1) = 0 = 2a_2 + 6a_3$, we deduce that $a_1 = 0$ and $a_2 = -3a_3$. Hence, $p(x) = a_0 + (-3a_3)x^2 + a_3x^3 = a_0 + a_3(x^3 - 3x^2)$. Thus, the set $B = \{1, x^3 - 3x^2\}$ spans W and is linearly independent, serving as a basis of W. Therefore, $\dim(W) = 2$.

Example 2.5.13. Let V be a vector space of dimension 120. If U and W are subspaces of V of dimensions 50 and 85, respectively, then we find minimum and maximum possible dimensions of $U \cap W$.

Since $U \cap W$ is a subspace of both U and W, $\dim(U \cap W) \leq 50$.

So, the maximum possible dimension of $U \cap W$ is 50.

Now,

$$\dim(U + W) = \dim(U) + \dim(W) - \dim(U \cap W)$$
$$\Rightarrow \dim(U \cap W) = \dim(U) + \dim(W) - \dim(U + W)$$
$$= 50 + 85 - \dim(U + W).$$

Thus, $\dim(U \cap W)$ is minimum if $\dim(U + W)$ is maximum. Since max $\dim(U + W) = \dim(V) = 120$, min $\dim(U \cap W) = 50 + 85 - 120 = 15$.

Exercises

(2.5.1) If W_1 and W_2 are subspaces of a vector space V, then show that $W_1 + W_2$ is a subspace of V.

(2.5.2) Let $W_1 = \{(x_1, x_2, x_3, x_4) \in \mathbb{R}^4 : x_1 + x_2 + x_3 + x_4 = 0\}$ and $W_2 = \{(x_1, x_2, x_3, x_4) \in \mathbb{R}^4 : x_1 - x_2 + x_3 - x_4 = 0\}$ be two subspaces of $\mathbb{R}^4$. Then find the dimension of $W_1 + W_2$ and show that $\mathbb{R}^4 \neq W_1 \oplus W_2$.

(2.5.3) For even $n \in \mathbb{N}$, let $W_1 = \{(x_1, x_2, \ldots, x_n) \in \mathbb{R}^n : x_1 + x_2 + \cdots + x_n = 0\}$ and $W_2 = \{(x_1, x_2, \ldots, x_n) \in \mathbb{R}^n : x_1 - x_2 + x_3 - x_4 \cdots - x_n = 0\}$ be two subspaces of $\mathbb{R}^n$. Then show that $\mathbb{R}^n \neq W_1 \oplus W_2$.

(2.5.4) For odd $n \in \mathbb{N}$, let $W_1 = \{(x_1, x_2, \ldots, x_n) \in \mathbb{R}^n : x_1 + x_2 + \cdots + x_n = 0\}$ and $W_2 = \{(x_1, x_2, \ldots, x_n) \in \mathbb{R}^n : x_1 - x_2 + x_3 - x_4 \cdots + x_n = 0\}$ be two subspaces of $\mathbb{R}^n$. Then show that $\mathbb{R}^n \neq W_1 \oplus W_2$.

(2.5.5) Let $U = \{p(x) \in P_5(x) : p(1) = p(-1) = 0\}$ and $W = \{p(x) \in P_5(x) : p(0) = p(1) = 0\}$ be subspaces of the polynomial space $P_5(x)$ over $\mathbb{R}$. Then find dimensions of $U \cap W$ and $U + W$.

(2.5.6) Let $U = \{p(x) \in P_4(x) : p'(1) = p''(-1) = 0\}$, where $p'(x) = \frac{dp(x)}{dx}$ and $p''(x) = \frac{d^2 p(x)}{dx^2}$ be a subset of $P_4(x)$. Then show that U is a subspace of the polynomial space $P_4(x)$ over the field $\mathbb{R}$.

(2.5.7) Let $U = \{p(x) \in P_5(x) : p(1) + p(-1) = 0 \text{ and } p(2) + p(-2) = 0\}$. Then show that U is a subspace of $P_5(x)$ over $\mathbb{R}$ and find a basis and dimension of U.

(2.5.8) If V is a vector space of dimension 150 and U and W are subspaces of V having dimensions 75 and 100, respectively, then find maximum and minimum possible dimensions of $U \cap W$.

(2.5.9) Let V be a vector space of dimension 10. If U is a subspace of V of dimension 9 and W is a subspace of dimension 6, which is not contained in U, then find the dimension of $U \cap W$.

(2.5.10) Show that the intersection of two distinct planes passing through origin is a line passing through origin.

2.6 Quotient space

Definition 2.6.1. Let W be a subspace of a vector space V over a field F. Then the set $(W, +)$ forms an Abelian subgroup of $(V, +)$, and consequently, it is a normal subgroup of $(V, +)$. The quotient group $\frac{V}{W}$ is regarded as a vector space over the field F, called the quotient space under the following vector addition and scalar multiplication defined as

$$(x + W) + (y + W) = (x + y) + W, \quad \text{and}$$
$$a.(x + W) = ax + W,$$

where $x + W, y + W \in \frac{V}{W}$ and $a \in F$.

Theorem 2.6.2. *Let V be a finite-dimensional vector space over a field F, and let W be its subspace. Then the dimension of the quotient space $\frac{V}{W}$ is given by* $\dim(\frac{V}{W}) = \dim(V) - \dim(W)$.

Proof. Let $\{x_1, x_2, \ldots, x_r\}$ be a basis of W. Since W is a subspace of V, the basis of W can be extended to form a basis $\{x_1, x_2, \ldots, x_r, y_1, y_2, \ldots, y_s\}$ of V. Specifically, this implies $\dim(W) = r$ and $\dim(V) = r + s$. To establish the result, we demonstrate that $\dim \frac{V}{W} = r + s - r = s$.

We claim that $B = y_1 + W, y_2 + W, \ldots, y_s + W$ serves as a basis of $\frac{V}{W}$.

Let $x + W \in \frac{V}{W}$. Given $x \in V$, we can represent x as a linear combination of the basis elements of V:

$$x = a_1 x_1 + a_2 x_2 + \cdots + a_r x_r + b_1 y_1 + b_2 y_2 + \cdots + b_s y_s,$$

where $a_i, b_j \in F$ for $1 \le i \le r$ and $1 \le j \le s$.

Since $a_1 x_1 + \cdots + a_r x_r \in W$, we have $a_1 x_1 + a_2 x_2 + \cdots + a_r x_r + W = W$. Consequently,

$$
\begin{aligned}
x + W &= (a_1 x_1 + a_2 x_2 + \cdots + a_r x_r) + (b_1 y_1 + b_2 y_2 + \cdots + b_s y_s) + W \\
&= (b_1 y_1 + b_2 y_2 + \cdots + b_s y_s) + W \\
&= (b_1 y_1 + W) + (b_2 y_2 + W) + \cdots + (b_s y_s + W) \\
&= b_1 (y_1 + W) + b_2 (y_2 + W) + \cdots + b_s (y_s + W).
\end{aligned}
$$

Thus, every element of $\frac{V}{W}$ can be expressed as a linear combination of elements in B, establishing $\frac{V}{W} = \langle B \rangle$.

Next, we demonstrate that B is linearly independent.

Assume that $a_1(y_1 + W) + a_2(y_2 + W) + \cdots + a_s(y_s + W) = W$ (the zero vector of $\frac{V}{W}$) for some $a_1, a_2, \ldots, a_s \in F$. This implies

$$(a_1 y_1 + W) + (a_2 y_2 + W) + \cdots + (a_s y_s + W) = W$$

or $(a_1 y_1 + a_2 y_2 + \cdots + a_s y_s) + W = W$.

Since $\{x_1, x_2, \ldots, x_r\}$ forms a basis of W, we can write

$$a_1 y_1 + a_2 y_2 + \cdots + a_s y_s = b_1 x_1 + b_2 x_2 + \cdots + b_r x_r \quad \text{for some } b_1, b_2, \ldots, b_r \in F.$$

Thus, $a_1 y_1 + a_2 y_2 + \cdots + a_s y_s - b_1 x_1 - b_2 x_2 - \cdots - b_r x_r = 0$.

Given that $\{x_1, x_2, \ldots, x_r, y_1, y_2, \ldots, y_s\}$ forms a linearly independent basis of V, we conclude that $a_1 = a_2 = \cdots = a_s = b_1 = b_2 = \cdots = b_r = 0$. Consequently, B is linearly independent.

Therefore, $\dim(\frac{V}{W}) = s = r + s - r = \dim(V) - \dim(W)$. $\qquad\square$

Example 2.6.3. Let V be a vector space over a field F. Then

$$\frac{V}{\{0\}} = \{x + \{0\} : x \in V\} = \{x : x \in V\} = V \quad \text{and}$$

$$\frac{V}{V} = \{x + V : x \in V\} = \{V\} = \{0\}, \quad \text{the null space.}$$

Example 2.6.4. Let $V = \mathbb{R}^2$ and $W = \{(0, x) : x \in \mathbb{R}\}$. Then

$$\frac{V}{W} = \{(r, s) + W : (r, s) \in \mathbb{R}^3\}.$$

For any $(r, s) \in \mathbb{R}^2$,

$$(r, s) + W = \{(r, s) + (0, x) : x \in \mathbb{R}\} = \{(r, s + x) : x \in \mathbb{R}\}$$
$$= \{(r, x') : x' \in \mathbb{R}\}, \quad \text{where } x' = s + x.$$

In other words, the coset $(r, s) + W$ represents the line passing through (r, s) and parallel to the y-axis.

Now, $(r, s) + W = (r', s') + W \Leftrightarrow (r, s) - (r', s') \in W = (r - r', s - s') \in W$. This means $r - r' = 0$. Hence, two elements (r, s) and (r', s') of V yield the same elements in $\frac{V}{W}$ if and only if $r = r'$. Consequently, $\frac{V}{W}$ can be identified with the set of vectors of the form $(r, 0)$, or equivalently, with the x-axis.

Furthermore, $\dim(\frac{V}{W}) = \dim(V) - \dim(W) = 2 - 1 = 1$.

Example 2.6.5. Let $W = \{(0, y, z) \in \mathbb{R}^3\}$ be the yz-plane in $\mathbb{R}^3$. Then W is a subspace of the vector space $V = \mathbb{R}^3$. The quotient space

$$\frac{V}{W} = \{(r, s, t) + W : (r, s, t) \in \mathbb{R}^3\}.$$

Each coset

$$(r, s, t) + W = \{(r, s, t) + (0, y, z) : y, z \in \mathbb{R}\}$$
$$= \{(r, y + s, z + t) : y, z \in \mathbb{R}\}$$
$$= \{(r, y', z') : y', z' \in \mathbb{R}\},$$

represents a plane parallel to yz-plane passing through the point $(r, s, t) \in \mathbb{R}^3$.

⚡ Exercises

(2.6.1) Let W be a subspace of a vector space V over a field F. Then show that the quotient set $\frac{V}{W}$ is a vector space under the operations $(x + W) + (y + W) = (x + y) + W$ and $a.(x + W) = ax + W$, where $a \in F$.

(2.6.2) Let $\{(x_1 + W), (x_2 + W), \ldots, (x_m + W)\}$ be a basis of the quotient space $\frac{V}{W}$ and let $\{y_1, y_2, \ldots, y_n\}$ be a basis of W. Then prove that $\{x_1, x_2, \ldots, x_m, y_1, y_2, \ldots, y_n\}$ is a basis of V.

(2.6.3) Let $V = \mathbb{R}^2$ and let W be any line in V passing through the origin. Then describe cosets of $\frac{V}{W}$ and find $\dim(\frac{V}{W})$.

(2.6.4) Let $V = P_{20}(x)$ and $W = \{p(x) \in V : p(1) = p(2) = 0\}$. Then find $\dim(\frac{V}{W})$.

3 Matrices and spaces of matrices

This chapter is devoted to several vector spaces obtained form the collection of various types of matrices under the matrix addition and scalar multiplication operations. Also, we discuss basis, dimension and some interesting properties of such vector spaces.

3.1 Matrix definition and matrix operations

Definition 3.1.1. A rectangular arrangement of mn elements of a field $F(\mathbb{R}$ or $\mathbb{C})$ in m rows and n columns is called a matrix of order $m \times n$. Thus, a matrix A of order $m \times n$ can be represented in the following form:

$$A = \begin{pmatrix} a_{11} & a_{12} & \cdots & a_{1n} \\ a_{21} & a_{22} & \cdots & a_{2n} \\ \vdots & \vdots & \ddots & \vdots \\ a_{m1} & a_{m2} & \cdots & a_{mn} \end{pmatrix}.$$

In brief, the above matrix is denoted by $A = [a_{ij}]_{m \times n}$, where a_{ij} is the entry at the intersection of i-th row and j-th column, called (i, j)-th element of A. Some special types of matrices are defined as follows.

Row matrix: A matrix having only one row is called a row matrix or a row vector.

Column matrix: A matrix is called a column matrix if it has only one column or column vector.

Square matrix: A matrix in which the number of rows and columns are equal is called a square matrix. A square matrix of order $n \times n$, denoted by A_n is called an n-square matrix.

Zero matrix: A matrix is said to be a zero matrix or a null matrix if all its elements are zeros. We use the notation $0_{m \times n}$ to denote the zero matrix of order $m \times n$.

Definition 3.1.2. Two matrices $A = [a_{ij}]_{m \times n}$ and $B = [b_{ij}]_{r \times s}$ are said to be equal if and only if (i) $m = r, n = s$ and (ii) $a_{ij} = b_{ij}$ for all i, j.

In other words, two matrices are said to be equal if and only if they have the same order and their corresponding entries are equal.

Definition 3.1.3. Let $A = [a_{ij}]_{m \times n}$ and $B = [b_{ij}]_{m \times n}$. Then the sum $A + B$ is a matrix $C = [c_{ij}]_{m \times n}$, where $c_{ij} = a_{ij} + b_{ij}$.

Note that the sum of two matrices can only be calculated when both matrices have the same orders. This operation is referred to as matrix addition.

Definition 3.1.4. Let $A = [a_{ij}]_{m \times n}$. The product of the matrix A by a scalar k is defined as $kA = [ka_{ij}]_{m \times n}$ and is referred to as scalar multiplication.

https://doi.org/10.1515/9783111516035-003

For example, if $A = \begin{bmatrix} 1 & -1 & 2 \\ 3 & 4 & -2 \end{bmatrix}$ and $k = 3$, then $3A = \begin{bmatrix} 3 & -3 & 6 \\ 9 & 12 & -6 \end{bmatrix}$.

Also, $(-1)A = -A = [-a_{ij}]$ is called the negative matrix of the matrix A.

Theorem 3.1.5. *Let A, B and C be matrices of order $m \times n$. Then:*

(i) $A + B = B + A$ *(commutative)*;

(ii) $(A + B) + C = A + (B + C)$ *(associative)*;

(iii) $k(A + B) = kA + kB$;

(iv) $(k + l)A = kA + lA$;

(v) $k(lA) = (kl)A$,

where k and l are scalars.

Proof. Let $A = [a_{ij}]_{m \times n}$, $B = [b_{ij}]_{m \times n}$ and $C = [c_{ij}]_{m \times n}$.

Since a_{ij}, b_{ij}, c_{ij}, k and l are elements of a field, $a_{ij} + b_{ij} = b_{ij} + a_{ij}$ and $k(a_{ij} + b_{ij}) = ka_{ij} + kb_{ij}$.

Hence,

(i) $A + B = [a_{ij}]_{m \times n} + [b_{ij}]_{m \times n} = [a_{ij} + b_{ij}]_{m \times n} = [b_{ij} + a_{ij}]_{m \times n} = [b_{ij}]_{m \times n} + [a_{ij}]_{m \times n} = B + A$.

(ii) $(A + B) + C = ([a_{ij}]_{m \times n} + [b_{ij}]_{m \times n}) + [c_{ij}]_{m \times n} = ([a_{ij} + b_{ij}]_{m \times n}) + [c_{ij}]_{m \times n} = [(a_{ij} + b_{ij}) + c_{ij}]_{m \times n} = [a_{ij} + (b_{ij} + c_{ij})]_{m \times n} = [a_{ij}]_{m \times n} + ([b_{ij} + c_{ij}]_{m \times n}) = A + (B + C)$.

(iii) $k(A + B) = k([a_{ij}]_{m \times n} + [b_{ij}]_{m \times n}) = k[(a_{ij} + b_{ij})]_{m \times n} = [k(a_{ij} + b_{ij})]_{m \times n} = [ka_{ij} + kb_{ij}]_{m \times n} = [ka_{ij}]_{m \times n} + [kb_{ij}]_{m \times n} = k[a_{ij}]_{m \times n} + k[b_{ij}]_{m \times n} = kA + kB$.

(iv) $(k + l)A = (k + l)[a_{ij}]_{m \times n} = [(k + l)a_{ij}]_{m \times n} = [ka_{ij} + la_{ij}]_{m \times n} = [ka_{ij}]_{m \times n} + [la_{ij}]_{m \times n} = k[a_{ij}]_{m \times n} + l[a_{ij}]_{m \times n} = kA + lA$.

(v) $k(lA) = k(l[a_{ij}]_{m \times n}) = k[la_{ij}]_{m \times n} = [kla_{ij}]_{m \times n} = kl[a_{ij}]_{m \times n} = klA$. □

Definition 3.1.6. Let $A = [a_{ij}]$ be an $m \times n$ matrix and $B = [b_{ij}]$ be an $n \times p$ matrix. Then the product AB is a matrix $C = [c_{ij}]$ of order $m \times p$, with entries $c_{ij} = a_{i1}b_{1j} + a_{i2}b_{2j} + \cdots + a_{in}b_{nj} = \sum_{k=1}^{n} a_{ik}b_{kj}$.

The product AB is defined only when the number of columns of A is equal to the number of rows of B.

For example, if

$$A = \begin{bmatrix} 1 & -1 & 0 \\ 3 & 2 & 1 \end{bmatrix} \quad \text{and} \quad B = \begin{bmatrix} 1 & 2 & 0 \\ 2 & 1 & 3 \\ 1 & 1 & -1 \end{bmatrix},$$

then

$$AB = \begin{bmatrix} 1.1 + (-1).2 + 0.1 & 1.2 + (-1).1 + 0.1 & 1.0 + (-1).3 + 0.(-1) \\ 3.1 + 2.2 + 1.1 & 3.2 + 2.1 + 1.1 & 3.0 + 2.3 + 1.(-1) \end{bmatrix}$$

$$= \begin{bmatrix} -1 & 1 & -3 \\ 8 & 9 & 5 \end{bmatrix}.$$

In this example, order of matrix A is 2×3 and order of matrix B is 3×3, therefore the product matrix AB is of order 2×3. However, since the number of columns of matrix B is not equal to the number of rows of matrix A, the product BA is not defined.

The following example demonstrates that matrix multiplication is not commutative.

Example 3.1.7. Let $A = \begin{bmatrix} 1 & 2 \\ 0 & 3 \end{bmatrix}$ and $B = \begin{bmatrix} 3 & -1 \\ 2 & 1 \end{bmatrix}$ be matrices of order 2×2. Then $AB = \begin{bmatrix} 7 & 1 \\ 6 & 3 \end{bmatrix}$ and $BA = \begin{bmatrix} 3 & 3 \\ 2 & 7 \end{bmatrix}$ are not equal.

The matrix multiplication satisfies the following properties.

Theorem 3.1.8. *Let A, B and C be matrices. Then:*
(i) $(AB)C = A(BC)$ *(associative)*;
(ii) $A(B + C) = AB + AC$ and $(B + C)A = BA + CA$ *(distributive)*;
(iii) *for any scalar k, $k(AB) = (kA)B = A(kB)$.*

Proof. (i)   Let $A = [a_{ij}]_{m \times n}$, $B = [b_{ij}]_{n \times p}$ and $C = [c_{ij}]_{p \times q}$. Then (i,j)-th element of the product AB is denoted by $[AB]_{ij}$. To prove that the matrices $(AB)C$ and $A(BC)$ are equal, we need to show that their corresponding entries are equal. That is, we have to show that $[(AB)C]_{ij} = [A(BC)]_{ij}$. Since order of AB is $m \times p$,

$$[(AB)C]_{ij} = \sum_{k=1}^{p} [AB]_{ik} c_{kj}$$
$$= \sum_{k=1}^{p} \left(\sum_{l=1}^{n} a_{il} b_{lk} \right) c_{kj}$$
$$= \sum_{k=1}^{p} \sum_{l=1}^{n} (a_{il} b_{lk}) c_{kj}$$
$$= \sum_{k=1}^{p} \sum_{l=1}^{n} a_{il} (b_{lk} c_{kj})$$
$$= \sum_{l=1}^{n} a_{il} \left(\sum_{k=1}^{p} b_{lk} c_{kj} \right)$$
$$= \sum_{l=1}^{n} a_{il} [BC]_{lj}$$
$$= [A(BC)]_{ij}.$$

(ii)   Let $A = [a_{ij}]_{m \times n}$, $B = [b_{ij}]_{n \times p}$ and $C = [c_{ij}]_{n \times p}$. Then

$$[A(B + C)]_{ij} = \sum_{l=1}^{n} a_{il} [B + C]_{lj}$$
$$= \sum_{l=1}^{n} a_{il} (b_{lj} + c_{lj})$$

$$= \sum_{l=1}^{n} a_{il}b_{lj} + \sum_{l=1}^{n} a_{il}c_{lj}$$
$$= [AB]_{ij} + [AC]_{ij}$$
$$= [AB + AC]_{ij}.$$

Hence, $A(B + C) = AB + AC$.

Similarly, $(B+C)A = BA+CA$, where matrices B and C are of order $m \times n$ and matrix A is of order $n \times p$.

(iii) Let $A = [a_{ij}]_{m \times n}$ and $B = [b_{ij}]_{n \times p}$. Then

$$k[AB]_{ij} = k\left(\sum_{l=1}^{n} a_{il}b_{lj}\right)$$
$$= \sum_{l=1}^{n} (ka_{il})b_{lj} \ (= [(kA)B]_{ij})$$
$$= \sum_{l=1}^{n} a_{il}(kb_{lj}) \ (= [A(kB)]_{ij}).$$

This proves that (i,j)-th element of matrices $k(AB)$, $(kA)B$ and $A(kB)$ are equal. Hence, $k(AB) = (kA)B = A(kB)$. □

Definition 3.1.9. Let $A = [a_{ij}]$ be an $m \times n$ matrix with entries in a field F. Then the transpose of A denoted by A^t is an $n \times m$ matrix $[b_{ji}]$, where $b_{ji} = a_{ij}$. That is, A^t is obtained by changing rows of A into corresponding columns.

For example, if $A = \left[\begin{smallmatrix} 1 & 2 & 3 \\ 0 & 2 & -1 \end{smallmatrix}\right]$, then $A^t = \left[\begin{smallmatrix} 1 & 0 \\ 2 & 2 \\ 3 & -1 \end{smallmatrix}\right]$.

Theorem 3.1.10. *When the relevant sums and products are defined, then for the matrices A, B and $k \in F$ we have:*

(i) $(A + B)^t = A^t + B^t$;

(ii) $(A^t)^t = A$;

(iii) $(kA)^t = kA^t$;

(iv) $(AB)^t = B^t A^t$.

Proof. (i)   Let $A = [a_{ij}]_{m \times n}$ and $B = [b_{ij}]_{m \times n}$. Then

$$(A + B)^t = ([a_{ij}]_{m \times n} + [b_{ij}]_{m \times n})^t$$
$$= [a_{ij} + b_{ij}]_{m \times n}^t$$
$$= [a_{ji} + b_{ji}]_{n \times m}$$
$$= [a_{ji}]_{n \times m} + [b_{ji}]_{n \times m}$$
$$= [a_{ij}]_{m \times n}^t + [b_{ij}]_{m \times n}^t$$
$$= A^t + B^t.$$

(ii) Let $A = [a_{ij}]_{m \times n}$. Then $A^t = [a_{ji}]_{n \times m}$, and hence

$$\left(A^t\right)^t = [a_{ij}]_{m \times n} = A.$$

(iii) Let $A = [a_{ij}]_{m \times n}$. Then $kA = [ka_{ij}]_{m \times n}$, and hence

$$(kA)^t = [ka_{ji}]_{n \times m} = k[a_{ji}]_{n \times m} = kA^t.$$

(iv) Let $A = [a_{ij}]_{m \times n}$ and $B = [b_{ij}]_{n \times p}$.

If $[(AB)^t]_{ji}$ denotes the (j, i)-th entry of the matrix $(AB)^t$, then

$$[(AB)^t]_{ji} = [AB]_{ij} = \sum_{l=1}^{n} a_{il} b_{lj} = \sum_{l=1}^{n} b_{lj} a_{il}$$

$$= \sum_{l=1}^{n} [B^t]_{jl} [A^t]_{li}$$

$$= [B^t A^t]_{ji}.$$

This shows that (j, i)-th entry of the matrix $(AB)^t$ is equal to (j, i)-th entry of the matrix $B^t A^t$ for all i, j. Hence, $(AB)^t = B^t A^t$. $\qquad \square$

Definition 3.1.11. Let $A = [a_{ij}]$ be an n-square matrix. Then the trace of A denoted by $\mathrm{tr}(A)$ is defined as

$$\mathrm{tr}(A) = a_{11} + a_{22} + \cdots + a_{nn} = \sum_{i=1}^{n} a_{ii}.$$

Thus, the sum of the diagonal elements of a square matrix is called the trace of the matrix.

For example, if $A = \begin{bmatrix} 1 & 2 & 3 \\ -1 & 4 & 0 \\ 2 & 3 & 5 \end{bmatrix}$, then $\mathrm{tr}(A) = 1 + 4 + 5 = 10$.

Theorem 3.1.12. *Let A and B be n-square matrices and let $k \in F$. Then:*
(i) $\mathrm{tr}(A + B) = \mathrm{tr}(A) + \mathrm{tr}(B)$;
(ii) $\mathrm{tr}(kA) = k(\mathrm{tr}(A))$;
(iii) $\mathrm{tr}(A^t) = \mathrm{tr}(A)$;
(iv) $\mathrm{tr}(AB) = \mathrm{tr}(BA)$;
(v) $\mathrm{tr}(AA^t) = \mathrm{tr}(A^t A) = \sum_{i,j} a_{ij}^2$, *where a_{ij} denotes the (i, j)-th entry of A.*

Proof. We prove parts (iv) and (v) of the theorem and remaining are left as an exercise.

(iv)

$$\text{tr}(AB) = \sum_{i=1}^{n} [AB]_{ii}$$

$$= \sum_{i=1}^{n} \left(\sum_{j=1}^{n} a_{ij} b_{ji} \right)$$

$$= \sum_{i=1}^{n} \left(\sum_{j=1}^{n} b_{ji} a_{ij} \right)$$

$$= \sum_{j=1}^{n} \sum_{i=1}^{n} b_{ji} a_{ij}$$

$$= \sum_{j=1}^{n} [BA]_{jj}$$

$$= \text{tr}(BA).$$

(v) Using the property $a_{ji} \in A^t \Rightarrow a_{ij} \in A$, we have

$$\text{tr}(AA^t) = \sum_{i=1}^{n} [AA^t]_{ii}$$

$$= \sum_{i=1}^{n} \left(\sum_{j=1}^{n} a_{ij} a_{ji} \right)$$

$$= \sum_{i=1}^{n} \sum_{j=1}^{n} a_{ij} a_{ij}$$

$$= \sum_{i,j} a_{ij}^2.$$

Hence, $\text{tr}(AA^t) = \sum_{i,j} a_{ij}^2$.
Similarly, $\text{tr}(A^t A) = \sum_{i,j} a_{ij}^2$.
From (iv), we have $\text{tr}(AA^t) = \text{tr}(A^t A)$. Hence,

$$\text{tr}(AA^t) = \text{tr}(A^t A) = \sum_{i,j} a_{ij}^2.$$

□

🄸 Exercises

(3.1.1) Let A and B be n-square matrices and let $k \in F$. Then prove that:
  (i) $\text{tr}(A + B) = \text{tr}(A) + \text{tr}(B)$;
  (ii) $\text{tr}(kA) = k(\text{tr}(A))$;
  (iii) $\text{tr}(A^t) = \text{tr}(A)$.

(3.1.2)   Let $A = [a_{ij}]$, where $a_{ij} = i + j$ and

$$B = [b_{ij}], \quad \text{where } b_{ij} = \begin{cases} 1, & \text{if } i + j \text{ is odd,} \\ 0, & \text{otherwise} \end{cases}$$

be matrices of order 4×4. Then find:
(i) $2A + 5B$;
(ii) $3A - 5B$;
(iii) AB and BA;
(iv) A^2B and AB^2;
(v) A^t and B^t.

(3.1.3)   Let $A = [a_{ij}]$, where $a_{ij} = (-1)^{i+j}$ and

$$B = [b_{ij}], \quad \text{where } b_{ij} = \begin{cases} 1, & \text{if } i + j \text{ is even,} \\ 0, & \text{otherwise} \end{cases}$$

be matrices of order 4×4. Then find:
(i) $A - B$;
(ii) $-A + 5B$;
(iii) AB and BA;
(iv) A^2B and AB^2;
(v) A^t and B^t.

(3.1.4)   Let $A = \begin{bmatrix} 1 & 2 & 3 & 4 \\ 0 & 2 & -1 & -3 \\ -2 & 2 & -5 & 1 \end{bmatrix}$ and $B = \begin{bmatrix} 4 & 3 & 2 & 1 \\ 2 & 0 & -1 & -3 \\ 4 & -2 & 3 & 6 \end{bmatrix}$ be 3×4 matrices. Then show that
(i) $(A + B)^t = A^t + B^t$;
(ii) $(A^t)^t = A$;
(iii) $(7A)^t = 7A^t$.

(3.1.5)   For the matrices $A = \begin{bmatrix} 1 & 2 & 3 \\ 0 & 2 & -1 \\ -2 & 2 & -5 \end{bmatrix}$ and $B = \begin{bmatrix} 4 & 3 & 2 \\ 2 & 0 & -1 \\ 4 & -2 & 3 \end{bmatrix}$, show that:
(i) $(AB)^t = B^tA^t$;
(ii) $\text{tr}(A + B) = \text{tr}(A) + \text{tr}(B)$;
(iii) $\text{tr}(5A) = 5(\text{tr}(A))$;
(iv) $\text{tr}(A^t) = \text{tr}(A)$;
(v) $\text{tr}(AB) = \text{tr}(BA)$;
(vi) $\text{tr}(AA^t) = \text{tr}(A^tA) = \sum_{i,j} a_{ij}^2$, where a_{ij} denotes the (i,j)-th entry of A.

3.2 Some special matrices and their properties

Definition 3.2.1 (Identity matrix). An n-square matrix I_n is called an identity matrix or unit matrix if its all diagonal entries are one and all off diagonal entries are zero.

For example, $I_3 = \begin{bmatrix} 1 & 0 & 0 \\ 0 & 1 & 0 \\ 0 & 0 & 1 \end{bmatrix}$.
-    If A is an n-square matrix, then $AI_n = I_nA = A$.
-    If A is an $m \times n$ matrix, then $AI_n = I_mA = A$.

Definition 3.2.2 (Diagonal matrix). An n-square matrix $A = [a_{ij}]$ is called a diagonal matrix if $a_{ij} = 0$ for all $i \neq j$. That is, a square matrix in which all its off diagonal entries are zero is called a diagonal matrix. Diagonal matrix A is denoted by $\text{diag}(a_{11}, a_{22}, \ldots, a_{nn})$.

For example, $\text{diag}(1, -1, 2) = \begin{bmatrix} 1 & 0 & 0 \\ 0 & -1 & 0 \\ 0 & 0 & 2 \end{bmatrix}$.

Identity matrix I_n is a diagonal matrix.

- If $A = \text{diag}(a_{11}, a_{22}, \ldots, a_{nn})$ is a diagonal matrix, B is an $n \times m$ matrix and C is an $m \times n$ matrix, then

    AB = the matrix obtained by multiplying by i-th row of B by a_{ii} for all $i = 1, 2, \ldots, n$, and

    CA = the matrix obtained by multiplying by i-th column of C by a_{ii} for all $i = 1, 2, \ldots, n$.

For example, let

$$A = \begin{bmatrix} 2 & 0 & 0 \\ 0 & 3 & 0 \\ 0 & 0 & 5 \end{bmatrix}, \quad B = \begin{bmatrix} a_1 & a_2 & a_3 & a_4 \\ b_1 & b_2 & b_3 & b_4 \\ c_1 & c_2 & c_3 & c_4 \end{bmatrix} \quad \text{and} \quad C = \begin{bmatrix} a_1 & a_2 & a_3 \\ b_1 & b_2 & b_3 \\ c_1 & c_2 & c_3 \\ d_1 & d_2 & d_3 \end{bmatrix}.$$

Then

$$AB = \begin{bmatrix} 2a_1 & 2a_2 & 2a_3 & 2a_4 \\ 3b_1 & 3b_2 & 3b_3 & 3b_4 \\ 5c_1 & 5c_2 & 5c_3 & 5c_4 \end{bmatrix} \quad \text{and} \quad CA = \begin{bmatrix} 2a_1 & 3a_2 & 5a_3 \\ 2b_1 & 3b_2 & 5b_3 \\ 2c_1 & 3c_2 & 5c_3 \\ 2d_1 & 3d_2 & 5d_3 \end{bmatrix}.$$

- If $A = \text{diag}(a_{11}, a_{22}, \ldots, a_{nn})$ and $B = \text{diag}(b_{11}, b_{22}, \ldots, b_{nn})$ be two diagonal matrices, then $AB = BA = \text{diag}(a_{11}b_{11}, a_{22}b_{22}, \ldots, a_{nn}b_{nn})$ is also a diagonal matrix. Hence, $A^n = \text{diag}(a_{11}^n, a_{22}^n, \ldots, a_{nn}^n)$.

Definition 3.2.3 (Scalar matrix). A diagonal matrix is called a scalar matrix if all its diagonal entries are equal. For any scalar k, the scalar matrix is of the form kI_n.

For example, $A = \begin{bmatrix} 5 & 0 & 0 \\ 0 & 5 & 0 \\ 0 & 0 & 5 \end{bmatrix} = 5I_3$ is a scalar matrix. Since every scalar matrix is a diagonal matrix, it satisfies all the properties of a diagonal matrix.

Definition 3.2.4 (Triangular matrix). A square matrix A is called an upper (lower) triangular matrix if all the entries below (above) the main diagonal are zero. Thus, an n-square matrix $A = [a_{ij}]$ is upper triangular if all $a_{ij} = 0$ for $i > j$ and lower triangular if all $a_{ij} = 0$ for $i < j$.

For example, $A = \begin{bmatrix} 1 & 2 & 3 \\ 0 & 4 & 5 \\ 0 & 0 & 6 \end{bmatrix}$ is an upper triangular matrix, and $B = \begin{bmatrix} 1 & 0 & 0 \\ 2 & 4 & 0 \\ 3 & 5 & 6 \end{bmatrix}$ is a lower triangular matrix.

- Transpose of an upper (lower) triangular matrix is a lower (upper) triangular matrix. In the above example, $A^t = B$ and $B^t = A$.

– Product of upper (lower) triangular matrices is an upper (lower) triangular matrix. For example, let $A = \begin{bmatrix} 1 & 2 \\ 0 & 3 \end{bmatrix}$ and $B = \begin{bmatrix} -1 & 2 \\ 0 & 1 \end{bmatrix}$ be upper triangular matrices. Then $AB = \begin{bmatrix} -1 & 4 \\ 0 & 3 \end{bmatrix}$ is an upper triangular matrix.
Similarly, if $C = \begin{bmatrix} 1 & 0 \\ 3 & 2 \end{bmatrix}$ and $D = \begin{bmatrix} -1 & 0 \\ 2 & 4 \end{bmatrix}$ are lower triangular matrices, then $CD = \begin{bmatrix} -1 & 0 \\ 1 & 8 \end{bmatrix}$ is a lower triangular matrix.
– If $A = [a_{ij}]_{n\times n}$ and $B = [b_{ij}]_{n\times n}$ are upper (lower) triangular matrices, then diagonal elements of upper (lower) triangular matrix AB are $\{a_{11}b_{11}, a_{22}b_{22}, \ldots, a_{nn}b_{nn}\}$.
For example, see the diagonals of AB and CD matrices given as above.

Definition 3.2.5 (Invertible or nonsingular matrix). An n-square matrix A is called invertible or nonsingular if there exists an n-square matrix B such that $AB = BA = I_n$.
The matrix B is called the inverse of A and we write $A^{-1} = B$. Clearly, $(A^{-1})^{-1} = A$.

Theorem 3.2.6. *Inverse of an invertible matrix is unique.*

Proof. Let A be an invertible matrix of order $n \times n$. Suppose B_1 and B_2 are n-square matrices such that $AB_1 = B_1A = I_n$ and $AB_2 = B_2A = I_n$. Then $B_1 = B_1I_n = B_1(AB_2) = (B_1A)B_2 = I_nB_2 = B_2$. Thus, both the inverses B_1 and B_2 of A are equal. □

Theorem 3.2.7. *The product of invertible matrices is invertible.*

Proof. Let A and B be n-square invertible matrices. Then we shall show that $(AB)^{-1} = B^{-1}A^{-1}$:

$$(AB)(B^{-1}A^{-1}) = A(BB^{-1})A^{-1} = AI_nA^{-1} = AA^{-1} = I_n \quad \text{and}$$
$$(B^{-1}A^{-1})(AB) = B^{-1}(A^{-1}A)B = B^{-1}I_nB = B^{-1}B = I_n$$

prove that $B^{-1}A^{-1}$ is the inverse of the matrix AB.
Hence, AB is invertible and $(AB)^{-1} = B^{-1}A^{-1}$. □

Theorem 3.2.8. *Let A be an n-square invertible matrix. Then A^t is invertible and $(A^t)^{-1} = (A^{-1})^t$.*

Proof. From the properties of transpose of a matrix, we have

$$A^t(A^{-1})^t = (A^{-1}A)^t = (I_n)^t = I_n, \quad \text{and}$$
$$(A^{-1})^tA^t = (AA^{-1})^t = (I_n)^t = I_n.$$

This proves that $(A^{-1})^t$ is the inverse of the matrix A^t. Hence, $(A^t)^{-1} = (A^{-1})^t$. □

Example 3.2.9.
(a) Every identity matrix is invertible.
(b) Let $D = \text{diag}(a_{11}, a_{22}, \ldots, a_{nn})$ be a diagonal matrix, where $a_{ii} \neq 0 \ \forall 1 \leq i \leq n$. Then $D^{-1} = \text{diag}(a_{11}^{-1}, a_{22}^{-1}, \ldots, a_{nn}^{-1})$ is the inverse of D. Since $DD^{-1} = D^{-1}D = \text{diag}(a_{11}a_{11}^{-1}, a_{22}a_{22}^{-1}, \ldots, a_{nn}a_{nn}^{-1}) = I_n$.

Thus, a diagonal matrix is invertible if and only if all its diagonal entries are nonzero.

Definition 3.2.10 (Symmetric matrix). Let $A = [a_{ij}]$ be an n-square matrix. Then A is called symmetric if $A^t = A$. If $A^t = -A$, then matrix A is called skew-symmetric.

In a skew-symmetric matrix $A = [a_{ij}]$, $a_{ij} = -a_{ji}$, $\forall 1 \le i,j \le n$, and hence $a_{ii} = -a_{ii} \Rightarrow 2a_{ii} = 0 \Rightarrow a_{ii} = 0$, $\forall 1 \le i \le n$. Thus, in a skew-symmetric matrix all the diagonal entries are zero.

For example, $A = \begin{bmatrix} 1 & -1 & 2 \\ -1 & 2 & 5 \\ 2 & 5 & 3 \end{bmatrix}$ is a symmetric matric and $B = \begin{bmatrix} 0 & 1 & 2 \\ -1 & 0 & 3 \\ -2 & -3 & 0 \end{bmatrix}$ is a skew-symmetric matrix.

Theorem 3.2.11. *Let A be an n-square matrix. Then:*
(i) *AA^t and A^tA are symmetric matrices,*
(ii) *If A is symmetric and invertible, then A^{-1} is symmetric.*

Proof. (i) From the properties of transpose of a matrix, we have the result: $(AA^t)^t = (A^t)^t A^t = AA^t$ and $(A^tA)^t = A^t(A^t)^t = A^tA$.
(ii) From Theorem 3.2.8, we have that $(A^t)^{-1} = (A^{-1})^t$. Since A is symmetric, $A^t = A$, and hence $(A^{-1})^t = (A^t)^{-1} = (A)^{-1} = A^{-1}$. This proves that A^{-1} is a symmetric matrix. □

Theorem 3.2.12. *Every square matrix can be expressed uniquely as the sum of a symmetric matrix and a skew-symmetric matrix.*

Proof. Let A be an n-square matrix. Then

$$A = \frac{(A + A^t)}{2} + \frac{(A - A^t)}{2},$$

where

$$\left(\frac{(A + A^t)}{2} \right)^t = \frac{1}{2}(A^t + (A^t)^t) = \frac{1}{2}(A^t + A) = \frac{1}{2}(A + A^t) \quad \text{and}$$

$$\left(\frac{(A - A^t)}{2} \right)^t = \frac{1}{2}(A^t - (A^t)^t) = \frac{1}{2}(A^t - A) = -\frac{1}{2}(A - A^t),$$

prove that $\frac{A+A^t}{2}$ is symmetric and $\frac{A-A^t}{2}$ is skew-symmetric.

Now, we shall that above expression is unique. Suppose $A = B + C$, where B is symmetric and C is skew-symmetric. Then $A^t = B^t + C^t = B - C$. Hence, $A + A^t = 2B \Rightarrow B = \frac{A+A^t}{2}$ and $A - A^t = 2C \Rightarrow C = \frac{A-A^t}{2}$. □

Example 3.2.13. Let $A = \begin{bmatrix} 1 & -1 & 2 \\ 3 & 4 & -2 \\ 6 & 2 & 3 \end{bmatrix}$, and hence $A^t = \begin{bmatrix} 1 & 3 & 6 \\ -1 & 4 & 2 \\ 2 & -2 & 3 \end{bmatrix}$. If

$$B = \frac{1}{2}(A + A^t) = \frac{1}{2} \begin{bmatrix} 2 & 2 & 8 \\ 2 & 8 & 0 \\ 8 & 0 & 6 \end{bmatrix} = \begin{bmatrix} 1 & 1 & 4 \\ 1 & 4 & 0 \\ 4 & 0 & 3 \end{bmatrix} \quad \text{and}$$

$$C = \frac{1}{2}(A - A^t) = \frac{1}{2}\begin{bmatrix} 0 & -4 & -4 \\ 4 & 0 & -4 \\ 4 & 4 & 0 \end{bmatrix} = \begin{bmatrix} 0 & -2 & -2 \\ 2 & 0 & -2 \\ 2 & 2 & 0 \end{bmatrix},$$

then $A = B + C$, where B is a symmetric matrix and C is a skew symmetric matrix.

Definition 3.2.14 (Orthogonal matrix). An n-square matrix A is said to be orthogonal if $AA^t = A^tA = I_n$. Thus, an orthogonal matrix is invertible and $A^t = A^{-1}$. In other words, an invertible matrix A is orthogonal if $A^{-1} = A^t$.

Let $A = \begin{bmatrix} \cos x & \sin x \\ -\sin x & \cos x \end{bmatrix}$. Then $AA^t = A^tA = I_2$. Hence, A is an orthogonal matrix.

Theorem 3.2.15. *The product of two orthogonal matrices is an orthogonal matrix.*

Proof. Let A and B be $n \times n$ orthogonal matrices. Then $(AB)(AB)^t = ABB^tA^t = AI_nA^t = AA^t = I_n$ and $(AB)^t(AB) = B^tA^tAB = B^tI_nB = B^tB = I_n$. Hence, AB is an orthogonal matrix. □

Definition 3.2.16 (Nilpotent matrix). A square matrix A of order $n \times n$ is called a nilpotent matrix if $A^m = 0$ for some $m \geq 1$.

Example 3.2.17.
(a) Let $A = \begin{bmatrix} 0 & 1 & 0 \\ 0 & 0 & 1 \\ 0 & 0 & 0 \end{bmatrix}$. Then $A^3 = \begin{bmatrix} 0 & 0 & 0 \\ 0 & 0 & 0 \\ 0 & 0 & 0 \end{bmatrix}$. Hence, A is a nilpotent matrix.
(b) If we take $A = \begin{bmatrix} 0 & 1 \\ 0 & 0 \end{bmatrix}$ and $B = \begin{bmatrix} 0 & 0 \\ 1 & 0 \end{bmatrix}$, then A and B are nilpotent matrices but $A + B = \begin{bmatrix} 0 & 1 \\ 1 & 0 \end{bmatrix}$ and $AB = \begin{bmatrix} 1 & 0 \\ 0 & 0 \end{bmatrix}$ are not nilpotent matrices.
(c) Every nilpotent matrix is a singular matrix.

Theorem 3.2.18. *Let A be an n-square nilpotent matrix such that $A^m = 0$, $m \geq 1$. Then the matrix $I_n + A$ is nonsingular.*

Proof. To prove the result, we shall show that $I_n + A$ is invertible and

$$(I_n + A)^{-1} = I_n - A + A^2 - \cdots + (-1)^{m-1}A^{m-1}.$$

We have

$$(I_n + A)(I_n - A + A^2 - \cdots + (-1)^{m-2}A^{m-2} + (-1)^{m-1}A^{m-1})$$
$$= I_n(I_n - A + A^2 - \cdots + (-1)^{m-2}A^{m-2} + (-1)^{m-1}A^{m-1})$$
$$\quad + A(I_n - A + A^2 - \cdots + (-1)^{m-2}A^{m-2} + (-1)^{m-1}A^{m-1})$$
$$= (I_n - A + A^2 - \cdots + (-1)^{m-2}A^{m-2} + (-1)^{m-1}A^{m-1})$$
$$\quad + (A - A^2 + A^3 - A^4 + \cdots + (-1)^{m-2}A^{m-1} + (-1)^{m-1}A^m)$$
$$= I_n + (-1)^{m-1}A^m$$
$$= I_n.$$

This proves that $I_n + A$ is invertible, and hence nonsingular. □

Definition 3.2.19 (Similar matrices). Let A and B be n-square matrices. Then A is said to be similar to B if there is a nonsingular matrix P such that $A = P^{-1}BP$.

Theorem 3.2.20. *Similarity is an equivalence relation on $M_{n,n}(F)$, the set of all n-square matrices with entries in the field F.*

Proof. Let $A, B, C \in M_{n,n}(F)$. Then $A = I_n^{-1}AI_n$ implies that A is similar to itself. Hence, similarity relation is reflexive.

Now, suppose that A is similar to B. Then there exist $P \in M_{n,n}(F)$ such that $A = P^{-1}BP$. Next, $A = P^{-1}BP \Rightarrow PAP^{-1} = B \Rightarrow (P^{-1})^{-1}AP^{-1} = B$.

This shows that there exist a nonsingular matrix P^{-1} such that $B = (P^{-1})^{-1}AP^{-1}$. That is, B is similar to A. Hence, similarity is a symmetric relation.

Finally, we shall show that the similarity relation is transitive.

Let A be similar to B and B be similar to C. Then there exist nonsingular matrices P and Q such that $A = P^{-1}BP$ and $B = Q^{-1}CQ$.

Since (QP) is the product of invertible matrices, (QP) is invertible, and hence $(QP)^{-1}C(QP) = P^{-1}Q^{-1}CQP = P^{-1}BP = A$ gives that A is similar to C. Thus, similarity is an equivalence relation on $M_{n,n}(F)$. □

Definition 3.2.21 (Complex matrices). A matrix A with complex entries is called a complex matrix. That is, a matrix $A = [a_{ij}]$ is a complex matrix if $a_{ij} \in \mathbb{C}, \forall i, j$.

Complex matrices satisfy the following additional definitions and properties along with all definitions and properties of matrices discussed previously:
- Let $A = [a_{ij}]_{m \times n}$ be a complex matrix. Then the matrix $\bar{A} = [\bar{a}_{ij}]_{m \times n}$ is called the complex conjugate of A, where $\bar{a}_{ij}$ is the conjugate of the complex number a_{ij}.
- The matrix $A^* = (\bar{A})^t = \bar{A}^t$ is called conjugate transpose or tranjugate of A.
- An n-square complex matrix A is called Hermitian if $A^* = A$, skew-Hermitian if $A^* = -A$ and unitary if $A^*A = AA^* = I_n$ (i. e., $A^* = A^{-1}$).

Note that in a Hermitian matrix $\bar{a}_{ii} = a_{ii}$ and in a skew-Hermitian matrix $\bar{a}_{ii} = -a_{ii}$. It means all the diagonal entries of a Hermitian matrix are real and all the diagonal entries of a skew-Hermitian matrix are purely imaginary or zero.

Example 3.2.22. Let $A = \begin{bmatrix} 2+3i & i & 1-i \\ -i & 3+4i & 6 \end{bmatrix}$ be a complex matrix of order 2×3. Then

$$\bar{A} = \begin{bmatrix} 2-3i & -i & 1+i \\ i & 3-4i & 6 \end{bmatrix} \quad \text{and} \quad A^* = (\bar{A})^t = \begin{bmatrix} 2-3i & i \\ -i & 3-4i \\ 1+i & 6 \end{bmatrix}.$$

Some important properties of complex matrices are listed in the form of the following theorem without proof. The proof is simple and left as an exercise.

Theorem 3.2.23. *When the relevant sums and products are defined, then for the complex matrices A, B and scalar $k \in \mathbb{C}$, we have the following:*

(i) $\overline{A + B} = \overline{A} + \overline{B}$;
(ii) $\overline{AB} = \overline{A}\,\overline{B}$;
(iii) $\overline{\overline{A}} = A$;
(iv) $\overline{kA} = \overline{k}\,\overline{A}$;
(v) $(A + B)^* = A^* + B^*$;
(vi) $(AB)^* = B^*A^*$;
(vii) $(A^*)^* = A$;
(viii) $(kA)^* = \overline{k}A^*$;
(ix) If A is invertible, then $(A^*)^{-1} = (A^{-1})^*$;
(x)   If A is an invertible Hermitian matrix, then A^{-1} is also Hermitian.

Theorem 3.2.24. *Every n-square complex matrix can be expressed uniquely as the sum of a Hermitian matrix and a skew-Hermitian matrix.*

Proof. Let A be an n-square complex matrix. Then $A = \frac{A+A^*}{2} + \frac{A-A^*}{2}$, where $\frac{A+A^*}{2}$ is Hermitian and $\frac{A-A^*}{2}$ is skew-Hermitian. For uniqueness and a more detailed proof, follow the steps of the corresponding theorem for symmetric and skew-symmetric matrices. □

Example 3.2.25. Let

$$A = \begin{bmatrix} 1 + 2i & -1 & 2 - 4i \\ 4 - i & 4 + 4i & -2 - 2i \\ 6 + 3i & 2 - 3i & 3 - 4i \end{bmatrix}.$$

Then

$$A^* = \begin{bmatrix} 1 - 2i & 4 + i & 6 - 3i \\ -1 & 4 - 4i & 2 + 3i \\ 2 + 4i & -2 + 2i & 3 + 4i \end{bmatrix}.$$

If

$$B = \frac{1}{2}(A + A^*) = \begin{bmatrix} 1 & \frac{3}{2} + \frac{1}{2}i & 4 - \frac{7}{2}i \\ \frac{3}{2} - \frac{1}{2}i & 4 & \frac{1}{2}i \\ 4 + \frac{7}{2}i & -\frac{1}{2}i & 3 \end{bmatrix} \quad \text{and}$$

$$C = \frac{1}{2}(A - A^*) = \begin{bmatrix} 2i & -\frac{5}{2} - \frac{1}{2}i & -2 - \frac{1}{2}i \\ \frac{5}{2} - \frac{1}{2}i & 4i & -2 - \frac{5}{2}i \\ 2 - \frac{1}{2}i & 2 - \frac{5}{2}i & -4i \end{bmatrix},$$

then $A = B + C$, where B is a Hermitian matrix and C is a skew-Hermitian matrix.

Theorem 3.2.26. *If A is a Hermitian (skew-Hermitian) matrix, then iA is a skew-Hermitian (Hermitian) matrix.*

Proof. Let A be a Hermitian matrix. Then $A^* = A$.

Hence, $(iA)^* = \bar{i}A^* = -iA$, proves that iA is skew-Hermitian.

If A is a skew-Hermitian matrix, then $A^* = -A$.

Hence, $(iA)^* = \bar{i}A^* = -i(-A) = iA$, proves that iA is Hermitian. ☐

ℹ **Exercises**

(3.2.1)   Let D be a diagonal matrix. Then for any polynomial $f(x)$ show that $f(D)$ is a diagonal matrix.

(3.2.2)   If A is a triangular (upper or lower) matrix, then show that $f(A)$ is a triangular matrix, where $f(x)$ is a polynomial.

(3.2.3)   Prove that the transpose of an upper triangular matrix is a lower triangular matrix and vice versa.

(3.2.4)   Prove that the product of two lower triangular matrices is a lower triangular matrix and the product of two upper triangular matrices is an upper triangular matrix.

(3.2.5)   Let

$$A = \begin{bmatrix} \frac{1}{3} & \frac{1}{3} & \frac{1}{3} \\ \frac{1}{3} & \frac{1}{3} & \frac{1}{3} \\ \frac{1}{3} & \frac{1}{3} & \frac{1}{3} \end{bmatrix} \quad \text{and} \quad B = \begin{bmatrix} 1 & 0 \\ 0 & -1 \end{bmatrix}.$$

Then find
(i) A^{100} (ii) A^{103} (iii) B^{100} (iv) B^{105}.

(3.2.6)   Express the following matrices as the sum of symmetric and skew-symmetric matrices:

$$A = \begin{bmatrix} 4 & 2 & 6 \\ 6 & -7 & 0 \\ -3 & 0 & 1 \end{bmatrix}, \quad B = \begin{bmatrix} -7 & 12 & 9 \\ 3 & -3 & 12 \\ 8 & 6 & 5 \end{bmatrix}$$

(3.2.7)   Let $A = \begin{bmatrix} a & b \\ c & d \end{bmatrix}$ be a matrix such that $(5A + I_2)^{-1} = \begin{bmatrix} 1 & 3 \\ 3 & 1 \end{bmatrix}$. Then find the relation between (i) a and b and (ii) c and d.

(3.2.8)   Let A and B be n-square matrices. Then show that $B^t AB$ is (a) symmetric if A is symmetric and (b) skew-symmetric if A is skew-symmetric.

(3.2.9)   Let A be an n-square matrix. Then show that $A^t A$ is a symmetric matrix.

(3.2.10)   Let $A = \begin{bmatrix} 4 & 2 & 6 \\ 6 & -7 & 0 \\ -3 & 0 & 1 \end{bmatrix}$. Then find a matrix B such that the matrix BA becomes a symmetric matrix.

(3.2.11)   Let A and B be n-square matrices. Then show that (a) $B^t(A + A^t)B$ is symmetric, (b) $B^t(A - A^t)B$ is skew-symmetric and (c) $B^t(A^t A)B$ is symmetric.

(3.2.12)   If A and B are similar matrices, then show that $\text{tr}(A) = \text{tr}(B)$.

(3.2.13)   If B is similar to A, then show that B^n is similar to A^n.

(3.2.14)   Let A be a 2-square matrix with real entries such that $\det(A+I_2) = 1+\det A$. Then show that $\text{tr}(A) = 0$.

(3.2.15)   Show that the matrices

$$A = \begin{bmatrix} 2 & 2-2i & 4 \\ 2+2i & 6 & 2i \\ 4 & -2i & 0 \end{bmatrix} \quad \text{and} \quad B = \begin{bmatrix} 1 & -2+3i & 1-2i \\ -2-3i & 0 & 4+3i \\ 1+2i & 4-3i & 2 \end{bmatrix}$$

are Hermitian and iA, iB are skew-Hermitian.

(3.2.16) Show that the matrices

$$A = \begin{bmatrix} i & 1+i & 3-2i \\ -1+i & i & 4+i \\ -3-2i & -4+i & 0 \end{bmatrix} \quad \text{and} \quad B = \begin{bmatrix} i & -1+\frac{3}{2}i & -2i \\ 1+\frac{3}{2}i & \frac{5}{2}i & -3 \\ -2i & 3 & \frac{3}{2}i \end{bmatrix}$$

are skew-Hermitian and iA and iB are Hermitian.

(3.2.17) Express the following matrices as the sum of Hermitian and skew-Hermitian matrices:

$$A = \begin{bmatrix} 2 & 2-2i & 4 \\ 2+2i & 6 & 2i \\ 4 & -2i & 0 \end{bmatrix}, \quad B = \begin{bmatrix} i & 1+i & 3-2i \\ -1+i & i & 4+i \\ -3-2i & -4+i & 0 \end{bmatrix},$$

$$C = \begin{bmatrix} 2+3i & 2-2i & 4-i \\ 2+2i & -6i & 2i \\ 4 & -2i & 6-4i \end{bmatrix}, \quad D = \begin{bmatrix} i & 1-2i & 4-2i \\ -1+4i & 8-2i & 4+i \\ 3+4i & -4+i & 0 \end{bmatrix}.$$

(3.2.18) Let A and B be n-square complex matrices. Then show that B^*AB is (a) Hermitian if A is Hermitian and (b) skew-Hermitian if A is skew-Hermitian.

(3.2.19) Let A and B be n-square Hermitian matrices. Then show that $AB + BA$ is Hermitian and $AB - BA$ is skew-Hermitian.

(3.2.20) Let A be an n-square invertible complex matrix. If $A + A^*$ is invertible, then show that $(A + A^*)^{-1}$ is Hermitian.

(3.2.21) Let A and B be n-square matrices. Then show that $B^*(A + A^*)B$ is Hermitian and $B^*(A - A^*)B$ is skew-Hermitian.

(3.2.22) Show that an n-square matrix A can be expressed as $H_1 + iH_2$ uniquely, where H_1, H_2 are Hermitian matrices. Hint: $H_1 + iH_2 = \{\frac{1}{2}(A + A^*)\} + i\{\frac{1}{2i}(A - A^*)\}$.

3.3 Vector spaces formed by the collection of matrices

In Section 3.1, we defined matrix addition and scalar multiplication. Using the properties of these operations, we can prove the following result.

Theorem 3.3.1. *Let $M_{m,n}(F)$ represent the set of all $m \times n$ matrices with entries from a field F. Then $M_{m,n}(F)$ forms a vector space over F with the operations addition and scalar multiplication defined as follows: For $A = [a_{ij}], B = [b_{ij}] \in M_{m,n}(F)$ and $k \in F$, $A + B = [a_{ij} + b_{ij}]$ and $kA = [ka_{ij}]$.*

Proof. We will first demonstrate that $M_{m,n}(F)$ is an Abelian group.

Let $A, B, C \in M_{m,n}(F)$. The properties $(A + B) + C = A + (B + C)$ and $A + B = B + A$ show that matrix addition is associative and commutative.

Clearly, the zero matrix $0_{m \times n}$ of order $m \times n$ serves as the additive identity.

Additionally, for $A = [a_{ij}] \in M_{m,n}(F)$, there exists $-A = [-a_{ij}] \in M_{m,n}(F)$ such that $A + (-A) = A - A = 0_{m \times n}$.

Thus, $M_{m,n}(F)$ is an Abelian group under matrix addition.

Next, for $k, l \in F$ and $A, B \in M_{m,n}(F)$, according to Theorem 3.1.5, we have

$$k(A + B) = kA + kB,$$
$$(k + l)A = kA + lA,$$
$$(kl)A = k(lA),$$

and

$$1 \cdot A = 1 \cdot [a_{ij}] = [1 \cdot a_{ij}] = [a_{ij}] = A.$$

This proves that $M_{m,n}(F)$ is a vector space over the field F. □

Corollary 3.3.2. *The set $M_{n,n}(F)$ of all n-square matrices with entries from a field F is a vector space over F under the operations matrix addition and scalar multiplication.*

Example 3.3.3. Let $M_{2,3}(\mathbb{R})$ be the vector space of all 2×3 real matrices. We shall show that the collection $B = \{e_{11}, e_{12}, e_{13}, e_{21}, e_{22}, e_{23}\}$ is a basis of the vector space $M_{2,3}(\mathbb{R})$ over $\mathbb{R}$, where

$$e_{11} = \begin{bmatrix} 1 & 0 & 0 \\ 0 & 0 & 0 \end{bmatrix}, \quad e_{12} = \begin{bmatrix} 0 & 1 & 0 \\ 0 & 0 & 0 \end{bmatrix}, \quad e_{13} = \begin{bmatrix} 0 & 0 & 1 \\ 0 & 0 & 0 \end{bmatrix},$$

$$e_{21} = \begin{bmatrix} 0 & 0 & 0 \\ 1 & 0 & 0 \end{bmatrix}, \quad e_{22} = \begin{bmatrix} 0 & 0 & 0 \\ 0 & 1 & 0 \end{bmatrix}, \quad e_{23} = \begin{bmatrix} 0 & 0 & 0 \\ 0 & 0 & 1 \end{bmatrix}.$$

Let $A = \begin{bmatrix} a & b & c \\ d & e & f \end{bmatrix}$ be any matrix in $M_{2,3}(\mathbb{R})$. Then

$$\begin{bmatrix} a & b & c \\ d & e & f \end{bmatrix} = a \begin{bmatrix} 1 & 0 & 0 \\ 0 & 0 & 0 \end{bmatrix} + b \begin{bmatrix} 0 & 1 & 0 \\ 0 & 0 & 0 \end{bmatrix} + c \begin{bmatrix} 0 & 0 & 1 \\ 0 & 0 & 0 \end{bmatrix} + d \begin{bmatrix} 0 & 0 & 0 \\ 1 & 0 & 0 \end{bmatrix}$$

$$+ e \begin{bmatrix} 0 & 0 & 0 \\ 0 & 1 & 0 \end{bmatrix} + f \begin{bmatrix} 0 & 0 & 0 \\ 0 & 0 & 1 \end{bmatrix}$$

$$= \begin{bmatrix} a & 0 & 0 \\ 0 & 0 & 0 \end{bmatrix} + \begin{bmatrix} 0 & b & 0 \\ 0 & 0 & 0 \end{bmatrix} + \begin{bmatrix} 0 & 0 & c \\ 0 & 0 & 0 \end{bmatrix} + \begin{bmatrix} 0 & 0 & 0 \\ d & 0 & 0 \end{bmatrix}$$

$$+ \begin{bmatrix} 0 & 0 & 0 \\ 0 & e & 0 \end{bmatrix} + \begin{bmatrix} 0 & 0 & 0 \\ 0 & 0 & f \end{bmatrix}.$$

That is,

$$A = ae_{11} + be_{12} + ce_{13} + de_{21} + ee_{22} + fe_{23}.$$

Thus, any matrix $A \in M_{2,3}(\mathbb{R})$ can be written as the linear combination of members of B.

Also,

$$a_1 e_{11} + a_2 e_{12} + a_3 e_{13} + a_4 e_{21} + a_5 e_{22} + a_6 e_{23} = 0_{2 \times 3}$$

$$\Rightarrow \begin{bmatrix} a_1 & a_2 & a_3 \\ a_4 & a_5 & a_6 \end{bmatrix} = \begin{bmatrix} 0 & 0 & 0 \\ 0 & 0 & 0 \end{bmatrix}$$

$$\Rightarrow a_1 = a_2 = a_3 = a_4 = a_5 = a_6 = 0.$$

This gives that B is linearly independent.

Hence, B is a basis of the vector space $M_{2,3}(\mathbb{R})$.

In view of this example, we have the following result.

Theorem 3.3.4. *The dimension of the vector spaces $M_{m,n}(F)$ and $M_{n,n}(F)$ are mn and n^2, respectively.*

Proof. Let $B = \{e_{ij} \in M_{m,n}(F) : e_{ij}$ is a matrix of order $m \times n$ whose ij-th entry is 1 and the rest of the entries are zero$\}$.

Now, we claim that B is a basis of $M_{m,n}(F)$.

Suppose $\sum_{ij} a_{ij} e_{ij} = 0_{m \times n}$, where $a_{ij} \in F$. Then

$$\sum_{ij} a_{ij} e_{ij} = [a_{ij}]_{m \times n} = 0_{m \times n}.$$

By the equality of matrices, we have $a_{ij} = 0 \ \forall i, j$.

Hence, B is linearly independent.

Also, any $A = [a_{ij}] \in M_{m,n}(F)$ can be written as $[a_{ij}] = \sum_{ij} a_{ij} e_{ij}$. That is, B generates $M_{m,n}(F)$. Hence, B is a basis of the vector space $M_{m,n}(F)$. Since B contains mn matrices, the dimension of $M_{m,n}(F)$ is mn. Similarly, dimension of the vector space $M_{n,n}(F)$ is n^2. □

Theorem 3.3.5. *Let $UT(M_{n,n}(F))$ and $LT(M_{n,n}(F))$ denote the set of all upper triangular matrices and lower triangular matrices of order $n \times n$, respectively. Then $UT(M_{n,n}(F))$ and $LT(M_{n,n}(F))$ are subspaces of $M_{n,n}(F)$ of dimension $\frac{n(n+1)}{2}$.*

Proof. Let $A = [a_{ij}], B = [b_{ij}] \in UT(M_{n,n}(F))$.

Then $a_{ij} = b_{ij} = 0 \ \forall i > j$.

Now, for $\alpha, \beta \in F$, $\alpha A + \beta B = [\alpha a_{ij}] + [\beta b_{ij}] = [\alpha a_{ij} + \beta b_{ij}]$.

Since $a_{ij} = b_{ij} = 0, \forall i > j$, $\alpha a_{ij} + \beta b_{ij} = 0, \forall i > j$. That is, $[\alpha a_{ij} + \beta b_{ij}]$ is an upper triangular matrix.

Thus, $(\alpha A + \beta B) \in UT(M_{n,n}(F))$ for all $\alpha, \beta \in F$ and $A, B \in UT(M_{n,n}(F))$. Hence, $UT(M_{n,n}(F))$ is a subspace of $M_{n,n}(F)$.

Next, we shall show that the set $B_U = \{e_{ij} \in M_{n,n}(F), i \leq j : e_{ij}$ is a matrix of order $m \times n$ whose ij-th entry is 1 and rest of the entries are zero$\}$ is a basis of the vector space $UT(M_{n,n}(F))$.

In the previous theorem, it is shown that the set B in $M_{n,n}(F)$ is linearly independent, and hence B_U being the subset of B is linearly independent. It remains to show that B_U generates $UT(M_{n,n}(F))$.

Let $A = [a_{ij}] \in UT(M_{n,n}(F))$. Then $a_{ij} = 0\ \forall i > j$. It means a_{ij} may be zero or may not be zero for $i \le j$. Hence, $[a_{ij}] = \sum_{i \le j} a_{ij} e_{ij}$.

Total number of e_{ij} in B_U

$$= \frac{(\text{total number of } e_{ij} \text{ in } M_{n,n}(F) - \text{total number of } e_{ii} \text{ in } M_{n,n}(F))}{2}$$

$$+ \text{total number of } e_{ii} \text{ in } M_{n,n}(F)$$

$$= \frac{n^2 - n}{2} + n$$

$$= \frac{n(n+1)}{2}.$$

Hence, the dimension of $UT(M_{n,n}(F)) = \frac{n(n+1)}{2}$.

Similarly, it can be shown that $B_L = \{e_{ij} \in M_{n,n}(F) : i \ge j\}$ is a basis of $LT(M_{n,n}(F))$ and $\dim(LT(M_{n,n}(F))) = \frac{n(n+1)}{2}$. $\qquad\square$

The following example helps to visualize the proof of the above theorem.

Example 3.3.6. Let $UT(M_{3,3}(\mathbb{R}))$ be the vector space of all 3×3 upper triangular real matrices. Then $B_U = \{e_{11}, e_{12}, e_{13}, e_{22}, e_{23}, e_{33}\}$ is the basis of $UT(M_{3,3}(\mathbb{R}))$.

Let $A = \begin{bmatrix} a & b & c \\ 0 & d & e \\ 0 & 0 & f \end{bmatrix}$ be an upper triangular real matrix. Then

$$ae_{11} + be_{12} + ce_{13} + de_{22} + ee_{23} + fe_{33}$$

$$= a\begin{bmatrix} 1 & 0 & 0 \\ 0 & 0 & 0 \\ 0 & 0 & 0 \end{bmatrix} + b\begin{bmatrix} 0 & 1 & 0 \\ 0 & 0 & 0 \\ 0 & 0 & 0 \end{bmatrix} + c\begin{bmatrix} 0 & 0 & 1 \\ 0 & 0 & 0 \\ 0 & 0 & 0 \end{bmatrix} + d\begin{bmatrix} 0 & 0 & 0 \\ 0 & 1 & 0 \\ 0 & 0 & 0 \end{bmatrix}$$

$$+ e\begin{bmatrix} 0 & 0 & 0 \\ 0 & 0 & 1 \\ 0 & 0 & 0 \end{bmatrix} + f\begin{bmatrix} 0 & 0 & 0 \\ 0 & 0 & 0 \\ 0 & 0 & 1 \end{bmatrix}$$

$$= \begin{bmatrix} a & b & c \\ 0 & d & e \\ 0 & 0 & f \end{bmatrix} = A.$$

Thus, B_U generates $UT(M_3(\mathbb{R}))$.

Also, $\dim(UT(M_{3,3}(\mathbb{R}))) = \frac{3(3+1)}{2} = 6$, which is equal to the number of members of basis B_U.

Similarly, we can show that $B_L = \{e_{11}, e_{21}, e_{22}, e_{31}, e_{32}, e_{33}\}$ is the basis of $LT(M_{3,3}(\mathbb{R}))$.

The following example is helpful to understand and visualize the proof of the next theorem in general form.

Example 3.3.7. Let $S(M_{3,3}(\mathbb{R}))$ be the vector space of all real symmetric matrices of order 3×3. Then $B_S = \{s_{11}, s_{12}, s_{13}, s_{22}, s_{23}, s_{33}\}$ is the basis of $S(M_3(\mathbb{R}))$, where

$$s_{11} = \begin{bmatrix} 1 & 0 & 0 \\ 0 & 0 & 0 \\ 0 & 0 & 0 \end{bmatrix}, \quad s_{12} = \begin{bmatrix} 0 & 1 & 0 \\ 1 & 0 & 0 \\ 0 & 0 & 0 \end{bmatrix}, \quad s_{13} = \begin{bmatrix} 0 & 0 & 1 \\ 0 & 0 & 0 \\ 1 & 0 & 0 \end{bmatrix},$$

$$s_{22} = \begin{bmatrix} 0 & 0 & 0 \\ 0 & 1 & 0 \\ 0 & 0 & 0 \end{bmatrix}, \quad s_{23} = \begin{bmatrix} 0 & 0 & 0 \\ 0 & 0 & 1 \\ 0 & 1 & 0 \end{bmatrix}, \quad s_{33} = \begin{bmatrix} 0 & 0 & 0 \\ 0 & 0 & 0 \\ 0 & 0 & 1 \end{bmatrix}.$$

Let $A = \begin{bmatrix} a & b & c \\ b & d & e \\ c & e & f \end{bmatrix}$ be a symmetric matrix. Then $A = as_{11} + bs_{12} + cs_{13} + ds_{22} + es_{23} + fs_{33}$.

Also, $\dim(S(M_{3,3}(\mathbb{R}))) = $ Number of elements in $B_S = 6 = \frac{3(3+1)}{2}$.

Theorem 3.3.8. *Let $S(M_{n,n}(F))$ denotes the set of all n-square symmetric matrices with entries in a field F. Then $S(M_{n,n}(F))$ is a subspace of $M_{n,n}(F)$ and $\dim(S(M_{n,n}(F))) = \frac{n(n+1)}{2}$.*

Proof. Let $A, B \in S(M_{n,n}(F))$ and $\alpha, \beta \in F$. Then $A^t = A$ and $B^t = B$.

From $(\alpha A + \beta B)^t = \alpha A^t + \beta B^t = \alpha A + \beta B$, we have that $\alpha A + \beta B$ is a symmetric matrix, and hence $S(M_{n,n}(F))$ is a subspace of $M_{n,n}(F)$.

Consider the set $B_S = \{s_{ij} \in M_{n,n}(F) : i \le j, s_{ij}$ is an n-square matrix whose ij-th and ji-th entries are 1 and rest of the entries are zero$\}$.

Now, we shall show that B_S is a basis of $S(M_{n,n}(F))$.

Let $\sum_{i \le j} a_{ij} s_{ij} = 0_{n \times n}$, where $a_{ij} \in F$. Then

$$\sum_{i \le j} a_{ij} s_{ij} = [a_{ij}]_{n \times n} = 0_{n \times n} \quad \text{implies that} \quad a_{ij} = 0 \quad \forall i, j.$$

Hence, B_S is linearly independent.

Let $A = [a_{ij}] \in S(M_{n,n}(F))$. Then $a_{ij} = a_{ji} \forall i, j$.

Since s_{ij} is a symmetric matrix, $a_{ij} s_{ij}$ is a symmetric matrix with ij-th and ji-th entries a_{ij} and rest of the entries zero. Thus, $A = [a_{ij}] = \sum_{i \le j} a_{ij} s_{ij}$, shows that $S(M_{n,n}(F))$ is generated by B_S. Hence, B_S is a basis of $S(M_{n,n}(F))$.

The number of elements in B_S can be counted by using the formula given in previous theorem, which is $\frac{n^2-n}{2} + n = \frac{n(n+1)}{2}$. Hence, $\dim(S(M_{n,n}(F))) = \frac{n(n+1)}{2}$. □

Theorem 3.3.9. *Let $S_k(M_{n,n}(F))$ denotes the set of all skew-symmetric matrices of order $n \times n$ with entries in a field F. Then $S_k(M_{n,n}(F))$ is a subspace of $M_{n,n}(F)$.*

Proof. Let $A, B \in S_k(M_{n,n}(F))$. Then $A^t = -A$ and $B^t = -B$.

For $\alpha, \beta \in F$, $(\alpha A + \beta B)^t = \alpha A^t + \beta B^t = \alpha(-A) + \beta(-B) = -(\alpha A + \beta B)$ shows that $\alpha A + \beta B$ is a skew-symmetric matrix. Hence, $S_k(M_{n,n}(F))$ is a subspace of $M_{n,n}(F)$. □

Theorem 3.3.10. $M_{n,n}(F)$ *is direct sum of* $S(M_{n,n}(F))$ *and* $S_k(M_{n,n}(F))$, *i. e.,* $M_{n,n}(F) = S(M_{n,n}(F)) \oplus S_k(M_{n,n}(F))$ *and* $\dim(S_k(M_{n,n}(F))) = \frac{n(n-1)}{2}$.

Proof. $M_{n,n}(F) = S(M_{n,n}(F)) \oplus S_k(M_{n,n}(F))$ follows from the result that every square matrix can be uniquely expressed as the sum of a symmetric matrix and a skew-symmetric matrix.

Hence,

$$\dim(M_{n,n}(F)) = \dim(S(M_{n,n}(F))) + \dim(S_k(M_{n,n}(F))) \quad \text{or}$$

$$n^2 = \frac{n(n+1)}{2} + \dim(S_k(M_{n,n}(F))), \quad \text{or}$$

$$\dim(S_k(M_{n,n}(F))) = n^2 - \frac{n^2+n}{2} = \frac{n(n-1)}{2}.$$

$\square$

Example 3.3.11. The set of all diagonal matrices of order $n \times n$ forms a vector space of dimension n. Similarly, the set of all scalar matrices of order $n \times n$ forms a vector space of dimension 1.

i Exercises

(3.3.1) Let $W_1 = \{[a_{ij}] \in M_{m,n}(\mathbb{C}) : \sum_{j=1}^{n} a_{ij} = 0, i = 1, 2, \dots, m\}$ and $W_2 = \{[a_{ij}] \in M_{m,n}(\mathbb{C}) : \sum_{i=1}^{m} a_{ij} = 0, j = 1, 2, \dots, n\}$. Then prove that:

 (i) W_1 and W_2 are subspaces of the vector space $M_{m,n}(\mathbb{C})$ over $\mathbb{C}$ and find bases and dimensions of W_1, W_2 and $W_1 \cap W_2$.

 (ii) Find bases and dimensions of subspaces W_1, W_2 and $W_1 \cap W_2$ over the field $\mathbb{R}$.

(3.3.2) Find the basis and dimension of the subspace $W = \{A \in M_{m,n}(\mathbb{C}) : \text{tr}(A) = 0\}$ over the field $\mathbb{R}$.

(3.3.3) Show that the following subsets of $M_{n,n}(F)$ are not subspaces of the vector space $M_{n,n}(F)$.

 (i)   The set of all nilpotent matrices.

 (ii) The set of all invertible matrices.

 (iii) The set of all orthogonal matrices.

(3.3.4) Let

$$W = \left\{ A \in M_{2,2}(\mathbb{C}) : A = \begin{bmatrix} z_1 & z_2 \\ z_3 & -z_1 \end{bmatrix} \right\}.$$

Then show that W is a subspace of $M_{2,2}(\mathbb{C})$ and find its basis and dimension.

(3.3.5) Let

$$W = \left\{ A \in M_{3,3}(\mathbb{C}) : A = \begin{bmatrix} a_{11} & a_{12} & a_{13} \\ a_{21} & a_{11} & a_{23} \\ a_{31} & a_{32} & -2a_{11} \end{bmatrix} \right\}.$$

Then show that W is a subspace of $M_{3,3}(\mathbb{C})$ and find its basis and dimension.

(3.3.6) Let A and B be n-square matrices with entries in a field F. Then show that $AB - BA \neq I_n$.

(3.3.7) Let A and B be n-square matrices with entries in a field F. Then classify $AB - BA$ between symmetric and skew-symmetric for the following cases:

    (i)   if A and B are symmetric;

    (ii)   if A and B are skew-symmetric;

    (iii) if A is symmetric and B is skew-symmetric;

    (iv) if A is skew-symmetric and B is symmetric.

(3.3.8) Let $M_{n,n}(\mathbb{C})$ be the vector space of all n-square complex matrices. Then show that the set of Hermitian and the set of skew-Hermitian matrices are not a subspace of $M_{n,n}(\mathbb{C})$.

(3.3.9) Show that the set $W = \{A \in M_{n,n}(\mathbb{R}) : A^t = 5A\}$ is a subspace of the vector space $M_{n,n}(\mathbb{R})$ over $\mathbb{R}$.

4 Linear transformations

Linear transformations play a crucial role in numerous branches of mathematics and applied sciences. Understanding the concept of linear transformations is essential for progress in disciplines such as linear algebra, functional analysis and more. In this chapter, we explore examples, properties and fundamental results related to linear transformations.

4.1 Definition and examples of linear transformations

A linear transformation is a function from one vector space to another vector space that preserves vector addition and scalar multiplication. We have the following definition, which is more precise.

Definition 4.1.1. Let V and W be vector spaces over the same field F. Then a map $T :
V \to W$ is called a linear transformation if it satisfies the following properties:
(i) $T(v_1 + v_2) = T(v_1) + T(v_2)$, $\forall v_1, v_2 \in V$;
(ii) $T(av) = aT(v)$, $\forall a \in F$ and $\forall v \in V$.

A bijective linear transformation is called an isomorphism. If $T : V \to W$ is an isomorphism, then V and W are considered isomorphic, denoted as $V \cong W$.

From an application standpoint, the subsequent definition of a linear transformation is more useful.

A map $T : V \to W$ from a vector space V to a vector space W is called a linear transformation if $T(av_1 + bv_2) = aT(v_1) + bT(v_2)$ for all $a, b \in F$ and $v_1, v_2 \in V$.

Hereafter, a map will refer to a map between two vector spaces over a field F.

Remark. We deduce that $(V, +)$ and $(W, +)$ are groups from the vector spaces V and W. Thus, condition (i) in the above definition indicates that $T : V \to W$ is a group homomorphism. Consequently, a linear transformation is also known as a vector space homomorphism.

The following theorem results from the group homomorphism property.

Theorem 4.1.2. Let $T : V \to W$ be a linear transformation. Then:
(i) $T(0) = 0$, and
(ii) $T(-v) = -T(v)$ $\forall v \in V$.

Example 4.1.3. The map $T : V \to W$ defined by $T(v) = 0$ $\forall v \in V$ is a linear transformation, called zero linear transformation.

https://doi.org/10.1515/9783111516035-004

Example 4.1.4. The identity map $I : V \to V$ defined by $I(v) = v \; \forall v \in V$ is a linear transformation. Since $I(av_1 + bv_2) = av_1 + bv_2 = aI(v_1) + bI(v_2)$ for all $a, b \in F$ and $v_1, v_2 \in V$.

Example 4.1.5. Let $T : \mathbb{R}^3 \to \mathbb{R}^3$ be a map defined by $T(x, y, z) = (x + y, y + z, z)$. Then for all $(x_1, y_1, z_1), (x_2, y_2, z_2) \in \mathbb{R}^3$ and $a, b \in \mathbb{R}$,

$$
\begin{aligned}
T\big(a(x_1, &y_1, z_1) + b(x_2, y_2, z_2)\big) \\
&= T(ax_1 + bx_2, ay_1 + by_2, az_1 + bz_2) \\
&= (ax_1 + bx_2 + ay_1 + by_2, ay_1 + by_2 + az_1 + bz_2, az_1 + bz_2) \\
&= \big((a(x_1 + y_1) + b(x_2 + y_2), a(y_1 + z_1) + b(y_2 + z_2), az_1 + bz_2\big) \\
&= a(x_1 + y_1, y_1 + z_1, z_1) + b(x_2 + y_2, y_2 + z_2, z_2) \\
&= aT(x_1, y_1, z_1) + bT(x_2, y_2, z_2).
\end{aligned}
$$

Thus, T is a linear transformation.

Example 4.1.6. The map $T : \mathbb{R}^3 \to \mathbb{R}^3$ defined by $T(x, y, z) = (x+1, y+1, z)$ is not a linear transformation since $T(0, 0, 0) = (1, 2, 0) \neq (0, 0, 0)$.

Example 4.1.7. Let F be any field. Then the i-th projection $p_i : F^n \to F$ defined by $p_i(x_1, x_2, \ldots, x_n) = x_i \; \forall 1 \le i \le n$ is a linear transformation.

Example 4.1.8. Let $P_n(x)$ be the vector space of at most n-degree polynomials over a field F. Then the map $T : P_n(x) \to P_n(x)$ defined by $T(f(x)) = f'(x)$, where $f'(x) = \frac{df(x)}{dx}$ is a linear transformation.

This follows from the calculus that

$$
T(af(x) + bg(x)) = \frac{d(af(x) + bg(x))}{dx} = a\frac{df(x)}{dx} + b\frac{dg(x)}{dx} = af'(x) + bg'(x)
$$
$$
= aT(f(x)) + bT(g(x)), \quad \forall f(x), g(x) \in P_n(x) \text{ and } \forall a, b \in F.
$$

T is called differentiation transformation and it is clear that the image of $P_n(x)$ under differentiation transformation is $P_{n-1}(x)$.

Example 4.1.9. Let $T : C(\mathbb{R}^{[0,1]}) \to \mathbb{R}$ be a map defined by

$$
T(f(x)) = \int_0^1 f(x)dx.
$$

Then for all $f(x), g(x) \in C(\mathbb{R}^{[0,1]})$ and $a, b \in \mathbb{R}$,

$$T(af(x) + bg(x)) = \int_0^1 (af(x) + bg(x))dx$$

$$= a\int_0^1 f(x)dx + b\int_0^1 g(x)dx$$

$$= aT(f(x)) + bT(g(x)),$$

shows that T is a linear transformation.

Example 4.1.10. Let $L_A : M_{n,p}(F) \to M_{m,p}(F)$ be a map defined by $L_A(B) = AB$, where A is a fixed matrix of order $m \times n$ with entries in the field F.

Then $L_A(aB + bC) = A(aB + bC) = aAB + bAC = aL_A(B) + bL_A(C)$ $\forall a, b \in F, B, C \in M_{np}(F)$.

This proves that L_A is a linear transformation.

To be more specific, the map $L_A : F^n \to F^m$ defined by $L_A(X) = AX$, where $X \in F^n$ and vectors in F^n and F^m are considered as column vectors, is a linear transformation if A is a matrix of order $m \times n$ with entries in F.

This linear transformation is called the left multiplication transformation.

Example 4.1.11. Let $T : M_{n,n}(F) \to M_{n,n}(F)$ be a map defined by $T(A) = A^t$.

Then for $a, b, \in F$ and $B, C \in M_{n,n}(F)$,

$$T(aB + bC) = (aB + bC)^t = aB^t + bC^t = aT(B) + bT(C)$$

shows that T is a linear transformation.

Example 4.1.12. Let $T : \mathbb{R}^3 \to \mathbb{R}^3$ be a map defined by

$$T(x, y, z) = (x \cos\theta - y \sin\theta, x \sin\theta + y \cos\theta, z).$$

Then for $a, b \in \mathbb{R}$ and $(x_1, y_1, z_1), (x_2, y_2, z_2) \in \mathbb{R}^3$,

$T(a(x_1, y_1, z_1) + b(x_2, y_2, z_2))$

$= T(ax_1 + bx_2, ay_1 + by_2, az_1 + bz_2)$

$= ((ax_1 + bx_2)\cos\theta - (ay_1 + by_2)\sin\theta, (ax_1 + bx_2)\sin\theta + (ay_1 + by_2)\cos\theta, az_1 + bz_2)$

$= (ax_1 \cos\theta - ay_1 \sin\theta, ax_1 \sin\theta + ay_1 \cos\theta, az_1)$

$\quad + (bx_2 \cos\theta - by_2 \sin\theta, bx_2 \sin\theta + by_2 \cos\theta, bz_2)$

$= aT(x_1, y_1, z_1) + bT(x_2, y_2, z_2).$

Thus, T is a linear transformation.

Example 4.1.13. Consider the map $T : \mathbb{R}^3 \to \mathbb{R}^3$, defined by $T(x, y, z) = (x \cos\theta - \sin\theta, y \sin\theta + \cos\theta, z)$.

Since $T(0,0,0) = (-\sin\theta, \cos\theta, 0) \neq (0,0,0)\ \forall\theta$, T is not a linear transformation.

Example 4.1.14. Let $T : \mathbb{R}^3 \to \mathbb{R}^2$ be a map defined by $T(x,y,z) = (x,yz)$. Then $T(0,0,0) = (0,0)$ but for any $a \in \mathbb{R} - \{0,1\}$,

$$T(a(x,y,z)) = T(ax,ay,az) = (ax, a^2yz) \neq aT(x,y,z) = a(x,yz).$$

Hence, T is not a linear transformation.

Example 4.1.15. Maps $T_1 : \mathbb{R}^2 \to \mathbb{R}^3$ defined by $T_1(x_1,x_2) = (x_1,x_2,0)$, and $T_2 : \mathbb{R}^3 \to \mathbb{R}^2$ defined by $T_2(x_1,x_2,x_3) = (x_1,x_2)$ are linear transformations.
More generally, for $m \leq n$,

$$T_1 : \mathbb{R}^m \to \mathbb{R}^n \quad \text{defined by } T_1(x_1,x_2,\dots,x_m) = (x_1,x_2,\dots,x_m,0,\dots,0) \quad \text{and}$$
$$T_2 : \mathbb{R}^n \to \mathbb{R}^m \quad \text{defined by } T_2(x_1,x_2,\dots,x_n) = (x_1,x_2,\dots,x_m)$$

are linear transformations.
T_1 is called natural inclusion and T_2 is called natural projection.

Example 4.1.16. Let $T_1 : V \to W$ and $T_2 : W \to U$ be linear transformations. Then the composite map $T_2 \circ T_1 : V \to U$ is a linear transformation.
Since $T_2 \circ T_1(ax+by) = T_2(T_1(ax+by)) = T_2(aT_1(x)+bT_1(y)) = aT_2(T_1(x))+bT_2(T_1(y)) = aT_2 \circ T_1(x) + bT_2 \circ T_1(y)$ for all $a,b \in F$ and $x,y \in V$.

Definition 4.1.17. Let $T : V \to W$ be a linear transformation. Then T is said to be singular if there exists a nonzero vector $v \in V$ such that $T(v) = 0$. If $T(v) = 0$ implies $v = 0$ for all $v \in V$, then T is called a nonsingular linear transformation. Thus, T is nonsingular $\Leftrightarrow$ T is injective.

Definition 4.1.18. A linear transformation $T : V \to W$ is called invertible if there exists a linear transformation $S : W \to V$ such that $ST = I_V$ and $TS = I_W$, where I_V and I_W are the identity maps on V and W, respectively. The linear transformation S is called the inverse of T and is denoted by $T^{-1} = S$.

Theorem 4.1.19. *The inverse of a linear transformation is unique.*

Proof. Let $T : V \to W$ be an invertible linear transformation. Suppose T_1 and T_2 are linear transformations from W to V such that

$$TT_1 = I_W = TT_2 \quad \text{and} \quad T_1T = I_V = T_2T.$$

Then

$$T_1 = T_1I_W = T_1(TT_2) = (T_1T)T_2 = I_VT_2 = T_2.$$

This proves that both inverses T_1 and T_2 of T are equal. □

Theorem 4.1.20. *A linear transformation is invertible if and only if it is bijective. In other words, a linear transformation is invertible if and only if it is an isomorphism.*

Proof. Let $T : V \to W$ be an invertible linear transformation. Then

$$TT^{-1} = I_W \quad \text{and} \quad T^{-1}T = I_V.$$

Now, we will show that T is bijective.

Suppose $T(v_1) = T(v_2)$. Then

$$T(v_1) - T(v_2) = 0 \Rightarrow T(v_1 - v_2) = 0 \in W.$$

Since T^{-1} is a linear transformation,

$$0 = T^{-1}(0) = T^{-1}(T(v_1 - v_2)) = T^{-1}T(v_1 - v_2) = I_V(v_1 - v_2) = v_1 - v_2.$$

Hence, $v_1 = v_2$, so T is injective.

To show that T is surjective, let $w \in W$. Then $T^{-1}(w) \in V$.

We have $TT^{-1}(w) = I_W(w) = w$. This implies that if we take $v = T^{-1}(w)$, then $T(v) = w$. Hence, T is surjective.

Conversely, suppose that $T : V \to W$ is a bijective linear transformation. Since T is a bijective function, there exists a $T^{-1} : W \to V$ such that $TT^{-1} = I_W$, $T^{-1}T = I_V$ and $(T^{-1})^{-1} = T$. To complete the proof, we need to show that T^{-1} is a linear transformation.

Let $w_1, w_2 \in W$ and $\alpha, \beta \in F$. Then, since T is bijective, there exist $v_1, v_2 \in V$ such that $T(v_1) = w_1$ and $T(v_2) = w_2$ or $v_1 = T^{-1}(w_1)$ and $v_2 = T^{-1}(w_2)$.

Now,

$$T(\alpha v_1 + \beta v_2) = \alpha T(v_1) + \beta T(v_2) = \alpha w_1 + \beta w_2$$
$$\Rightarrow T^{-1}(\alpha w_1 + \beta w_2) = \alpha v_1 + \beta v_2 = \alpha T^{-1}(w_1) + \beta T^{-1}(w_2).$$

This proves that T^{-1} is a linear transformation, and hence T is invertible. $\quad\square$

Theorem 4.1.21. *If T_1 and T_2 are invertible linear transformations on a vector space V, then the composition $T_2 \circ T_1$ is invertible and $(T_2 \circ T_1)^{-1} = T_1^{-1} \circ T_2^{-1}$.*

Proof. Since T_1 and T_2 are invertible, T_1^{-1} and T_2^{-1} exist. Then

$$(T_2 \circ T_1)(T_1^{-1} \circ T_2^{-1}) = T_2 \circ T_1 T_1^{-1} \circ T_2^{-1} = T_2 \circ I_V \circ T_2^{-1} = I_V.$$

Also, $(T_1^{-1} \circ T_2^{-1})(T_2 \circ T_1) = T_1^{-1} \circ T_2^{-1}T_2 \circ T_1 = T_1^{-1}I_V \circ T_1 = I_V$.
This proves that $T_1^{-1} \circ T_2^{-1}$ is the inverse of $(T_2 \circ T_1)^{-1}$.
Hence, $(T_2 \circ T_1)^{-1} = T_1^{-1} \circ T_2^{-1}$. $\quad\square$

Exercises

(4.1.1) Let $\mathbb{C}$ be a vector space over $\mathbb{R}$. Then show that the map $T : \mathbb{C} \to \mathbb{C}$ defined by $T(z) = \bar{z}$ is a linear transformation. If $\mathbb{C}$ is a vector space over $\mathbb{C}$. Then show that the map $T : \mathbb{C} \to \mathbb{C}$ defined by $T(z) = \bar{z}$ is a nonlinear transformation.

(4.1.2) Let $T : \mathbb{R}^3 \to \mathbb{R}^3$ be a map defined by $T(x, y, z) = (x + y, y + z, z + x)$. Then prove that T is a linear transformation.

(4.1.3) Determine whether the following maps on $\mathbb{R}^3$ are linear transformations:
   (i) $T(x, y, z) = (x - y, y - z, z - x)$.
   (ii) $T(x, y, z) = (z, y, x)$.
   (iii) $T(x, y, z) = (x, y^2, z^2)$.
   (iv) $T(x, y, z) = (x, y, 0)$.
   (v) $T(x, y, z) = (x - y + 3z, 3x + y, -x - 3y + 3z)$.
   (vi) $T(x, y, z) = (x - 1, y + 2, z + 3)$.
   (vii) $T(x, y, z) = (|x|, y, z - x)$.

(4.1.4) Show that the following maps are linear transformations:
   (i) $T : M_{n,n}(\mathbb{R}) \to \mathbb{R}$ defined by $T(A) = \mathrm{tr}(A)$.
   (ii) $T : M_{n,n}(\mathbb{R}) \to M_{n,n}(\mathbb{R})$ defined by $T(A) = \frac{A + A^t}{2}$.
   (iii) $T : M_{n,n}(\mathbb{R}) \to M_{n,n}(\mathbb{R})$ defined by $T(A) = kA, k \in \mathbb{R}$.
   (iv) $T : C(\mathbb{R}^{[0,1]}) \to \mathbb{R}$ defined by $T(f(x)) = \int_0^1 f(t^2)dt$.

(4.1.5) Show that the following maps are nonlinear transformations:
   (i) $T : M_{n,n}(\mathbb{R}) \to M_{n,n}(\mathbb{R})$ defined by $T(A) = A^t A$.
   (ii) $T : M_{n,n}(\mathbb{R}) \to M_{n,n}(\mathbb{R})$ defined by $T(A) = A^2$.
   (iii) $T : M_{n,n}(\mathbb{R}) \to M_{n,n}(\mathbb{R})$ defined by $T(A) = A + I_n$.

(4.1.6) Let $P_n(x)$ denotes the vector space of all polynomials with real coefficients of degree at most $n \in \mathbb{N}$. Then prove that the following maps are linear transformations:
   (i) $T : P_n(x) \to P_{n-2}(x)$ defined by $T(f(x)) = f''(x)$.
   (ii) $T : P_n(x) \to P_{n-1}(x)$ defined by $T(f(x)) = f''(x) + f'(x)$.
   (iii) $T : P_n(x) \to P_{n+1}(x)$ defined by $T(f(x)) = \int_0^x f(t)dt$.
   (iv) $T : P_n(x) \to P_{n+1}(x)$ defined by $T(f(x)) = f'(x) - \int_0^x f(t)dt$.

(4.1.7) Show that $T : P_2(x) \to P_3(x)$ defined by $T(f(x)) = xf(x) + f'(x)$ is a linear transformation where $f'(x) = \frac{df(x)}{dx}$.

(4.1.8) Let V be a vector space over a field F. Then prove that all the linear transformations $T : F \to V$ are of the form of $T(k) = kw$ for some $w \in V$.
   Hint: Since $k = k.1 \Rightarrow T(k) = T(k.1) = kT(1)$, take $T(1) = w \in V$.

4.2 Rank and nullity of a linear transformation

In this section, we will prove the fundamental theorem of linear transformations and the rank-nullity theorem. Additionally, we will explore some related results and examples concerning the rank and nullity of linear transformations.

Definition 4.2.1. Let $T : V \to W$ be a linear transformation. Then $T^{-1}(\{0\}) = \{v \in V : T(v) = 0\}$ is called the kernel of T and is denoted by $\mathrm{Ker}(T)$. That is, $\mathrm{Ker}(T) = \{v \in V : T(v) = 0\}$.

Theorem 4.2.2. *Let $T : V \rightarrow W$ be a linear transformation. Then the image of T, denoted by $\mathrm{Im}(T)$, is a subspace of W and the kernel of T, denoted by $\mathrm{Ker}(T)$, is a subspace of V.*

Proof. We have

$$\mathrm{Im}(T) = T(V) = \{w \in W : \text{there exists } v \in V \text{ with } T(v) = w\}.$$

Since $T(0) = 0$, both $\mathrm{Im}(T)$ and $\mathrm{Ker}(T)$ are nonempty.

Let $w_1, w_2 \in \mathrm{Im}(T)$. Then there exist v_1 and v_2 in V such that $T(v_1) = w_1$ and $T(v_2) = w_2$.

Now, for $a, b \in F$,

$$aw_1 + bw_2 = aT(v_1) + bT(v_2) = T(av_1 + bv_2).$$

This implies that $aw_1 + bw_2 \in \mathrm{Im}(T)$. Hence, $\mathrm{Im}(T)$ is a subspace of W.

To prove that $\mathrm{Ker}(T)$ is a subspace of V, we need to show that $av_1 + bv_2 \in \mathrm{Ker}(T)$ where $v_1, v_2 \in \mathrm{Ker}(T)$ and $a, b \in F$. That is, we need to show that $T(av_1 + bv_2) = 0$.

Since T is a linear transformation,

$$T(av_1 + bv_2) = aT(v_1) + bT(v_2) = a0 + b0 = 0.$$

This completes the proof. □

Note that $\mathrm{Im}(T)$ is called the rank space of T and $\mathrm{Ker}(T)$ is called the null space of T.

Definition 4.2.3. Let $T : V \rightarrow W$ be a linear transformation from a finite-dimensional vector space V to a vector space W. Then the dimension of the rank space $\mathrm{Im}(T)$ is called the rank of T, denoted by $\mathrm{rank}(T)$ and the dimension of the null space $\mathrm{Ker}(T)$ is called the nullity of T, denoted by $\mathrm{null}(T)$.

Theorem 4.2.4. *Let $T : V \rightarrow W$ be a linear transformation. Then T is injective if and only if $\mathrm{Ker}(T) = \{0\}$.*

Proof. Let T be an injective linear transformation. Then

$$x \in \mathrm{Ker}(T) \Rightarrow T(x) = 0 = T(0).$$

Since T is injective, $x = 0$. Hence, $\mathrm{Ker}(T) = \{0\}$.

Conversely, suppose that $\mathrm{Ker}(T) = \{0\}$. Then

$$T(x) = T(y) \Rightarrow T(x) - T(y) = 0$$
$$\Rightarrow T(x - y) = 0$$
$$\Rightarrow (x - y) \in \mathrm{Ker}(T)$$
$$\Rightarrow x - y = 0$$
$$\Rightarrow x = y.$$

Hence, T is injective. □

Theorem 4.2.5 (Fundamental theorem of linear transformation). *Let $T : V \to W$ be a linear transformation. Then $\frac{V}{\mathrm{Ker}(T)}$ is isomorphic to $\mathrm{Im}(T)$, i. e., $\frac{V}{\mathrm{Ker}(T)} \cong T(V)$.*

Proof. Let $f : \frac{V}{\mathrm{Ker}(T)} \to T(V)$ be a map defined by $f(v + \mathrm{Ker}(T)) = T(v)$. Then f is well-defined. This follows from below:

$$v_1 + \mathrm{Ker}(T) = v_2 + \mathrm{Ker}(T) \Rightarrow v_1 - v_2 \in \mathrm{Ker}(T)$$
$$\Rightarrow T(v_1 - v_2) = 0$$
$$\Rightarrow T(v_1) - T(v_2) = 0$$
$$\Rightarrow T(v_1) = T(v_2)$$
$$\Rightarrow f(v_1 + \mathrm{Ker}(T)) = f(v_2 + \mathrm{Ker}(T)).$$

Now, we shall show that f is a bijective linear transformation.
For $a, b \in F$,

$$f(a(v_1 + \mathrm{Ker}(T)) + b(v_2 + \mathrm{Ker}(T)))$$
$$= f(av_1 + \mathrm{Ker}(T) + bv_2 + \mathrm{Ker}(T))$$
$$= f(av_1 + bv_2 + \mathrm{Ker}(T))$$
$$= T(av_1 + bv_2)$$
$$= aT(v_1) + bT(v_2)$$
$$= af(v_1 + \mathrm{Ker}(T)) + bf(v_2 + \mathrm{Ker}(T)),$$

show that f is a linear transformation.
    f is injective, for

$$v + \mathrm{Ker}(T) \in \mathrm{Ker}(f) \Rightarrow f(v + \mathrm{Ker}(T)) = 0 \Rightarrow T(v) = 0$$
$$\Rightarrow v \in \mathrm{Ker}(T) \Rightarrow v + \mathrm{Ker}(T) = \mathrm{Ker}(T),$$

which is the zero of $\frac{V}{\mathrm{Ker}(T)}$. Thus, $\mathrm{Ker}(f) = \{0\}$, and hence f is injective.
    Next, for any $T(v) \in T(V), f(v + \mathrm{Ker}(T)) = T(v)$, shows that f is surjective.
    Hence, f is an isomorphism and so $\frac{V}{\mathrm{Ker}(T)} \cong T(V)$.
    In case, $T : V \to W$, a surjective linear transformation $T(V) = W$, and hence $\frac{V}{\mathrm{Ker}(T)} \cong W$.   □

Theorem 4.2.6. *Let $T : V \to W$ be a linear transformation from a finite-dimensional vector space V to a vector space W. Then $\mathrm{rank}(T) + \mathrm{null}(T) = \dim(V)$.*

Proof. Let $\dim(V) = n$. Since $\mathrm{Ker}(T)$ is a subspace of V, suppose $\dim \mathrm{Ker}(T) = m \leq n$.
    Let $B_1 = \{x_1, x_2, \ldots, x_m\}$ be a basis of $\mathrm{Ker}(T)$. Then it can be extended to a basis $B_2 = \{x_1, x_2, \ldots, x_m, y_1, y_2, \ldots, y_{n-m}\}$ of V.
    Now, we claim that $\{T(y_1), T(y_2), \ldots, T(y_{n-m})\}$ is a basis of $T(V)$, which proves that $\dim(T(V)) = n - m$.

First, we prove that $\{T(y_1), T(y_2), \ldots, T(y_{n-m})\}$ spans $T(V)$.

Let $w \in T(V)$. Then there exists $v \in V$ such that $T(v) = w$. Since B_2 is a basis of V, v can be expressed as

$$v = a_1 x_1 + a_2 x_2 + \cdots + a_m x_m + b_1 y_1 + b_2 y_2 + \cdots + b_{n-m} y_{n-m},$$

for some scalars $a_1, a_2, \ldots, a_m, b_1, b_2, \ldots, b_{n-m}$.

Hence,

$$w = T(v) = T(a_1 x_1 + a_2 x_2 + \cdots + a_m x_m + b_1 y_1 + b_2 y_2 + \cdots + b_{n-m} y_{n-m})$$
$$= a_1 T(x_1) + a_2 T(x_2) + \cdots + a_m T(x_m) + b_1 T(y_1) + b_2 T(y_2) + \cdots + b_{n-m} T(y_{n-m}).$$

Since $x_i \in \mathrm{Ker}(T)$, $T(x_i) = 0 \ \forall 1 \le i \le m$. Hence, from the above equation, we have

$$w = b_1 T(y_1) + b_2 T(y_2) + \cdots + b_{n-m} T(y_{n-m}).$$

This means $T(V)$ is generated by $\{T(y_1), T(y_2), \ldots, T(y_{n-m})\}$.

Now, we shall show that $\{T(y_1), T(y_2), \ldots, T(y_{n-m})\}$ is linearly independent.

Let

$$c_1 T(y_1) + c_2 T(y_2) + \cdots + c_{n-m} T(y_{n-m}) = 0,$$

for some scalars $c_1, c_2, \ldots, c_{n-m}$. Then by the property of linear transformation we have that

$$T(c_1 y_1 + c_2 y_2 + \cdots + c_{n-m} y_{n-m}) = 0$$
$$\Rightarrow c_1 y_1 + c_2 y_2 + \cdots + c_{n-m} y_{n-m} \in \mathrm{Ker}(T).$$

From the basis B_1 of $\mathrm{Ker}(T)$, we have the following expression:

$$c_1 y_1 + c_2 y_2 + \cdots + c_{n-m} y_{n-m} = d_1 x_1 + d_2 x_2 + \cdots + d_m x_m,$$

for some scalars $d_1, d_2, \ldots, d_m$.

Then

$$c_1 y_1 + c_2 y_2 + \cdots + c_{n-m} y_{n-m} - d_1 x_1 - d_2 x_2 - \cdots - d_m x_m = 0.$$

Since

$$B_2 = \{x_1, x_2, \ldots, x_m, y_1, y_2, \ldots, y_{n-m}\}$$

is a basis, and thus linearly independent, we have $c_1 = c_2 = \cdots = c_{n-m} = d_1 = d_2 = \cdots = d_m = 0$. This proves that $\{T(y_1), T(y_2), \ldots, T(y_{n-m})\}$ is linearly independent.

Hence,

$$\dim(T(V)) = n - m = \dim(V) - \dim(\mathrm{Ker}(T)) \quad \text{or}$$
$$\mathrm{rank}(T) + \mathrm{null}(T) = \dim(V).$$

This significant result is also known as the rank-nullity theorem. □

Theorem 4.2.7. *Let V and W be vector spaces over the same field F and let $S = \{v_1, v_2, \ldots, v_n\}$ be a basis of V. Then every function $f : S \to W$ can be uniquely extended to a linear transformation $T : V \to W$ such that $T(v_i) = f(v_i) \; \forall 1 \le i \le n$.*

Proof. Let $v \in V$. Then v can be uniquely expressed as

$$v = a_1 v_1 + a_2 v_2 + \cdots + a_n v_n, \quad \text{for some } a_1, a_2, \ldots, a_n \in F.$$

Now, we claim that the map $T : V \to W$ defined by

$$T(v) = T(a_1 v_1 + a_2 v_2 + \cdots + a_n v_n) = a_1 f(v_1) + a_2 f(v_2) + \cdots + a_n f(v_n)$$

is a linear transformation.

Let $v = a_1 v_1 + a_2 v_2 + \cdots + a_n v_n$ and $u = b_1 v_1 + b_2 v_2 + \cdots + b_n v_n$ be vectors in V. Then for $\alpha, \beta \in F$,

$$
\begin{aligned}
T(\alpha u + \beta v) &= T((\alpha a_1 + \beta b_1)v_1 + (\alpha a_2 + \beta b_2)v_2 + \cdots + (\alpha a_n + \beta b_n)v_n) \\
&= (\alpha a_1 + \beta b_1)f(v_1) + (\alpha a_2 + \beta b_2)f(v_2) + \cdots + (\alpha a_n + \beta b_n)f(v_n) \\
&= \alpha a_1 f(v_1) + \alpha a_2 f(v_2) + \cdots + \alpha a_n f(v_n) + \beta b_1 f(v_1) + \beta b_2 f(v_2) + \cdots + \beta b_n f(v_n) \\
&= \alpha(a_1 f(v_1) + a_2 f(v_2) + \cdots + a_n f(v_n)) + \beta(b_1 f(v_1) + b_2 f(v_2) + \cdots + b_n f(v_n)) \\
&= \alpha T(a_1 v_1 + a_2 v_2 + \cdots + a_n v_n) + \beta T(b_1 v_1 + b_2 v_2 + \cdots + b_n v_n) \\
&= \alpha T(v) + \beta T(u).
\end{aligned}
$$

This shows that T is a linear transformation.
Further,

$$
\begin{aligned}
T(v_i) &= T(0v_1 + 0v_2 + \cdots + v_i + 0v_{i+1} + \cdots + 0v_n) \\
&= 0f(v_1) + 0f(v_2) + \cdots + f(v_i) + 0f(v_{i+1}) + \cdots + 0f(v_n) \\
&= f(v_i) \quad \forall 1 \le i \le n.
\end{aligned}
$$

Finally, we shall show that the extended linear transformation is unique. Suppose $T_1 : V \to W$ is a linear transformation such that $T_1(v_i) = f(v_i) \; \forall i$.
Then

$$
\begin{aligned}
T_1(v) &= T_1(a_1 v_1 + a_2 v_2 + \cdots + a_n v_n) \\
&= a_1 T_1(v_1) + a_2 T_1(v_2) + \cdots + a_n T_1(v_n)
\end{aligned}
$$

$$= a_1 f(v_1) + a_2 f(v_2) + \cdots + a_n f(v_n)$$
$$= T(a_1 v_1 + a_2 v_2 + \cdots + a_n v_n)$$
$$= T(v) \quad \forall v \in V.$$

Hence, $T_1 = T$. $\qquad \square$

Theorem 4.2.8. *Let V be an m-dimensional vector space and W be an n-dimensional vector space over a field F. Then V is isomorphic to W if and only if $m = n$.*

Proof. Let $V \cong W$. Then there exists an isomorphism $T : V \to W$. If $\{v_1, v_2, \ldots, v_m\}$ is a basis of V, then we claim that $\{T(v_1), T(v_2), \ldots, T(v_m)\}$ is a basis of W.

Since T is an isomorphism, for each $w \in W$ there exists a $v \in V$ such that $T(v) = w$. Suppose $v = \sum_{i=1}^{m} a_i v_i$, where $a_1, a_2, \ldots, a_m \in F$. Then $w = T(v) = \sum_{i=1}^{m} a_i T(v_i)$ shows that w is generated by $\{T(v_1), T(v_2), \ldots, T(v_m)\}$.

For linear independence, let $b_1 T(v_1) + b_2 T(v_2) + \cdots + b_m T(v_m) = 0$, where $b_1, b_2, \ldots, b_m \in F$. Then

$$T(b_1 v_1 + b_2 v_2 + \cdots + b_m v_m) = 0 \Rightarrow b_1 v_1 + b_2 v_2 + \cdots + b_m v_m \in \mathrm{Ker}(T).$$

Since T is injective, $\mathrm{Ker}(T) = \{0\}$. Hence, $b_1 v_1 + b_2 v_2 + \cdots + b_m v_m = 0$.

Since $v_1, v_2, \ldots, v_m$ being basis vectors are linearly independent, $b_1 = b_2 = \cdots = b_m = 0$. This proves that the set $\{T(v_1), T(v_2), \ldots, T(v_m)\}$ is linearly independent, and hence a basis of W. That is, $\dim(W) = m = \dim(V)$.

Conversely, suppose that $\dim(W) = \dim(V) = m$. Then we have to prove that $V \cong W$.

Let $\{v_1, v_2, \ldots, v_m\}$ and $\{w_1, w_2, \ldots, w_m\}$ be bases of vector spaces V and W, respectively. Then we shall show that the linear transformation $T : V \to W$, which is the unique extension of the function $f : \{v_1, v_2, \ldots, v_m\} \to W$ defined by $f(v_i) = w_i \, \forall 1 \leq i \leq m$ is an isomorphism. That is, $T(v_i) = f(v_i) = w_i \, \forall 1 \leq i \leq m$ and T is injective and surjective both.

Let $v = \sum_{i=1}^{m} a_i v_i \in \mathrm{Ker}(T)$. Then $T(v) = \sum_{i=1}^{m} a_i T(v_i) = 0 \Rightarrow \sum_{i=1}^{m} a_i w_i = 0$.

Since $\{w_1, w_2, \ldots, w_m\}$ is a basis, $a_i = 0 \, \forall i$ and so $v = 0$. Thus, $\mathrm{Ker}(T) = \{0\}$, and hence T is injective.

Let $w \in W$. Then for some $a_1, a_2, \ldots, a_m \in F$, w can be expressed as $w = \sum_{i=1}^{m} a_i w_i$.

Now, $w = \sum_{i=1}^{m} a_i w_i = \sum_{i=1}^{m} a_i T(v_i) = T(\sum_{i=1}^{m} a_i v_i) = T(v)$, where $v = \sum_{i=1}^{m} a_i v_i \in V$. That is, there exists $v \in V$ such that $T(v) = w$. This proves that T is surjective.

Hence, T is a bijective linear transformation and so $V \cong W$. $\qquad \square$

Corollary 4.2.9. *Let V be an n-dimensional vector space over a field F. Then $V \cong F^n$.*

Proof. Let $W = F^n$. Then W is a vector space of dimension n over the field F. Hence, from the above theorem we have $V \cong F^n$. $\qquad \square$

Example 4.2.10. Let $V = M_{m,n}(\mathbb{R})$. Then V is an mn-dimensional vector space over the field $\mathbb{R}$. Hence, from above corollary we have $M_{m,n}(\mathbb{R}) \cong \mathbb{R}^{mn}$.

Theorem 4.2.11. *Let $T : V \to W$ be a linear transformation and let $S = \{v_1, v_2, \ldots, v_n\}$ be a generator of V. Then $\{T(v_1), T(v_2), \ldots, T(v_n)\}$ is a generator of $T(V)$.*

Proof. Let $w \in T(V)$. Then there exists $v \in V$ such that $T(v) = w$. Since S is a generator of V for some $a_1, a_2, \ldots, a_n \in F$, $v = a_1 v_1 + a_2 v_2 + \cdots + a_n v_n$.

Then $w = T(v) = T(a_1 v_1 + a_2 v_2 + \cdots + a_n v_n) = a_1 T(v_1) + a_2 T(v_2) + \cdots + a_n T(v_n)$ proves the result. $\qquad\square$

Example 4.2.12. Let $T : \mathbb{R}^3 \to \mathbb{R}^3$ be a linear transformation defined by $T(x, y, z) = (x - y + 2z, x - y, -x - 2y + 2z)$. Then $\mathrm{Ker}(T) = \{(x, y, z) \in \mathbb{R}^3 : T(x, y, z) = (0, 0, 0)\}$.

Let $(x, y, z) \in \mathrm{Ker}(T)$. Then

$$T(x, y, z) = (0, 0, 0)$$
$$\Rightarrow (x - y + 2z, x - y, -x - 2y + 2z) = (0, 0, 0)$$
$$\Rightarrow x - y + 2z = 0, x - y = 0, -x - 2y + 2z = 0.$$

After the solution of above equations, we get $x = y = z = 0$. Hence, $\mathrm{Ker}(T) = \{(0, 0, 0)\}$.

From the rank-nullity theorem,

$$\mathrm{rank}(T) + \mathrm{null}(T) = \dim(\mathbb{R}^3)$$
$$\Rightarrow \mathrm{rank}(T) + \mathrm{null}(T) = 3$$
$$\Rightarrow \mathrm{rank}(T) + 0 = 3$$
$$\Rightarrow \mathrm{rank}(T) = 3.$$

That is, $\dim(T(\mathbb{R}^3)) = 3$.

Example 4.2.13. Let $T : \mathbb{R}^4 \to \mathbb{R}^4$ be a linear transformation defined by $T(x_1, x_2, x_3, x_4) = (x_1 + x_3, 2x_1 + x_2 + 3x_3, 2x_2 + 2x_3, x_4)$. Then $\mathrm{Ker}(T) = \{(x_1, x_2, x_3, x_4) \in \mathbb{R}^4 : T(x_1, x_2, x_3, x_4) = (0, 0, 0, 0)\}$.

Let $(x_1, x_2, x_3, x_4) \in \mathrm{Ker}(T)$. Then

$$T(x_1, x_2, x_3, x_4) = (0, 0, 0, 0)$$
$$\Rightarrow (x_1 + x_3, 2x_1 + x_2 + 3x_3, 2x_2 + 2x_3, x_4) = (0, 0, 0, 0)$$
$$\Rightarrow x_1 + x_3 = 0, 2x_1 + x_2 + 3x_3 = 0, 2x_2 + 2x_3 = 0, x_4 = 0.$$

Now, $x_1 + x_3 = 0$ and $2x_2 + 2x_3 = 0 \Rightarrow x_1 = x_2 = -x_3$.

Thus, from above equations we get that

$$(x_1, x_2, x_3, x_4) = (-x_3, -x_3, x_3, 0) = x_3(-1, -1, 1, 0).$$

Hence, $\mathrm{Ker}(T) = \{x_3(-1, -1, 1, 0) : x_3 \in \mathbb{R}\}$ and so $\mathrm{null}(T) = 1$.

By the rank-nullity theorem,

$$\text{rank}(T) + \text{null}(T) = \dim(\mathbb{R}^4)$$
$$\Rightarrow \text{rank}(T) + \text{null}(T) = 4$$
$$\Rightarrow \text{rank}(T) + 1 = 4$$
$$\Rightarrow \text{rank}(T) = 3.$$

Example 4.2.14. For any $n \in \mathbb{N}$, let $P_n(x)$ denotes the vector space of all polynomials with real coefficients and degree at most n. Then we find the rank and nullity of the linear transformation $T : P_3(x) \to P_4(x)$ defined by $T(p(x)) = \dot{p}(x) - \int_0^x p(t)dt$.

Let $p(x) = a_0 + a_1 x + a_2 x^2 + a_3 x^3 \in \text{Ker } T$. Then

$$T(p(x)) = \dot{p}(x) - \int_0^x p(t)dt = 0$$

$$\Rightarrow a_1 + 2a_2 x + 3a_3 x^2 - \int_0^x (a_0 + a_1 t + a_2 t^2 + a_3 t^3)dt = 0$$

$$\Rightarrow a_1 + 2a_2 x + 3a_3 x^2 - a_0 x - a_1 x^2/2 - a_2 x^3/3 - a_3 x^4/4 = 0$$

$$\Rightarrow a_1 + (2a_2 - a_0)x + (3a_3 - a_1/2)x^2 - a_2 x^3/3 - a_3 x^4/4 = 0$$

$$\Rightarrow a_1 = 0, 2a_2 - a_0 = 0, 3a_3 - a_1/2 = 0, -a_2/3 = 0, -a_3/4 = 0$$

$$\Rightarrow a_0 = a_1 = a_2 = a_3 = 0$$

$$\Rightarrow p(x) = 0.$$

Hence,

$$\text{Ker}(T) = \{0\} \quad \text{or} \quad \text{null}(T) = 0 \quad \text{and}$$
$$\text{rank}(T) = \dim(P_3(x)) - \text{null}(T) = 4.$$

Alternate method:

Let $\{1, x, x^2, x^3\}$ be a basis of $P_3(x)$. Then from Theorem 4.2.11, $\{T(1), T(x), T(x^2), T(x^3)\}$ generates $P_3(x)$, and hence contains maximal linearly independent set. Now we have

$$T(1) = 0 - \int_0^x dt = x,$$

$$T(x) = 1 - \int_0^x t dt = 1 - x^2/2,$$

$$T(x^2) = 2x - \int_0^x t^2 dt = 2x - x^3/3,$$

$$T(x^3) = 3x^2 - \int_0^x t^3 dt = 3x^2 - x^4/4.$$

Let

$$a_1 T(1) + a_2 T(x) + a_3 T(x^2) + a_4 T(x^3) = 0$$

$$\Rightarrow a_1 x + a_2(1 - x^2/2) + a_3(2x - x^3/3) + a_4(3x^2 - x^4/4) = 0$$

$$\Rightarrow a_2 + (a_1 + 2a_3)x + (-a_2/2 + 3a_4)x^2 + (-a_3/3)x^3 - (a_4/4)x^4 = 0$$

$$\Rightarrow a_2 = 0, a_1 + 2a_3 = 0, -a_2/2 + 3a_4 = 0, -a_3/3 = 0, -a_4/4 = 0$$

$$\Rightarrow a_1 = a_2 = a_3 = a_4 = 0.$$

Then the set $\{T(1), T(x), T(x^2), T(x^3)\}$ is linearly independent, and hence a basis of $\mathrm{Im}(T)$.

Thus $\mathrm{rank}(T) = 4$ and $\mathrm{null}(T) = \dim(P_3(x)) - \mathrm{rank}(T) = 4 - 4 = 0$.

Example 4.2.15. Let us find a linear map $T : \mathbb{R}^3 \to \mathbb{R}^4$ whose image is generated by $\{(1, 0, 2, 3), (-3, 2, 4, 0)\}$.

Let $B = \{e_1 = (1, 0, 0), e_2 = (0, 1, 0), e_3 = (0, 0, 1)\}$ be the standard basis of $\mathbb{R}^3$. Define the linear transformation T by $T(e_1) = (1, 0, 2, 3)$, $T(e_2) = (-3, 2, 4, 0)$, $T(e_3) = (0, 0, 0, 0)$.

From Theorems 4.2.7 and 4.2.11, such a unique linear transformation exists and

$$\{T(e_1), T(e_2), T(e_3)\} = \{(1, 0, 2, 3), (-3, 2, 4, 0), (0, 0, 0, 0)\},$$

spans $\mathrm{Im}(T)$, and hence $\{(1, 0, 2, 3), (-3, 2, 4, 0)\}$ spans $\mathrm{Im}(T)$.

Let $(x_1, x_2, x_3) \in \mathbb{R}^3$. Then $(x_1, x_2, x_3) = x_1 e_1 + x_2 e_2 + x_3 e_3$.

Hence,

$$\begin{aligned}
T(x_1, x_2, x_3) &= T(x_1 e_1 + x_2 e_2 + x_3 e_3) \\
&= x_1 T(e_1) + x_2 T(e_2) + x_3 T(e_3) \\
&= x_1(1, 0, 2, 3) + x_2(-3, 2, 4, 0) + x_3(0, 0, 0, 0) \\
&= (x_1 - 3x_2, 2x_2, 2x_1 + 4x_2, 3x_1).
\end{aligned}$$

Example 4.2.16. Let V and W be vector spaces of dimensions 2 and 4, respectively, over the same field F. If $T : V \to W$ is a linear transformation, then by the rank-nullity theorem linearly independent sets $B_1 = \{w_1, w_2, w_3, w_4\}$ and $B_2 = \{w_1, w_2, w_3\}$ cannot be basis of $\mathrm{Im}(T)$. But for $B_3 = \{w_1, w_2\}$ and $B_4 = \{w_1\}$ there always exist linear transformations $T_1 : V \to W$ and $T_2 : V \to W$ such that B_3 and B_4 are bases of $\mathrm{Im}(T_1)$ and $\mathrm{Im}(T_2)$, respectively.

If $B = \{v_1, v_2\}$ is a basis of V, then T_1 and T_2 can be defined as follows:

$$T_1(v_1) = w_1, \quad T_1(v_2) = w_2 \quad \text{and} \quad T_2(v_1) = w_1, \quad T_2(v_2) = 0.$$

More generally, if V and W are vector spaces of dimensions m and n, respectively, over the same field F, then for $m < n$, there does not exist any linear transformation $T : V \to W$ with $\text{rank}(T) > m$, but always there is a possibility of a linear transformation $T : V \to W$ with $\text{rank}(T) \le m$.

Example 4.2.17. Let us find the solution of the following question:

Is it possible to construct a linear transformation $T : F^2 \to F^4$ such that $\text{Im}(T) = \{(x, y, z, w) \in F^4 : x + y + z - w = 0\}$?

We have the dimension of $\text{Im}(T) = 3$ (prove). Then by the rank-nullity theorem,

$$\text{rank}(T) + \text{null}(T) = 2 \Rightarrow 3 + \text{null}(T) = 2.$$

Since $\text{null}(T) \ge 0, 3 + \text{null}(T) \ne 2$. Hence, there does not exist such a linear transformation.

Example 4.2.18. In this example, we obtain the linear transformation from the given its image set. Let $T : \mathbb{R}^2 \to \mathbb{R}^3$ be a linear transformation such that $\text{Im}(T) = \{(x, y, z) \in \mathbb{R}^3 : x + y - z = 0\}$. Then $\text{Im}(T) = \{(x, y, x + y) : x, y \in \mathbb{R}\}$.

From the standard basis $B = \{e_1 = (1, 0, 0), e_2 = (0, 1, 0), e_3 = (0, 0, 1)\}$ of $\mathbb{R}^3$, we have

$$(x, y, x + y) = xe_1 + ye_2 + (x + y)e_3$$
$$= x(e_1 + e_3) + y(e_2 + e_3)$$
$$= x(1, 0, 1) + y(0, 1, 1).$$

This shows that $\text{Im}(T)$ is generated by $\{(1, 0, 1), (0, 1, 1)\}$.

Define $T(1, 0) = (1, 0, 1)$ and $T(0, 1) = (0, 1, 1)$.

Then

$$T(x, y) = T(x(1, 0) + y(0, 1))$$
$$= xT(1, 0) + yT(0, 1)$$
$$= x(1, 0, 1) + y(0, 1, 1)$$
$$= (x, y, x + y).$$

Theorem 4.2.19. *Let V_1, V_2, V_3 be vector spaces over a field F and let $T_1 : V_1 \to V_2$, $T_2 : V_2 \to V_3$ be linear transformations. Then we have the following:*
(i) $\text{rank}(T_2 \circ T_1) \le \min\{\text{rank}(T_1), \text{rank}(T_2)\}$,
(ii) $\text{null}(T_1) \le \text{null}(T_2 \circ T_1)$,
(iii) *if T_1 is invertible, then* $\text{rank}(T_2 \circ T_1) = \text{rank}(T_2)$.

Proof. (i)   We have that $T_2 \circ T_1 : V_1 \to V_3$ is a linear transformation. Since $T_1(V_1)$ is a subspace of V_2, $T_2(T_1(V_1))$ is a subspace of $T_2(V_2)$. Hence, $\text{rank}(T_2 \circ T_1) \le \text{rank}(T_2)$. Again, since $T_1(V_1)$ is a subspace of V_2, the restriction map,

$$T_2/T_1(V_1) : T_1(V_1) \to V_3 \text{ is a linear transformation.}$$

Then by the rank-nullity theorem,

$$\dim(T_1(V_1)) = \dim(T_2(T_1(V_1))) + \text{null}(T_2/T_1(V_1)) \quad \text{or}$$
$$\text{rank}(T_2 \circ T_1) = \text{rank}(T_1) - \text{null}(T_2/T_1(V_1)) \le \text{rank}(T_1).$$

This proves that $\text{rank}(T_2 \circ T_1) \le \min\{\text{rank}(T_1), \text{rank}(T_2)\}$.

(ii)   We have that if $\text{Ker}(T_1) \subseteq \text{Ker}((T_2 \circ T_1))$, then

$$\dim(\text{Ker}(T_1)) \le \dim(\text{Ker}(T_2 \circ T_1)) \quad \text{and so} \quad \text{null}(T_1) \le \text{null}(T_2 \circ T_1).$$

Thus, to prove the result it is sufficient to show that $\text{Ker}(T_1) \subseteq \text{Ker}(T_2 \circ T_1)$. Let $x \in \text{Ker}(T_1)$. Then

$$T_1(x) = 0 \Rightarrow T_2(T_1(x)) = T_2(0) = 0$$
$$\Rightarrow T_2 \circ T_1(x) = 0 \Rightarrow x \in \text{Ker}(T_2 \circ T_1).$$

Hence, $\text{Ker}(T_1) \subseteq \text{Ker}(T_2 \circ T_1)$.

(iii)   From (i), we have

$$\text{rank}(T_2) = \text{rank}(T_2 \circ T_1 \circ T_1^{-1}) \le \text{rank}(T_2 \circ T_1) \le \text{rank}(T_2).$$

Hence, $\text{rank}(T_2 \circ T_1) = \text{rank}(T_2)$.   □

Example 4.2.20.
(a) Let $T : V \to V$ be a linear transformation on a finite-dimensional vector space V, such that $\text{rank}(T^2) = \text{rank}(T)$. Then $V = I_m(T) \oplus \text{Ker}(T)$.
Since $T^2 = T \circ T : V \to V$ is a linear transformation, by the rank-nullity theorem, we have $\text{rank}(T^2) + \text{null}(T^2) = \dim(V)$.
This implies

$$\text{null}(T^2) = \dim(V) - \text{rank}(T^2)$$
$$= \dim(V) - \text{rank}(T) = \text{null}(T).$$

Thus,

$$\text{null}(T^2) = \text{null}(T) \Rightarrow \dim(\text{Ker}(T^2)) = \dim(\text{Ker}(T))$$
$$\Rightarrow \text{Ker}(T^2) \cong \text{Ker}(T).$$

Also,

$$x \in \text{Ker}(T) \Rightarrow T(x) = 0 \Rightarrow T(T(x)) = T(0) \Rightarrow T^2(x) = 0$$
$$\Rightarrow x \in \text{Ker}(T^2) \quad \text{proves that} \quad \text{Ker}(T) \subseteq \text{Ker}(T^2).$$

In view of $\text{Ker}(T^2) \cong \text{Ker}(T)$ and $\text{Ker}(T) \subseteq \text{Ker}(T^2)$, we have that $\text{Ker}(T^2) = \text{Ker}(T)$. Now, $T(V) \subseteq V \Rightarrow T^2(V) \subseteq T(V)$ and

$$\text{rank}(T^2) = \text{rank}(T) \Rightarrow \dim(I_m(T^2)) \cong \dim(I_m(T)).$$

Hence, $I_m(T^2) = I_m(T)$.
Finally, we shall show that $I_m(T) \cap \text{Ker}(T) = \{0\}$.
Let $x \in I_m(T) \cap \text{Ker}(T)$. Then $x \in I_m(T)$ and $x \in \text{Ker}(T)$.

$$x \in I_m T \Rightarrow \exists y \in V : T(y) = x \text{ and } x \in \text{Ker } T \Rightarrow T(x) = 0.$$
$$x = T(y) \Rightarrow T(x) = T^2(y) \Rightarrow 0 = T^2(y) \Rightarrow y \in \text{Ker}(T^2) = \text{Ker}(T)$$
$$\Rightarrow T(y) = 0.$$

Hence, $x = 0$.
Thus, $I_m(T) \cap \text{Ker}(T) = \{0\}$.
Hence, $\dim(V) = \text{rank}(T) + \text{null}(T)$ with $I_m(T) \cap \text{Ker}(T) = \{0\}$ proves that $V = I_m(T) \oplus \text{Ker}(T)$.

(b) Let V be a finite-dimensional vector space and let $T : V \to V$ be a linear transformation. Then $V = I_m T^r \oplus \text{Ker } T^r$ for some positive integer r.
Since $T(V) \subseteq V$, $T^2(V) \subseteq T(V)$, we can find the following chain:

$$V \supseteq I_m(T) \supseteq I_m(T^2) \supseteq \cdots \supseteq I_m(T^r) \supseteq I_m(T^{r+1}) \supseteq \cdots \supseteq I_m(T^{r+k})$$
$$\supseteq \cdots \supseteq \{0\}, \quad \text{for all } k, r \geq 1.$$

Hence,

$$\infty > \dim(V) \geq \text{rank}(T) \geq \text{rank}(T^2) \geq \cdots \geq \text{rank}(T^r) \geq \text{rank}(T^{r+1})$$
$$\geq \cdots \geq \text{rank}(T^{r+k}) \geq \cdots \geq 0, \quad \text{for all } k, r \geq 1.$$

Thus, there exists some positive integer r such that $\text{rank}(T^r) = \text{rank}(T^{r+k})$ for all positive integers k. If we take $r = k$, then $\text{rank}(T^r) = \text{rank}(T^{r+r}) = \text{rank}(T^{2r}) = \text{rank}((T^r)^2)$. Hence, from part (a) we have

$$V = I_m(T^r) \oplus \text{Ker}(T^r).$$

Exercises

(4.2.1) Let V and W be n-dimensional vector spaces over the same field F. Then show that the linear transformation $T : V \to W$ is surjective if and only if T is injective.

(4.2.2) Using the fundamental theorem of vector space homomorphism, prove the rank-nullity theorem.

(4.2.3) Let V be an m-dimensional vector space and W be an n-dimensional vector space over the same field F. Then for the linear transformation $T : V \to W$, prove that
(i) if $m > n$, then T cannot be injective and (ii) if $m < n$, then T can not be surjective.

(4.2.4) Let $T : V \to W$ be an isomorphism. Then prove that $\{T(v_i)\}, 1 \le i \le n$ is a basis of W, whenever $\{v_i\}, 1 \le i \le n$ is a basis of V.

(4.2.5) Let $T : F^2 \to F^3$ be a linear transformation defined by $T(x,y) = (x, x+y, y)$. Then find the rank and nullity of T.

(4.2.6) Let $T : P_n(x) \to P_{n+1}(x)$ be a linear transformation defined by $T(p(x)) = p'(x) - \int_0^x p(t)dt$. Then find the rank$(T)$.

(4.2.7) Let V and W be vector spaces over the same field F of dimensions n and m, respectively. If $T : V \to W$ is a linear transformation and $B = \{v_1, v_2, \ldots, v_r\}$ is a subset of V such that $T(B) = \{T(v_1), T(v_2), \ldots, T(v_r)\}$ is a linearly independent set in W, then prove that B is a linearly independent set.

(4.2.8) Let $T : V \to W$ be an injective linear transformation from an n-dimensional vector space V to an m-dimensional vector space W. Then prove that if $\{v_1, v_2, \ldots, v_r\}$ is a linearly independent set in V, then $\{T(v_1), T(v_2), \ldots, T(v_r)\}$ is a linearly independent set in W.

(4.2.9) Find the rank and nullity of the linear transformations given in exercises 4.1.2, 4.1.4(i), 4.1.4(ii), 4.1.4(iii) and 4.1.6.

(4.2.10) Find the linear transformation $T : \mathbb{R}^3 \to \mathbb{R}^4$ whose image is generated by:
(i) $\{(1,2,3,4),(3,-2,0,0)\}$;
(ii) $\{(1,0,0,0),(0,1,0,0),(0,0,1,0)\}$;
(iii) $\{(1,0,0,0),(0,0,1,0),(0,0,0,1)\}$;
(iv) $\{(3,0,0,0),(2,-1,0,0),(-1,4,2,0)\}$;
(v) $\{(1,2,-1,6)\}$.

(4.2.11) Construct a linear transformation $T : \mathbb{R}^4 \to \mathbb{R}^5$ such that
(i) $\text{Im}(T) = \{(x_1,x_2,x_3,x_4,x_5) : x_1 + 2x_2 - x_3 = 0\}$;
(ii) $\text{Im}(T) = \{(x_1,x_2,x_3,x_4,x_5) : 2x_2 + 3x_3 + 4x_4 = 0\}$.

(4.2.12) Let V be a finite-dimensional vector space and let $T : V \to V$ be a linear map. Then show that the following statements are equivalent:
(i) T is an isomorphism.
(ii) $\text{Ker}(T) = \{0\}$.
(iii) $\text{Im}(T) = V$.

(4.2.13) Let V be a vector space of dimension $n \ge 2$. If $T : V \to V$ is a linear transformation such that $T^{m+1} = 0$ and $T^m \ne 0$ for some $m \ge 1$, then find the relation between rank(T^m) and null(T^m).

(4.2.14) Let $T : \mathbb{R}^3 \to \mathbb{R}^3$ be a linear transformation defined by $T(x,y,z) = (x+3y+2z, 3x+4y+z, 2x+y-z)$, $(x,y,z) \in \mathbb{R}^3$. Then find rank(T^2) and null(T^3).

(4.2.15) Let $S, T : \mathbb{R}^3 \to \mathbb{R}^3$ be two linear transformations defined by
$S(x_1,x_2,x_3) = (x_1, x_1+x_2, x_1-x_2-x_3)$ and $T(x_1,x_2,x_3) = (x_1+2x_3, x_2-x_3, x_1+x_2+x_3)$. Then show that S is invertible but not T.

(4.2.16) Let U, V and W be three finite-dimensional vector spaces such that $\dim(U) = \dim(W)$. If $T_1 : U \to V$ and $T_2 : V \to W$ are linear transformations such that $T_2 \circ T_1 : U \to W$ is injective, then show that T_1 is injective and T_2 is surjective.

4.3 Vector spaces of linear transformations

Consider the set $\text{Hom}(V, W) = \{T : T : V \to W \text{ is a linear transformation}\}$, where V and W are vector spaces over the same field F. Then for $T_1, T_2 \in \text{Hom}(V, W)$, $T_1 + T_2$ defined by $(T_1 + T_2)(x) = T_1(x) + T_2(x)$ is a linear transformation:

For $a, b \in F$ and $x, y \in V$,

$$(T_1 + T_2)(ax + by) = T_1(ax + by) + T_2(ax + by)$$
$$= aT_1(x) + bT_1(y) + aT_2(x) + bT_2(y)$$
$$= aT_1(x) + aT_2(x) + bT_1(y) + bT_2(y)$$
$$= a(T_1 + T_2)(x) + b(T_1 + T_2)(y).$$

Also, for $a \in F$, aT_1 defined by $(aT_1)(x) = aT_1(x)$ is a linear transformation:

$$\text{for } (aT_1)(ax + by) = aT_1(ax + by) = a\big(aT_1(x) + bT_1(y)\big)$$
$$= aaT_1(x) + abT_1(y) = a(aT_1)(x) + b(aT_1)(y).$$

The following theorem is the direct consequence of Example 2.1.7.

Theorem 4.3.1. *Let V and W be vector spaces over the same field F. Then the set $\text{Hom}(V, W)$ is a vector space over F with respect to addition and scalar multiplications defined as follows:*

$$(T_1 + T_2)(x) = T_1(x) + T_2(x) \quad \text{and}$$
$$(aT_1)(x) = aT_1(x), \quad \text{for all } T_1, T_2 \in \text{Hom}(V, W), x \in V \text{ and } a \in F.$$

Now, we shall show that if V and W are finite-dimensional vector spaces, then $\text{Hom}(V, W)$ is a finite-dimensional vector space.

Theorem 4.3.2. *Let V and W be finite-dimensional vector spaces over the same field F of dimensions m and n, respectively. Then $\dim(\text{Hom}(V, W)) = \dim(V) \cdot \dim(W) = mn$.*

Proof. Let $\{v_1, v_2, \ldots, v_m\}$ and $\{w_1, w_2, \ldots, w_n\}$ be ordered bases of V and W, respectively. For every $(i, j), 1 \le i \le m, 1 \le j \le n$, there exists a unique linear transformation $T_{ij} \in \text{Hom}(V, W)$ such that

$$T_{ij}(v_k) = \begin{cases} w_j, & \text{if } i = k, \\ 0, & \text{if } i \ne k. \end{cases}$$

Now, we claim that the set $\{T_{ij}, 1 \le i \le m, 1 \le j \le n\}$ is a basis of the vector space $\text{Hom}(V, W)$.

First, we shall show that $\text{Hom}(V, W)$ is generated by $\{T_{ij}, 1 \le i \le m, 1 \le j \le n\}$. That is, $T \in \text{Hom}(V, W)$ can be uniquely expressed as $T = \sum a_{ij} T_{ij}$, where $a_{ij} \in F$. Then it is sufficient to show that both the linear transformations T and $\sum a_{ij} T_{ij}$ are equal on the basis of V.

Since $\{w_1, w_2, \ldots, w_n\}$ is a basis of W, $T(v_k) \in W$ can be uniquely expressed as $T(v_k) = \sum_{j=1}^{n} a_{kj} w_j$, where $a_{kj} \in F$ and $1 \le k \le m$.

Now,

$$\left(\sum_{ij} a_{ij} T_{ij} \right)(v_k) = \sum_{ij} a_{ij} T_{ij}(v_k)$$

$$= \sum_{j} \left(\sum_{i} a_{ij} T_{ij}(v_k) \right)$$

$$= \sum_{j} (a_{1j} T_{1j}(v_k) + a_{2j} T_{2j}(v_k) + \cdots + a_{kj} T_{kj}(v_k) + \cdots + a_{mj} T_{mj}(v_k))$$

$$= \sum_{j} a_{kj} w_j, \quad \text{since } T_{ij}(v_k) = 0 \text{ if } i \ne k.$$

Thus, $T(v_k) = (\sum_{ij} a_{ij} T_{ij})(v_k) \; \forall 1 \le k \le m$. Hence, $T = \sum_{ij} a_{ij} T_{ij}$.

To complete the proof, it remains to show that the set $\{T_{ij}, 1 \le i \le m, 1 \le j \le n\}$ is linearly independent.

Suppose $\sum_{ij} a_{ij} T_{ij} = 0$, where $a_{ij} \in F$. Then

$$\left(\sum_{ij} a_{ij} T_{ij} \right)(v_k) = 0$$

$$\Rightarrow \sum_{ij} a_{ij} T_{ij}(v_k) = 0$$

$$\Rightarrow \sum_{j} \left(\sum_{i} a_{ij} T_{ij}(v_k) \right) = 0$$

$$\Rightarrow \sum_{j} a_{kj} T_{kj}(v_k) = 0$$

$$\Rightarrow \sum_{j} a_{kj} w_j = 0 \quad \forall k.$$

Since $\{w_1, w_2, \ldots, w_n\}$ is linearly independent,

$$\sum_{j} a_{kj} w_j - 0 \to a_{kj} = 0 \quad \forall j, \text{ and for all } k = 1, 2, \ldots, m, \text{ also.}$$

This shows that all $a_{ij} = 0$ for $1 \le i \le m, 1 \le j \le n$, and hence the set $\{T_{ij}, 1 \le i \le m, 1 \le j \le n\}$ is linearly independent. This proves that $\{T_{ij}, 1 \le i \le m, 1 \le j \le n\}$ is a basis of $\mathrm{Hom}(V, W)$. Hence, $\dim(\mathrm{Hom}(V, W)) = \dim(V) \cdot \dim(W) = mn$. □

Definition 4.3.3. Let V be a vector space over a field F. Since every field is a vector space over itself, the vector space $\mathrm{Hom}(V, F) = \{f : f : V \to F \text{ is a linear transformation}\}$, denoted by V^* is called the dual space of V. A linear transformation $f \in V^*$ is called linear functional.

If V is an m-dimensional vector space, then $\dim(V^*) = \dim(\text{Hom}(V,F)) = \dim(V) \cdot \dim(F) = m \cdot 1 = m$. Thus, for a finite-dimensional vector space V, $\dim(V^*) = \dim(V)$, and hence $V^* \cong V$.

Linear transformations given in Example 4.1.7 and Example 4.1.9 are linear functionals.

If $\{e_1, e_2, \ldots, e_m\}$ is a basis of the vector space V over a field F and $\{1\}$ is the basis of vector space $W = F$ over F, then $T_{ij} = T_{i1}, 1 \le i \le m, j = 1$ defined in the previous theorem is

$$T_{i1}(e_k) = \begin{cases} 1, & \text{if } i = k, \\ 0, & \text{otherwise.} \end{cases}$$

In the form of simple notation, let us denote $T_{i1} = e_i^*$, then

$$e_i^*(e_j) = \begin{cases} 1, & \text{if } i = j, \\ 0, & \text{otherwise.} \end{cases}$$

From the proof of the above theorem, we have the following definition.

Definition 4.3.4. Let V be an m-dimensional vector space over a field F and let $\{e_1, e_2, \ldots, e_m\}$ be a basis of V. Then the set $\{e_1^*, e_2^*, \ldots, e_m^*\}$ is a basis of V^* called the dual basis to $\{e_1, e_2, \ldots, e_m\}$, where $e_i^*(e_j) = 1$, if $i \ne j$ and 0, otherwise for all $1 \le i, j \le m$.

Definition 4.3.5. Let V be a vector space over a field F. Then $\text{Hom}(V^*, F) = (V^*)^*$ denoted by V^{**} is called the second dual of V.

Theorem 4.3.6. *Let V be a vector space over a field F. Then for $x \in V$, the map $x^{**} : V^* \to F$ given by $x^{**}(f) = f(x)$ is a linear functional on V^*, where $f \in V^*$.*

Proof. Let $f, g \in V^*$ and $a, b \in F$. Then $x^{**}(af + bg) = (af + bg)(x) = af(x) + bg(x) = ax^{**}(f) + bx^{**}(g)$. This completes the proof. □

Theorem 4.3.7. *Let $\{x_1, x_2, \ldots, x_n\}$ be a basis of a vector space V over a field F and let $\{x_1^*, x_2^*, \ldots, x_n^*\}$ be its dual basis in V^*. Then:*
(i) *any vector $x \in V$ can be expressed as $x = x_1^*(x)x_1 + x_2^*(x)x_2 + \cdots + x_n^*(x)x_n$ and*
(ii) *functional $f \in V^*$ can be expressed as $f = f(x_1)x_1^* + f(x_2)x_2^* + \cdots + f(x_n)x_n^*$.*

Proof. (i) Let $x = \sum_{i=1}^n a_i x_i$, $a_i \in F$. Then $x_1^*(x) = \sum_{i=1}^n a_i x_1^*(x_i) = a_1$, where $x_1^*(x_i) = 0$ if $i \ne 1$.
Similarly, for all $i = 2, 3, \ldots, n$, $x_i^*(x) = a_i$.
Hence, $x = \sum_{i=1}^n a_i x_i = \sum_{i=1}^n x_i^*(x)x_i$.

(ii) Let $x \in V$. Then from (i), we have $x = \sum_{i=1}^{n} x_i^*(x)x_i$.
Hence,

$$f(x) = f\left(\sum_{i=1}^{n} x_i^*(x)x_i\right)$$
$$= \sum_{i=1}^{n} x_i^*(x)f(x_i)$$
$$= \sum_{i=1}^{n} f(x_i)x_i^*(x)$$
$$= \left(\sum_{i=1}^{n} f(x_i)x_i^*\right)(x).$$

Thus, $f(x) = (\sum_{i=1}^{n} f(x_i)x_i^*)(x)$, for all $x \in V$. Hence, $f = \sum_{i=1}^{n} f(x_i)x_i^*$. □

Theorem 4.3.8. *Let V be an n-dimensional vector space over a field F. Then V is isomorphic to* V^{**}.

Proof. Since V and V^* are vector spaces over the same field F, $V^{**} = \text{Hom}(V^*, F)$ is a vector space over F.

Also, $\dim(V^{**}) = \dim(V^*) = n = \dim(V)$.

Hence, V and V^{**} are isomorphic vector spaces. □

Theorem 4.3.9. *Let V be an n-dimensional vector space over a field F. Then the map* $T : V \to V^{**}$ *defined by* $T(x) = x^{**}$ *is an isomorphism.*

Proof. Let $x, y \in V$ and $a, b \in F$. Then $T(ax + by) = (ax + by)^{**}$.

Now, for $f \in V^*$, $(ax + by)^{**}(f) = f(ax + by) = af(x) + bf(y) = ax^{**}(f) + by^{**}(f) = (ax^{**} + by^{**})(f)$ shows that $(ax + by)^{**} = ax^{**} + by^{**}$. Hence, $T(ax + by) = ax^{**} + by^{**}$ gives that T is a linear transformation.

Now, we shall show that T is bijective.

Let $x \in \text{Ker}(T)$. Then $T(x) = x^{**} = 0$, where 0 is the zero linear functional in V^*. Hence, $x^{**}(f) = 0 \Rightarrow f(x) = 0$ for all $f \in V^*$. This shows that $x = 0$. For if $x \neq 0$, the set $\{x\}$ being linearly independent, can be enlarged to a basis of V. Then we can define a linear functional f on V whose restriction on the basis is given by $f(x) = 1$ and 0 at all other members of the basis, which contradicts the fact that $f(x) = 0$ for all $f \in V^*$. Hence, $\text{Ker}(T) = \{0\}$ and so that T is injective.

Also, $\text{rank}(T) = \dim(V) - \text{null}(T) = n$ shows that T is surjective. Hence, T is an isomorphism. □

Example 4.3.10. Let $\{e_1 = (1, -1, 2), e_2 = (0, 1, -1), e_3 = (0, 3, 4)\}$ be a basis of $\mathbb{R}^3$. Suppose $\{e_1^*, e_2^*, e_3^*\}$ is the dual basis of the basis $\{e_1, e_2, e_3\}$.

The linear functionals e_1^*, e_2^*, e_3^* may be expressed in the form (by the Riesz representation theorem proved in the coming chapter):

$$e_1^*(x,y,z) = a_1 x + a_2 y + a_3 z,$$
$$e_2^*(x,y,z) = b_1 x + b_2 y + b_3 z,$$
$$e_3^*(x,y,z) = c_1 x + c_2 y + c_3 z.$$

By the definition $e_i^*(e_j) = 1$, if $i = j$ and 0 if $i \neq j$, we have

$$1 = e_1^*(e_1) = e_1^*(1,-1,2) = a_1 - a_2 + 2a_3,$$
$$0 = e_1^*(e_2) = e_1^*(0,1,-1) = a_2 - a_3,$$
$$0 = e_1^*(e_3) = e_1^*(0,3,4) = 3a_2 + 4a_3.$$

Solving above equations yield $a_1 = 1$, $a_2 = 0$, $a_3 = 0$.
Hence, $e_1^*(x,y,z) = x$.
Similarly, we get

$$e_2^*(x,y,z) = \frac{10}{7}x + \frac{4}{7}y + \frac{-3}{7}z \quad \text{and}$$
$$e_3^*(x,y,z) = \frac{-1}{7}x + \frac{-1}{7}y + \frac{2}{7}z.$$

Example 4.3.11. Let $V = P_2(x)$ be the vector space of polynomials over $\mathbb{R}$ of degree ≤ 2, and let e_1^*, e_2^*, e_3^* be the linear functionals on V defined as follows:

$$e_1^*(f(x)) = \int_0^1 f(x)dx, \quad e_2^*(f(x)) = \acute{f}(1), \quad e_3^*(f(x)) = f(0),$$

where $f(x) = a_0 + a_1 x + a_2 x^2 \in V$ and $\acute{f}(x)$ denotes the derivative of $f(x)$ with respect to x.

Since $\dim(V^*) = \dim(V) = 3$, the set e_1^*, e_2^*, e_3^* is a basis of V^* if it is linearly independent.

To show that e_1^*, e_2^*, e_3^* is linearly independent, let $\alpha, \beta, \gamma \in \mathbb{R}$ such that $\alpha e_1^* + \beta e_2^* + \gamma e_3^* = 0$, the zero functional.

Then, for $f(x) \in V$,

$$(\alpha e_1^* + \beta e_2^* + \gamma e_3^*)f(x) = 0,$$
$$\alpha e_1^*(f(x)) + \beta e_2^*(f(x)) + \gamma e_3^*(f(x)) = 0,$$

$$\alpha \int_0^1 f(x)dx + \beta \acute{f}(1) + \gamma f(0) = 0.$$

Let $f(x) = a_0 + a_1x + a_2x^2$. Then from the above equation, we have

$$\alpha \int_0^1 (a_0 + a_1x + a_2x^2)dx + \beta(a_1 + 2a_2) + \gamma a_0 = 0,$$

$$\alpha\left(a_0 + \frac{a_1}{2} + \frac{a_2}{3}\right) + \beta(a_1 + 2a_2) + \gamma a_0 = 0, \quad \text{for all } a_0, a_1, a_2 \in \mathbb{R},$$

$$(\alpha + \gamma)a_0 + \left(\frac{\alpha}{2} + \beta\right)a_1 + \left(\frac{\alpha}{3} + 2\beta\right)a_2 = 0$$

$$\Rightarrow \alpha + \gamma = 0, \quad \text{if } a_0 = 1, a_1 = a_2 = 0,$$

$$\frac{\alpha}{2} + \beta = 0, \quad \text{if } a_1 = 1, a_0 = a_2 = 0,$$

$$\frac{\alpha}{3} + 2\beta = 0, \quad \text{if } a_2 = 1, a_0 = a_1 = 0.$$

Solving above equations yield $\alpha = \beta = \gamma = 0$. Thus, the set $\{e_1^*, e_2^*, e_3^*\}$ is linearly independent, and hence a basis of V^*.

Now, we shall find the basis $\{e_1, e_2, e_3\}$ of V whose dual is $\{e_1^*, e_2^*, e_3^*\}$.

Let $e_1 = a_0 + a_1x + a_2x^2$, $e_2 = b_0 + b_1x + b_2x^2$ and $e_3 = c_0 + c_1x + c_2x^2$. Then by the definition of dual basis we have

$$1 = e_1^*(e_1) = \int_0^1 (a_0 + a_1x + a_2x^2)dx = a_0 + \frac{a_1}{2} + \frac{a_2}{3},$$

$$0 = e_1^*(e_2) = \int_0^1 (b_0 + b_1x + b_2x^2)dx = b_0 + \frac{b_1}{2} + \frac{b_2}{3},$$

$$0 = e_1^*(e_3) = \int_0^1 (c_0 + c_1x + c_2x^2)dx = c_0 + \frac{c_1}{2} + \frac{c_2}{3}.$$

Similarly,

$$0 = e_2^*(e_1) = a_1 + 2a_2, \quad 1 = e_2^*(e_2) = b_1 + 2b_2, \quad 0 = e_2^*(e_3) = c_1 + 2c_2 \quad \text{and}$$

$$0 = e_3^*(e_1) = a_0, \quad 0 = e_3^*(e_2) = b_0, \quad 1 = e_3^*(e_3) = c_0.$$

After solving the above nine equations, we have that $a_0 = 0$, $a_1 = 3$, $a_2 = \frac{-3}{2}$; $b_0 = 0$, $b_1 = \frac{-1}{2}$, $b_2 = \frac{3}{4}$ and $c_0 = 1$, $c_1 = -3$, $c_2 = \frac{3}{2}$.

Hence, the required basis is

$$\left\{e_1 = 3x - \frac{-3}{2}x^2, e_2 = \frac{-x}{2} + \frac{3}{4}x^2, e_3 = 1 - 3x + \frac{3}{2}x^2\right\}.$$

Definition 4.3.12. Let V be a vector space over a field F and let S be a subset of V. Then the set $S^\circ = \{f \in V^* : f(x) = 0, \forall x \in S\}$ is called the annihilator of S.

For $f_1, f_2 \in S^\circ$ and $a, b \in F$,

$$(af_1 + bf_2)(x) = af_1(x) + bf_2(x) = a \cdot 0 + b \cdot 0 = 0, \quad \forall x \in S.$$

This shows that the annihilator of a subset S of a vector space V is a subspace of the dual space V^* of V.

For example, $\{0\}^\circ = V^*$ and $V^* = \{0\}$.

Theorem 4.3.13. *Let W be a subspace of an n-dimensional vector space V. Then $n = \dim(V) = \dim(W) + \dim(W^\circ)$.*

Proof. Let $\dim(W) = r$. Then we have to show that $\dim(W^\circ) = n - r$.

Suppose $\{w_1, w_2, \ldots, w_r\}$ is a basis of W, which can be extended to a basis $\{w_1, w_2, \ldots, w_r, v_1, v_2, \ldots, v_{n-r}\}$ of V.

Let $\{w_1^*, w_2^*, \ldots, w_r^*, v_1^*, v_2^*, \ldots, v_{n-r}^*\}$ be a basis of V^*. Now, we claim that $\{v_1^*, v_2^*, \ldots, v_{n-r}^*\}$ is a basis of W°.

Since $\{v_1^*, v_2^*, \ldots, v_{n-r}^*\}$ is a subset of basis of V^*, it is linearly independent. Also, every element $w \in W$ can be expressed as $w = \sum_{i=1}^{r} a_i w_i$, where $a_i \in F$. Then by the definition of dual basis,

$$v_j^*(w) = v_j^*\left(\sum_{i=1}^{r} a_i w_i\right) = \sum_{i=1}^{r} a_i v_j^*(w_i) = \sum_{i=1}^{r} a_i \cdot 0 = 0$$

shows that $\{v_1^*, v_2^*, \ldots, v_{n-r}^*\}$ is a subset of W°.

Next, we shall show that $\{v_1^*, v_2^*, \ldots, v_{n-r}^*\}$ generates W°.

Let $f \in W^\circ$. Then $f \in V^*$. From Theorem 4.3.7,

$$f = f(w_1)w_1^* + f(w_1)w_1^* + f(w_2)w_2^* + \cdots + f(w_r)w_r^* + f(v_1)v_1^* + \cdots + f(v_{n-r})v_{n-r}^*.$$

Since $f \in W^\circ, f(w_i) = 0, \forall 1 \le i \le r, f = f(v_1)v_1^* + \cdots + f(v_{n-r})v_{n-r}^*$. This proves that $\{v_1^*, v_2^*, \ldots, v_{n-r}^*\}$ generates W°, and hence a basis of W°. Thus, $\dim(W^\circ) = n - r = n - \dim(W) \Rightarrow \dim(W) + \dim(W^\circ) = \dim(V)$. □

Corollary 4.3.14. *Let W be a subspace of an n-dimensional vector space V. Then $W^{\circ\circ} \cong W$.*

Proof. Let $\dim(W) = m$. Then

$$\dim(W) + \dim(W^\circ) = \dim(V) \Rightarrow \dim(W^\circ) = n - m.$$

Again, $\dim(V^*) = n$ and W° is a subspace of V^*,

$$\dim(W^{\circ\circ}) = \dim(V^*) - \dim(W^\circ) = n - (n - m) = m.$$

Thus, $\dim(W) = \dim(W^{\circ\circ})$ and hence $W^{\circ\circ} \cong W$. □

Example 4.3.15. Let W be a subspace of $\mathbb{R}^3$ generated by the set $\{w_1 = (1, 2, 3), w_2 = (0, 1, -1)\}$. In this example, we compute the basis of W°.

Since $\dim(W^\circ) = 3 - \dim(W) = 1$, there exists only one linear functional in the basis of W° such that $f(w_1) = f(w_2) = 0$.

Let $f(x, y, z) = ax + by + cz \in W^\circ$. Then

$$f(1, 2, 3) = a + 2b + 3c = 0 \quad \text{and} \quad f(0, 1, -1) = b - c = 0,$$
$$b - c = 0 \Rightarrow b = c.$$

And so,

$$a + 2b + 3c = 0 \Rightarrow a + 2c + 3c = 0 \Rightarrow a = -5c.$$

Hence, $f(x, y, z) = -5cx + cy + cz = c(-5x + y + z)$ gives that $f(x, y, z) = -5x + y + z$.
Let $w \in W$. Then $w = a_1 w_1 + a_2 w_2$, for some $a_1, a_2 \in \mathbb{R}$.
Hence, $f(w) = f(a_1 w_1 + a_2 w_2) = a_1 f(w_1) + a_2 f(w_2) = a_1 \cdot 0 + a_2 \cdot 0 = 0$ implies that $f \in W^\circ$.

Exercises

(4.3.1)   Prove Theorem 4.3.1.

(4.3.2)   Let $f : \mathbb{R}^n \to \mathbb{R}$ be a function defined by $f(x_1, x_2, \ldots, x_n) = a_1 x_1 + a_2 x_2 + \cdots + a_n x_n$, where $a_1, a_2, \ldots, a_n \in \mathbb{R}$. Then show that f is a linear functional.

(4.3.3)   Find the dual basis of each of the following bases of $\mathbb{R}^3$:
   (i) $\{(1, -1, 1), (4, 1, 0), (8, 1, 1)\}$;
   (ii) $\{(1, 3, 3), (1, 4, 3), (1, 3, 4)\}$;
   (iii) $\{(1, 1, 3), (1, 3, -3), (-2, -4, -4)\}$;
   (iv) $\{(1, 0, 0), (0, 1, 0), (0, 0, 1)\}$.

(4.3.4)   Let $V = P_1(x)$ be the vector space of all polynomials with real coefficients of degree at most one and let $e_1^*(f(x)) = \int_0^1 f(x) dx$, $e_2^*(f(x)) = f'(1)$, $e_3^*(f(x)) = f(0)$, are linear functionals on V. Then show that each of the following sets (i) $\{e_1^*, e_2^*\}$, (ii) $\{e_1^*, e_3^*\}$, (iii) $\{e_2^*, e_3^*\}$ is a basis of V^* and find the corresponding bases $\{e_i, e_j\}$ in V whose dual is $\{e_i^*, e_j^*\}$, where $1 \leq i, j \leq 3$.

(4.3.5)   Let $V = P_2(x)$ be the vector space of all polynomials with real coefficients of degree at most two. Then
   (a)   show that e_a^* defined by $e_a^*(f(x)) = f(a)$, $a \in \mathbb{R}$ is a linear functional on V;
   (b)   show that $\{e_{-1}^*, e_0^*, e_1^*\}$ is a basis of V^*;
   (c)   find the basis $\{e_{-1}, e_0, e_1\}$ of V such that $\{e_{-1}^*, e_0^*, e_1^*\}$ is its dual;
   (d)   for distinct $r, s, t \in \mathbb{R}$, show that $[c_r^*, c_s^*, c_t^*]$ is a basis of V^*.

(4.3.6)   If $\{e_1, e_2, \ldots, e_n\}$ is a basis of a vector space V and $\{e_1^*, e_2^*, \ldots, e_n^*\}$ is its dual basis, then for any $x \in V$ show that $x = e_1^*(x)e_1 + e_2^*(x)e_2 + \cdots + e_n^*(x)e_1$.

(4.3.7)   Let S be a subset of a vector space V. Then show that $S^\circ = [\langle S \rangle]^\circ$.

(4.3.8)   Let W be a subspace of $\mathbb{R}^3$. Then find the basis of W°, if:
   (i) W is generated by $\{w_1 = (1, 1, -1)\}$;
   (ii) W is generated by $\{w_1 = (0, 1, 2), w_2 = (3, 2, 1)\}$;
   (iii) W is generated by $\{w_1 = (-3, 2, 4), w_2 = (1, 0, 2)\}$.

(4.3.9)   Let W be a subspace of $\mathbb{R}^4$. Then find the basis of W° if W is generated by:

(i) $\{(1,-1,1,-1)\}$;
(ii) $\{(1,2,3,-1),(1,0,3,4)\}$;
(iii) $\{(0,-3,4,2),(-1,2,0,3)\}$;
(iv) $\{(1,0,2,-3),(0,2,3,4),(-5,2,3,0)\}$.

(4.3.10) If W_1 and W_2 are subspaces of a vector space V (may be finite- or infinite-dimensional), then prove that $(W_1+W_2)^\circ = W_1^\circ \cap W_2^\circ$.

4.4 Transpose of a linear transformation

The transpose of a linear transformation between two vector spaces given over the same field is an induced map between the dual spaces of the two vector spaces.

Definition 4.4.1. Let V and W be vector spaces over the same field F and let $T : V \to W$ be a linear transformation. Then the map $T^t : W^* \to V^*$ defined by $T^t(f) = f \circ T$, being the composition of linear transformations is a linear transformation, called the transpose of T.

$f \in W^*$ means $f : W \to F$ is a linear transformation, and hence the composition $f \circ T : V \to F$ is a linear transformations. Thus, $T^t(f) = f \circ T \in V^*$.

Theorem 4.4.2. Let $T_1 : V_1 \to V_2$ and $T_2 : V_2 \to V_3$ be linear transformations. Then $(T_2 \circ T_1)^t = T_1^t \circ T_2^t$.

Proof. Since $T_2 \circ T_1 : V_1 \to V_3$ is a linear transformation, $(T_2 \circ T_1)^t$ is a linear transformation from V_3^* to V_1^*. Then for $f \in V_3^*$,

$$(T_2 \circ T_1)^t(f) = f \circ (T_2 \circ T_1) = (f \circ T_2) \circ T_1 = T_2^t(f) \circ T_1.$$

Since $T_2^t(f) \in V_2^*$,

$$T_2^t(f) \circ T_1 = T_1^t(T_2^t(f)) = (T_1^t \circ T_2^t)(f).$$

Thus, $(T_2 \circ T_1)^t(f) = (T_1^t \circ T_2^t)(f)$ for all f.
Hence, $(T_2 \circ T_1)^t = T_1^t \circ T_2^t$. □

Theorem 4.4.3. Let V and W be vector spaces over a field F. Then: (i) $(T_1+T_2)^t = T_1^t + T_2^t$ (ii) $(aT)^t = aT^t$, for all $T_1, T_2 \in \mathrm{Hom}(V,W)$ and $a \in F$.

Proof. (i) We have that $T_1 + T_2 : V \to W$ is a linear transformation.
Hence,

$$(T_1+T_2)^t : W^* \to V^* \quad \text{and} \quad (T_1^t + T_2^t) : W^* \to V^*$$

are linear transformations.
Let $f \in W^*$. Then

$$(T_1 + T_2)^t(f) = f \circ (T_1 + T_2) = f \circ T_1 + f \circ T_2 = T_1^t(f) + T_2^t(f) = (T_1^t + T_2^t)(f).$$

Hence, $(T_1 + T_2)^t = T_1^t + T_2^t$.

(ii) Since $\alpha T : V \to W$ is a linear transformation, $(\alpha T)^t : W^* \to V^*$ is linear transformation.

For $f \in W^*$, $(\alpha T)^t(f) = f \circ \alpha T = \alpha f \circ T = \alpha(f \circ T) = \alpha T^t(f)$.

Hence, $(\alpha T)^t = \alpha T^t$. □

Theorem 4.4.4. *Let V and W be finite-dimensional vector spaces and let $T : V \to W$ be a linear transformation. Then* $\operatorname{rank}(T) = \operatorname{rank}(T^t)$.

Proof. Since $T : V \to W$ is a linear transformation, $T^t : W^* \to V^*$ is a linear transformation.

By the rank-nullity theorem,

$$\operatorname{rank}(T^t) + \operatorname{null}(T^t) = \dim(W^*) = \dim(W)$$
$$\Rightarrow \operatorname{rank}(T^t) = \dim(W) - \operatorname{null}(T^t).$$

Since $T(V) \subset W$, by Theorem 4.3.13,

$$\dim(T(V))^\circ + \dim(T(V)) = \dim(W)$$
$$\Rightarrow \dim(T(V))^\circ = \dim(W) - \dim(T(V)).$$

If $\dim(T(V))^\circ = \operatorname{null}(T^t)$, then from the above equation

$$\dim(W) - \dim(T(V)) = \operatorname{null}(T^t),$$

and hence from the above first equation,

$$\begin{aligned}
\operatorname{rank}(T^t) &= \dim(W) - \operatorname{null}(T^t) \\
&= \dim(W) - (\dim(W) - \dim(T(V))) \\
&= \dim(T(V)) \\
&= \operatorname{rank}(T).
\end{aligned}$$

To complete the proof, we have to show that $\dim(T(V))^\circ = \operatorname{null}(T^t)$.

$$\begin{aligned}
f \in \operatorname{Ker}(T^t) &\Leftrightarrow T^t(f) = 0 \quad \text{(zero linear functional)} \\
&\Leftrightarrow f \circ T = 0 \\
&\Leftrightarrow (f \circ T)(x) = 0 \in F, \quad \forall x \in V \\
&\Leftrightarrow f(T(x)) = 0 \in F, \quad \forall x \in V \\
&\Leftrightarrow f \in (T(V))^\circ.
\end{aligned}$$

Hence, $\operatorname{Ker}(T^t) = (T(V))^\circ \Rightarrow \operatorname{null}(T^t) = \dim(T(V))^\circ$. □

From the proof of the above theorem, we write the following corollary immediately.

Corollary 4.4.5. *Let $T : V \to V$ be a linear transformation on a finite-dimensional vector space V. Then T is nonsingular if and only if T^t is nonsingular.*

Example 4.4.6. Let $T : \mathbb{R}^3 \to \mathbb{R}^2$ be a linear transformation defined by $T(x,y,z) = (x+2y+z, y-z)$. Then $T^t : (\mathbb{R}^2)^* \to (\mathbb{R}^3)^*$. Let $f : \mathbb{R}^2 \to \mathbb{R}$ be a linear functional defined by $f(x,y) = x+3y$. That is, $f \in (\mathbb{R}^2)^*$. Then $(T^t(f))(x,y,z) = (f \circ T)(x,y,z) = f(T(x,y,z)) = f(x+2y+z, y-z) = (x+2y+z) + 3(y-z) = x + 5y - 2z$.

ℹ️ Exercises

(4.4.1) Let $T : V \to W$ be a linear transformation. Then prove that $\text{null}(T^t) = \dim(\text{Im } T)^{\circ}$.

(4.4.2) Let $f : \mathbb{R}^2 \to \mathbb{R}$ be a linear functional defined by $f(x,y) = 2x - y$. Then for each of the following linear transformations $T : \mathbb{R}^2 \to \mathbb{R}^2$, find $(T^t(f))(x,y)$:

    (a) $T(x,y) = (0,0)$;

    (b) $T(x,y) = (x,y)$;

    (c) $T(x,y) = (0,y)$;

    (d) $T(x,y) = (x+y,y)$;

    (e) $T(x,y) = (x-y,x+y)$.

(4.4.3) Let $f : \mathbb{R}^2 \to \mathbb{R}$ be a linear functional defined by $f(x,y) = 2x - 5y$. Then for each of the following linear transformations $T : \mathbb{R}^3 \to \mathbb{R}^2$, find $(T^t(f))(x,y,z)$:

    (a) $T(x,y,z) = (x+y,-z)$;

    (b) $T(x,y,z) = (x,y+z)$;

    (c) $T(x,y,z) = (x+y+z,x-y)$;

    (d) $T(x,y,z) = (2x-y,x+y+z)$;

    (e) $T(x,y,z) = (2x+3y-z,x+z)$.

---

5 Matrices and linear transformations

In the preceding chapters, we have studied matrices and linear transformations separately. But fortunately, there is a beautiful relationship between the linear transformation on a finite-dimensional vector space and a matrix. In this chapter, we shall see how a linear transformation between two finite-dimensional vector spaces can be represented by a matrix. By developing a vector space isomorphism between the vector space of matrices and the vector space of linear transformations, we can use the properties of one space to study the properties of the other.

5.1 The matrix representation of a linear transformation

We begin this section by introducing the concept of an ordered basis.

Definition 5.1.1. Let V be an n-dimensional vector space over the field F. Then the basis $B = \{v_1, v_2, \ldots, v_n\}$ of V is called an ordered basis if the order of elements of B is fixed.

For example, if $B_1 = \{v_1, v_2, v_3\}$ and $B_2 = \{v_2, v_1, v_3\}$ are two ordered bases of a vector space V, then $B_1 \neq B_2$. Thus, every basis of an n-dimensional vector space gives rise to $n!$ distinct ordered bases.

Example 5.1.2. In the vector space F^3, the standard basis,

$$\{e_1 = (1,0,0), e_2 = (0,1,0), e_3 = (0,0,1)\},$$

is the ordered basis for F^3.

In general, $\{e_1, e_2, \ldots, e_n\}$ is an ordered basis of F^n. Similarly, the basis $\{1, x, x^2, \ldots, x^n\}$ is an ordered basis of the polynomial space $P_n(x)$.

Now we proceed to the title of this section, which is the matrix representation of a linear transformation.

Let V and W be finite-dimensional vector spaces over the same field F with ordered bases $B_1 = \{v_1, v_2, \ldots, v_n\}$ and $B_2 = \{w_1, w_2, \ldots, w_m\}$, respectively. If $T : V \to W$ is a linear transformation, then for each $1 \leq j \leq n$, $T(v_j) \in W$ can be expressed uniquely as a linear combination of the vectors in the basis B_2.

That is,

$$T(v_1) = a_{11}w_1 + a_{21}w_2 + a_{31}w_3 + \cdots + a_{m1}w_m,$$
$$T(v_2) = a_{12}w_1 + a_{22}w_2 + a_{32}w_3 + \cdots + a_{m2}w_m,$$
$$\vdots$$
$$T(v_n) = a_{1n}w_1 + a_{2n}w_2 + a_{3n}w_3 + \cdots + a_{mn}w_m,$$

where $a_{ij} \in F, \forall 1 \leq i \leq m, 1 \leq j \leq n$.

https://doi.org/10.1515/9783111516035-005

Since a linear transformation defined on the basis of a vector space is completely determined on the whole space, the matrix $[a_{ij}]$ of unique ordered mn scalars $a_{ij} \in F$ represents the linear transformation T relative to the ordered bases B_1 and B_2.

Definition 5.1.3. Let $T : V \rightarrow W$ be a linear transformation, where V and W are finite-dimensional vector spaces with ordered bases $B_1 = \{v_1, v_2, \ldots, v_n\}$ and $B_2 = \{w_1, w_2, \ldots, w_m\}$, respectively. Then the matrix $[a_{ij}]$ of order $m \times n$ is called the matrix of the linear transformation T related to the ordered bases B_1 and B_2 where $a_{ij} \in F$ such that $T(v_j) = \sum_{i=1}^{m} a_{ij} w_i, 1 \leq j \leq n$.

We use the notation $[T]_{B_1,B_2}$ to denote the above matrix, i. e., $[T]_{B_1,B_2} = [a_{ij}]$. Thus, a linear transformation from an n-dimensional vector space to an m-dimensional vector space is represented by a matrix of order $m \times n$.

We will illustrate the matrix representation of several linear transformations in the next few examples.

Example 5.1.4. Let $T : F^3 \rightarrow F^2$ be a linear transformation defined by $T(x, y, z) = (2x + 3y - z, x + z)$. Then the matrix representing T related to standard bases $B_1 = \{(1, 0, 0), (0, 1, 0), (0, 0, 1)\}$ and $B_2 = \{(1, 0), (0, 1)\}$, will be of order 2×3. By taking the coefficients of the subsequent column vectors, we are able to obtain the matrix:

$$T(1, 0, 0) = (2, 1) = 2(1, 0) + 1(0, 1),$$
$$T(0, 1, 0) = (3, 0) = 3(1, 0) + 0(0, 1),$$
$$T(0, 0, 1) = (-1, 1) = -1(1, 0) + 1(0, 1).$$

Hence, the matrix is

$$[T]_{B_1,B_2} = \begin{bmatrix} 2 & 3 & -1 \\ 1 & 0 & 1 \end{bmatrix}.$$

Example 5.1.5. Let $P_4(x)$ and $P_3(x)$ be the polynomial spaces over the field F. Then the matrix of linear transformation $T : P_4(x) \rightarrow P_3(x)$, defined by $T(f(x)) = \frac{d}{dx} f(x)$, with respect to standard bases is of order 4×5.

Let $B_1 = \{1, x, x^2, x^3, x^4\}$ and $B_2 = \{1, x, x^2, x^3\}$ be the bases of $P_4(x)$ and $P_3(x)$, respectively. Then

$$T(1) = 0 = 0.1 + 0.x + 0.x^2 + 0.x^3,$$
$$T(x) = 1 = 1.1 + 0.x + 0.x^2 + 0.x_3,$$
$$T(x^2) = 2x = 0.1 + 2.x + 0.x^2 + 0.x^3,$$
$$T(x^3) = 3x^2 = 0.1 + 0.x + 3.x^2 + 0.x^3,$$
$$T(x^4) = 4x^3 = 0.1 + 0.x + 0.x^2 + 4.x^3,$$

and hence the matrix representing T is

$$[T]_{B_1,B_2} = \begin{bmatrix} 0 & 1 & 0 & 0 & 0 \\ 0 & 0 & 2 & 0 & 0 \\ 0 & 0 & 0 & 3 & 0 \\ 0 & 0 & 0 & 0 & 4 \end{bmatrix}.$$

Example 5.1.6. Let $T : \mathbb{R}^3 \to \mathbb{R}^3$ be a linear transformation defined by $T(x,y,z) = (x + y, y + z, z + x)$. Then the matrix $[T]_{B_1,B_1}$ where $B_1 = \{(1,1,1),(0,1,1),(0,0,1)\}$ is obtained as follows:

$$T(1,1,1) = (2,2,2) = 2(1,1,1) + 0(0,1,1) + 0(0,0,1),$$
$$T(0,1,1) = (1,2,1).$$

Suppose $(1,2,1) = a_1(1,1,1) + a_2(0,1,1) + a_3(0,0,1)$.
Then

$$(1,2,1) = (a_1, a_1 + a_2, a_1 + a_2 + a_3)$$
$$\Rightarrow a_1 = 1, a_1 + a_2 = 2, a_1 + a_2 + a_3 = 1.$$

On solving, $a_1 = 1, a_2 = 1, a_3 = -1$.
Hence, $T(0,1,1) = (1,2,1) = 1(1,1,1) + 1(0,1,1) - 1(0,0,1)$.
Next, $T(0,0,1) = (0,1,1) = 0(1,1,1) + 1(0,1,1) + 0(0,0,1)$.
Then

$$[T]_{B_1,B_2} = \begin{bmatrix} 2 & 1 & 0 \\ 0 & 1 & 1 \\ 0 & -1 & 0 \end{bmatrix}.$$

Example 5.1.7. Let $T : \mathbb{R}^3 \to P_2(x)$ be a linear transformation defined by $T(a,b,c) = (a - b)x^2 + cx + a + b + c$, where $(a,b,c) \in \mathbb{R}^3$. Then we determine the matrix $[T]_{B_1,B_2}$, where $B_1 = \{(1,0,0),(1,1,0),(1,1,1)\}$ and $B_2 = \{1, 1 + x, 1 + x + x^2\}$ are ordered bases of vector spaces $\mathbb{R}^3$ and $P_2(x)$ over $\mathbb{R}$, respectively.
$T(1,0,0) = x^2 + 1$.
Suppose, $x^2 + 1 = a_1(1) + a_2(1 + x) + a_3(1 + x + x^2)$.
Then

$$x^2 + 1 = a_1 + a_2 + a_3 + (a_2 + a_3)x + a_3x^2$$
$$\Rightarrow a_1 + a_2 + a_3 = 1, a_2 + a_3 = 0, a_3 = 1$$
$$\Rightarrow a_1 = 1, a_2 = -1, a_3 = 1.$$

Similarly,

$$T(1,1,0) = 2 = 2.1 + 0.(1 + x) + 0.(1 + x + x^2),$$

and

$$T(1,1,1) = x + 3 = 2.(1) + 1.(1 + x) + 0.(1 + x + x^2).$$

Hence, the matrix

$$[T]_{B_1,B_2} = \begin{bmatrix} 1 & 2 & 2 \\ -1 & 0 & 1 \\ 1 & 0 & 0 \end{bmatrix}.$$

Example 5.1.8. If $I_V : V \to V$ is an identity linear transformation, then $[I_V]_{B,B} = I_n$, the identity matrix of order n.

If V and W are finite-dimensional vector spaces over a field F of dimensions n and m, respectively, then the following theorem shows that the vector spaces $\mathrm{Hom}(V, W)$ and $M_{m,n}(F)$ are isomorphic.

Theorem 5.1.9. *Let $B_1 = \{v_1, v_2, \ldots, v_n\}$ and $B_2 = \{w_1, w_2, \ldots, w_m\}$ be ordered bases of vector spaces V and W over a field F, respectively. Then the map $f : \mathrm{Hom}(V, W) \to M_{m,n}(F)$ defined by $f(T) = [T]_{B_1,B_2}$ is a vector space isomorphism. That is, $\mathrm{Hom}(V, W) \cong M_{m,n}(F)$.*

Proof. First of all, we shall show that f is a linear transformation.

Let $T_1, T_2 \in \mathrm{Hom}(V, W)$ and let $[T_1]_{B_1,B_2} = [a_{ij}]$, $[T_2]_{B_1,B_2} = [b_{ij}]$.

Then for $a, b \in F$ we have to prove that $f(aT_1 + bT_2) = af(T_1) + bf(T_2)$, or we have to show that $[aT_1 + bT_2]_{B_1,B_2} = a[T_1]_{B_1,B_2} + b[T_2]_{B_1,B_2}$.

We have that

$$[T_1]_{B_1,B_2} = [a_{ij}] \Rightarrow T_1(v_j) = \sum_{i=1}^{m} a_{ij} w_i$$

and

$$[T_2]_{B_1,B_2} = [b_{ij}] \Rightarrow T_2(v_j) = \sum_{i=1}^{m} b_{ij} w_i.$$

Hence,

$$(aT_1 + bT_2)(v_j) = aT_1(v_j) + bT_2(v_j)$$
$$= a \sum_{i=1}^{m} a_{ij} w_i + b \sum_{i=1}^{m} b_{ij} w_i$$
$$= \sum_{i=1}^{m} aa_{ij} w_i + \sum_{i=1}^{m} bb_{ij} w_i$$
$$= \sum_{i=1}^{m} (aa_{ij} + bb_{ij}) w_i.$$

This gives that (i, j)-th element of the matrix $[aT_1 + bT_2]_{B_1,B_2} = aa_{ij} + bb_{ij}$, is equal to $a(i, j)$-th element of the matrix $[T_1]_{B_1,B_2} + b(i, j)$-th element of the matrix $[T_2]_{B_1,B_2}$.

Hence, $[aT_1 + bT_2]_{B_1,B_2} = a[T_1]_{B_1,B_2} + b[T_2]_{B_1,B_2}$.

Now, we shall show that f is bijective.

Suppose, $f(T_1) = f(T_2) = [a_{ij}]$. Then $T_1(v_j) = T_2(v_j) \ \forall v_j \in B_1$.

Since T_1 and T_2 are equal on the basis B_1, we have $T_1 = T_2$, and hence f is injective.

Next, for the given matrix $[a_{ij}] \in M_{m,n}(F)$, there exists an unique $T \in \text{Hom}(V, W)$ such that $T(v_j) = \sum_{i=1}^{n} a_{ij} w_i \ \forall v_j \in B_1$. Hence, $f(T) = [T]_{B_1,B_2} = [a_{ij}]$, and so f is surjective, also.

Thus, the map $f : \text{Hom}(V, W) \to M_{m,n}(F)$ is a vector space isomorphism, and hence $\text{Hom}(V, W) \cong M_{m,n}(F)$. □

Remark. In view of above notation, if $T_1, T_2 : V \to W$ are two linear transformations, then $[T_1 + T_2]_{B_1,B_2} = [T_1]_{B_1,B_2} + [T_2]_{B_1,B_2}$ and $[aT_1]_{B_1,B_2} = a[T_1]_{B_1,B_2}$, $a \in F$. That is, the sums and scalar multiplication of matrices are associated with the corresponding sums and scalar multiplication of linear transformations.

Since the composition of two linear transformations is a linear transformation, we obtain the matrix representation of a composition of linear transformations.

Theorem 5.1.10. *Let V, W and U be finite-dimensional vector spaces with the bases $B_1 = \{v_1, v_2, \ldots, v_n\}$, $B_2 = \{w_1, w_2, \ldots, w_m\}$ and $B_3 = \{u_1, u_2, \ldots, u_p\}$, respectively. Then for the linear transformations,*

$$T_1 : V \to W, \qquad T_2 : W \to U, \qquad [T_2 \circ T_1]_{B_1,B_3} = [T_2]_{B_2,B_3}[T_1]_{B_1,B_2}.$$

Proof. Let $[T_1]_{B_1,B_2} = [a_{ij}]_{m \times n}$ and $[T_2]_{B_2,B_3} = [b_{ki}]_{p \times m}$.

Then $T_1(v_j) = \sum_{i=1}^{m} a_{ij} w_i$ and $T_2(w_i) = \sum_{k=1}^{p} b_{ki} u_k$.

Since $T_2 \circ T_1$ is a linear transformation from V to U,

$$T_2 \circ T_1(v_j) = T_2(T_1(v_j))$$

$$= T_2\left(\sum_{i=1}^{m} a_{ij} w_i \right)$$

$$= \sum_{i=1}^{m} a_{ij} T_2(w_i)$$

$$= \sum_{i=1}^{m} a_{ij} \left(\sum_{k=1}^{p} b_{ki} u_k \right)$$

$$= \sum_{i=1}^{m} \sum_{k=1}^{p} a_{ij} b_{ki} u_k$$

$$= \sum_{k=1}^{p} \left(\sum_{i=1}^{m} b_{ki} a_{ij} \right) u_k$$

$$= \sum_{k=1}^{p} c_{kj} u_k,$$

where $c_{kj} = \sum_{i=1}^{m} b_{ki} a_{ij}$.

Thus the (k,j)-th element of the matrix $[T_2 \circ T_1]_{B_1,B_3}$ is equal to the (k,j)-th element of the matrix $[T_2]_{B_2,B_3} [T_1]_{B_1,B_2}$.

Hence, $[T_2 \circ T_1]_{B_1,B_3} = [T_2]_{B_2,B_3} [T_1]_{B_1,B_2}$. $\qquad\square$

Remark. Above matrix representation of a composition of linear transformations gives the idea behind the multiplication of matrices.

Theorem 5.1.11. *Let V and W be finite-dimensional vector spaces over the same field F with ordered bases B_1 and B_2, respectively. Then the linear transformation $T : V \to W$ is invertible if and only if $[T]_{B_1,B_2}$ is invertible and $[T^{-1}]_{B_2,B_1} = ([T]_{B_1,B_2})^{-1}$.*

Proof. Suppose T is invertible. Since V and W are finite-dimensional vector spaces, $\dim(V) = \dim(W)$.

Let $\dim(V) = \dim(W) = n$. Then $[T]_{B_1,B_2}$ is an n-square matrix.

Since T is invertible, there exists $T^{-1} : W \to V$ such that $TT^{-1} = I_W$ and $T^{-1}T = I_V$, where I_V and I_W denote the identity linear transformations on V and W, respectively.

Then $I_n = [I_V]_{B_1,B_1} = [T^{-1}T]_{B_1,B_1} = [T^{-1}]_{B_2,B_1} [T]_{B_1,B_2}$.

Similarly, $[T]_{B_1,B_2} [T^{-1}]_{B_2,B_1} = I_n$.

This show that $[T]_{B_1,B_2}$ is invertible and $[T_{B_1,B_2}]^{-1} = [T^{-1}]_{B_2,B_1}$.

Conversely, suppose that $[T]_{B_1,B_2} = [a_{ij}]$ is invertible. Then there exists an n-square matrix $B = [b_{ij}]$ such that $[T]_{B_1,B_2} B = B[T]_{B_1,B_2} = I_n$.

If $B_1 = \{v_1, v_2, \dots, v_n\}$ and $B_2 = \{w_1, w_2, \dots, w_n\}$ are ordered bases of V and W, respectively, then there exists a linear transformation $T_1 : W \to V$ such that $T_1(w_j) = \sum_{i=1}^{n} b_{ij} v_i \ \forall 1 \le j \le n$.

Then $[T_1]_{B_2,B_1} = B$. Now, $[T_1 T]_{B_1,B_1} = [T_1]_{B_2,B_1} [T]_{B_1,B_2} = B[T]_{B_1,B_2} = I_n = [I_V]_{B_1,B_2} \Rightarrow$ $T_1 T = I_V$ and similarly, $TT_1 = I_W$. This shows that $T^{-1} = T_1$, and hence T is invertible. $\qquad\square$

Example 5.1.12. Let $A = \begin{bmatrix} 3 & -2 & 1 & 0 \\ 1 & 6 & 2 & 1 \\ -3 & 0 & 7 & 1 \end{bmatrix}$ be the matrix of linear transformation $T : F^4 \to F^3$, relative to the standard ordered bases B_1 and B_2 of F^4 and F^3, respectively, where F is a field. In this example, we obtain the linear transformation T representing the matrix A.

Let

$$B_1 = \{(1,0,0,0), (0,1,0,0), (0,0,1,0), (0,0,0,1)\}, \quad \text{and}$$
$$B_2 = \{(1,0,0), (0,1,0), (0,0,1)\}.$$

By the matrix representation of a linear transformation T, we have

$$T(1,0,0,0) = 3(1,0,0) + 1(0,1,0) + (-3)(0,0,1) = (3,1,-3).$$

Similarly, $T(0,1,0,0,) = (-2,6,0)$, $T(0,0,1,0) = (1,2,7)$ and $T(0,0,0,1) = (0,1,1)$.

Hence,

$$
\begin{aligned}
T(x,y,z,w) &= T(x(1,0,1,0) + y(0,1,0,0) + z(0,0,1,0) + w(0,0,0,1))\\
&= xT(1,0,0,0) + yT(0,1,0,0) + zT(0,0,1,0) + wT(0,0,0,1)\\
&= x(3,1,-3) + y(-2,6,0) + z(1,2,7) + w(0,1,1)\\
&= (3x - 2y + z, x + 6y + 2Z + w, -3x + 7z + w).
\end{aligned}
$$

In the following theorem, we shall show that if A is a matrix representing the linear transformation T, then A^t is the matrix representing T^t.

Theorem 5.1.13. *Let V and W be finite-dimensional vector spaces over a field F with ordered bases $B_1 = \{v_1, v_2, \ldots, v_n\}$ and $B_2 = \{w_1, w_2, \ldots, w_m\}$, respectively. Then for the linear transformation $T : V \to W$, $([T]_{B_1,B_2})^t = [T^t]_{B_2,B_1}$.*

Proof. Let $[T]_{B_1,B_2} = [a_{ij}]_{m\times n}$. Then $T(v_j) = \sum_{i=1}^{m} a_{ij} w_i \ \forall 1 \le i \le n$.

Suppose $B_1^* = \{v_1^*, v_2^*, \ldots, v_n^*\}$ and $B_2^* = \{w_1^*, w_2^*, \ldots, w_m^*\}$ are the corresponding dual bases of dual spaces V^* and W^*, respectively.

Then for the linear transformation $T^t : W^* \to V^*$, suppose $T^t(w_k^*) = \sum_{j=1}^{n} b_{jk} v_j^*$.

From the definition of transpose of T, $T^t(w_k^*) = w_k^* \circ T \in V^*$, which implies that $w_k^* \circ T = \sum_{j=1}^{n} b_{jk} v_j^*$.

Then

$$
(w_k^* \circ T)(v_r) = \sum_{j=1}^{n} b_{jk} v_j^*(v_r)
$$

or

$$
w_k^*(T(v_r)) = b_{rk},
$$

where $v_r^*(v_r) = 1$, and $v_r^*(v_j) = 0$ if $r \ne j$.

Thus,

$$
b_{rk} = w_k^*(T(v_r)) = w_k^*\left(\sum_{i=1}^{m} a_{ir} w_i\right) = \sum_{i=1}^{m} a_{ir} w_k^*(w_i) = a_{kr}.
$$

This shows that $[a_{ij}]^t = [b_{ji}] = [T^t]_{B_2^*, B_1^*}$. □

Exercises

(5.1.1) Let $T : \mathbb{R}^3 \to \mathbb{R}^2$ be a linear transformation. Then determine the matrix $[T]_{B_1,B_2}$ if:
  (i) $T(x,y,z) = (2x - y, 2y - z)$, $B_1 = \{(1,0,0), (1,-1,0), (1,1,-1)\}$, and $B_2 = \{(1,1), (-1,0)\}$;
  (ii) $T(x,y,z) = (2x - y, 2y - z)$, $B_1 = \{(1,2,1), (-1,0,2), (2,1,-3)\}$ and $B_2 = \{(1,-1), (2,3)\}$;
  (iii) $T(x,y,z) = (2x + 3y - z, x + z)$, $B_1 = \{(1,1,2), (1,2,2), (2,2,3)\}$ and $B_2 = \{(1,1), (-1,0)\}$;
  (iv) $T(x,y,z) = (2x + 3y - z, x + z)$, $B_1 = \{(1,2,2), (1,1,2), (2,2,3)\}$ and $B_2 = \{(1,1), (-1,0)\}$.

(5.1.2) Let $T : \mathbb{R}^3 \to \mathbb{R}^3$ be a linear transformation. Then determine the matrix $[T]_{B_1,B_2}$ if:

(i) $T(x,y,z) = (2y + z, x - 4y, 3x)$, $B_1 = \{(1,0,0),(1,-1,0),(1,1,-1)\}$ and $B_2 = \{(1,2,1),(-1,0,2),(2,1,-3)\}$;

(ii) $T(x,y,z) = (2y + z, x - 4y, 3x)$, $B_1 = \{(1,2,1),(-1,0,2),(2,1,-3)\}$ and $B_2 = \{(1,0,0),(1,-1,0),(1,1,-1)\}$;

(iii) $T(x,y,z) = (x + 3y + 2z, 3x + 4y + z, 2x + y - z)$ and $B_1 = B_2 = \{(1,1,2),(1,2,2),(2,2,3)\}$;

(iv) $T(x,y,z) = (x + 3y + 2z, 3x + 4y + z, 2x + y - z)$, and $B_1 = B_2 = \{(1,0,0),(1,1,0),(1,1,-1)\}$.

(5.1.3) Find the matrix of the linear transformation $T_i, 1 \le i \le 5$ related to the standard bases:

(i) $T_1 : P_4(x) \to P_2(x)$ defined as $T(f(x)) = f'''(x)$;

(ii) $T_2 : P_3(x) \to P_2(x)$ defined as $T(f(x)) = f'(x) + f''(x)$;

(iii) $T_3 : P_2(x) \to P_3(x)$ defined as $T(f(x)) = \int_0^x f(t)dt$;

(iv) $T_4 : P_2(x) \to P_3(x)$ defined as $T(f(x)) = f'(x) - \int_0^x f(t)dt$;

(v) $T_5 : P_2(x) \to P_3(x)$ defined as $T(f(x)) = xf(x) + f'(x)$.

(5.1.4) For the linear transformations given in the above exercise, determine:

(i) $[T_1]_{B_1,B_2}$, where B_1 is the standard basis and $B_2 = \{1, 1+x, 1+x+x^2\}$;

(ii) $[T_2]_{B_1,B_2}$, where B_1 is the standard basis and $B_2 = \{1, 1+x, 1+x+x^2\}$;

(iii) $[T_3]_{B_1,B_2}$, where $B_1 = \{1, 1+x, 1+x+x^2\}$ and B_2 is the standard basis;

(iv) $[T_4]_{B_1,B_2}$, where B_1 is the standard basis and $B_2 = \{1, 1+x, 1+x+x^2, 1+x+x^2+x^3\}$;

(iv) $[T_5]_{B_1,B_2}$, where $B_1 = \{1, 1+x, 1+x+x^2\}$ and $B_2 = \{1, 1+x, 1+x+x^2, 1+x+x^2+x^3\}$.

(5.1.5) Let $T : M_{2,2}(\mathbb{R}) \to P_2(x)$ be a linear transformation defined by

$$T\left(\begin{bmatrix} a & b \\ c & d \end{bmatrix}\right) = a + c + 2dx + cx^2.$$

Then find the matrix $[T]_{B_1,B_2}$ where B_1 and B_2 are standard bases, i. e.,

$$B_1 = \left\{\begin{bmatrix} 1 & 0 \\ 0 & 0 \end{bmatrix}, \begin{bmatrix} 0 & 1 \\ 0 & 0 \end{bmatrix}, \begin{bmatrix} 0 & 0 \\ 1 & 0 \end{bmatrix}, \begin{bmatrix} 0 & 0 \\ 0 & 1 \end{bmatrix}\right\} \quad \text{and} \quad B_2 = \{1, x, x^2\}.$$

(5.1.6) Let $T : \mathbb{R}^4 \to P_2(x)$ be a linear transformation defined by $T(a,b,c,d) = a + c + 2dx + cx^2$. Then find the matrix of T related to the standard bases and compare with the matrix of previous question.

(5.1.7) Let $T : M_{2,2}(\mathbb{R}) \to M_{2,2}(\mathbb{R})$ be the linear transformation defined by $T(A) = \frac{(A+A^t)}{2}$. Then determine the matrix $[T]_{B_1,B_2}$, where B_1 is the standard basis of $M_{2,2}(\mathbb{R})$.

(5.1.8) Find the linear transformation $T : P_2(x) \to P_2(x)$ whose matrix representation relative to the standard basis is

$$A = \begin{bmatrix} 1 & 2 & 3 \\ 3 & 2 & 1 \\ 1 & -1 & 2 \end{bmatrix}.$$

(5.1.9) Find the linear transformation $T : \mathbb{R}^4 \to \mathbb{R}^3$ whose matrix representation is

$$[T]_{B_1,B_2} = \begin{bmatrix} -1 & 2 & 3 & 1 \\ 1 & 1 & 1 & 0 \\ 1 & 2 & -2 & -1 \end{bmatrix},$$

where

$$B_1 = \{(1,0,0,0),(1,1,0,0),(1,1,1,0),(1,1,1,1)\} \quad \text{and} \quad B_2 = \{(1,1,1),(1,1,0),(1,0,0)\}$$

are ordered bases of $\mathbb{R}^4$ and $\mathbb{R}^3$, respectively.

(5.1.10) For exercise (5.1.8), find T relative to the basis $\{1, 1+x, 1+x+x^2\}$.

5.2 Effect of change of bases on matrix representation

In this section, we prove the result, which gives the relation between the matrices of a linear transformation with the different ordered bases. We begin this section with the transition matrices and their properties.

Definition 5.2.1. Let V be a vector space with the ordered bases B_1 and B_2. Then the matrix $[I_V]_{B_1,B_2}$, of the identity linear transformation $I_V : V \to V$ is called the transition matrix from the basis B_1 to the basis B_2. Transition matrices are invertible.

Example 5.2.2. Let $I : \mathbb{R}^3 \to \mathbb{R}^3$ be the identity linear transformation. Then the transition matrix from the basis $B_1 = \{(1,0,0),(0,1,0),(0,0,1)\}$ to the basis $B_2 = \{(1,0,0),(1,-1,0),(1,1,-1)\}$ is given as follows:

$$I(1,0,0) = (1,0,0) = 1(1,0,0) + 0(1,-1,0) + 0(1,1,-1),$$
$$I(0,1,0) = (0,1,0) = 1(1,0,0) - 1(1,-1,0) + 0(1,1,-1),$$
$$I(0,0,1) = (0,0,1) = 2(1,0,0) - 1(1,-1,0) - 1(1,1,-1).$$

Hence,

$$[I]_{B_1,B_2} = \begin{bmatrix} 1 & 1 & 2 \\ 0 & -1 & -1 \\ 0 & 0 & -1 \end{bmatrix},$$

and

$$([I]_{B_1,B_2})^{-1} = [I^{-1}]_{B_2,B_1} = [I]_{B_2,B_1} = \begin{bmatrix} 1 & 1 & 1 \\ 0 & -1 & 1 \\ 0 & 0 & -1 \end{bmatrix}.$$

Theorem 5.2.3. Let $B_1 = \{v_1,v_2,\ldots,v_n\}$, $B_1' = \{v_1',v_2',\ldots,v_n'\}$ be the bases of the vector space V and $B_2 = \{w_1,w_2,\ldots,w_m\}$, $B_2' = \{w_1',w_2',\ldots,w_m'\}$ be the bases of the vector space W. If $T : V \to W$ is a linear transformation, then the matrix $[T]_{B_1',B_2'} = ([I_W]_{B_2',B_2})^{-1}[T]_{B_1,B_2}[I_V]_{B_1',B_1}$, where I_V and I_W are identity linear transformations. Conversely, if B_1 and B_2 are bases of vector spaces V and W, respectively, and for an $m \times n$ matrix A if there are nonsingular matrices P and Q such that $B = Q^{-1}AP$, then there exist a linear transformation $T : V \to W$ and bases B_1' and B_2' of V and W, respectively, such that $[T]_{B_1,B_2} = A$ and $[T]_{B_1',B_2'} = B$.

Proof. Let us denote the vector space with the basis B by V_B. Then from the composition of maps $V_{B_1'} \xrightarrow{I_V} V_{B_1} \xrightarrow{T} W_{B_2} \xrightarrow{I_W} W_{B_2'}$ we have the linear transformation $I_W \circ T \circ I_V = T$ from $V_{B_1'}$ to $W_{B_2'}$.

From Theorem 5.1.10,

$$[T]_{B_1', B_2'} = [I_W \circ T \circ T_V]_{B_1', B_2'}$$
$$= [I_W]_{B_2, B_2'} [T]_{B_1, B_2} [I_V]_{B_1' B_1}$$
$$= ([I_W^{-1}]_{B_2', B_2})^{-1} [T]_{B_1, B_2} [I_V]_{B_1', B_1}$$
$$= ([I_W]_{B_2', B_2})^{-1} [T]_{B_1, B_2} [I_V]_{B_1', B_1}.$$

Now, we prove the converse result.

Let $A = [a_{ij}]$. Then define $T : V \to W$ by $T(v_j) = \sum_{i=1}^{m} a_{ij} w_i$, and hence $[T]_{B_1, B_2} = A$. If $P = [p_{ij}]_{n \times n}$ and $Q = [q_{ij}]_{m \times m}$, then define $v_j' = \sum_{i=1}^{n} p_{ij} v_i$ and $w_j' = \sum_{i=1}^{m} q_{ij} w_i$.

Since P and Q are nonsingular matrices, $B_1' = \{v_1', v_2', \ldots, v_n'\}$ and $B_2' = \{w_1', w_2', \ldots, w_m'\}$ are bases of V and W, respectively, and so $Q = [I_W]_{B_2', B_2}$ and $P = [I_V]_{B_1', B_1}$ are transition matrices.

Given that $B = Q^{-1} A P$. Hence,

$$B = ([I_W]_{B_2', B_2})^{-1} [T]_{B_1, B_2} [I_V]_{B_1', B_1}$$
$$= [I_W^{-1}]_{B_2, B_2'} [T]_{B_1, B_2} [I_V]_{B_1', B_1}$$
$$= [I_W]_{B_2, B_2'} [T]_{B_1, B_2} [I_V]_{B_1', B_1}$$
$$= [I_W \circ T \circ I_V]_{B_1', B_2'}$$
$$= [T]_{B_1', B_2'}. \qquad \square$$

Corollary 5.2.4. *Let V be a finite-dimensional vector space with ordered bases B_1 and B_1' and let $T : V \to V$ be a linear transformation. Then $[T]_{B_1', B_1'} = ([I_V]_{B_1', B_1})^{-1} [T]_{B_1, B_1} [I_V]_{B_1', B_1}$. Conversely, if B_1 is an ordered basis of an n-dimensional vector space V and there is an n-square matrix B, which is similar to A, then there exist a linear transformation $T : V \to V$ and a basis B_1' of V, such that $[T]_{B_1, B_1} = A$ and $B = [T]_{B_1', B_1'}$.*

Proof. In the above theorem if we replace W by V, B_1, B_2 by B_1 and B_1', B_2' by B_1', we get $[T]_{B_1', B_1'} = ([I_V]_{B_1', B_1})^{-1} [T]_{B_1, B_1} [I_V]_{B_1', B_1}.$ $\qquad \square$

Alternate Proof. From Theorem 5.1.10, we have

$$[I_V]_{B_1', B_1} [T]_{B_1', B_1'} = [I_V T]_{B_1', B_1} = [T I_V]_{B_1', B_1} = [T]_{B_1, B_1} [I_V]_{B_1', B_1}.$$

Since the transition matrix $[I_V]_{B_1', B_1}$ is invertible,

$$[T]_{B_1', B_1'} = ([I_V]_{B_1', B_1})^{-1} [T]_{B_1, B_1} [I_V]_{B_1', B_1}.$$

Conversely, since B is similar to A, there exists a nonsingular matrix P such that $B = P^{-1}AP$.

On replacing W by V, Q^{-1} by P^{-1}, B_1, B_2 by B_1 and B_1', B_2' by B_1' in the above theorem, we get $A = [T]_{B_1,B_1}$ and $B = [T]_{B_1',B_1'}$. □

Remark. From the above corollary, it is clear that two matrices A and B are similar if and only if they are represented by the same linear transformation relative to different ordered bases.

The following examples illustrates Theorem 5.2.3.

Example 5.2.5. Let $T : \mathbb{R}^3 \to \mathbb{R}^2$ be the linear transformation defined by $T(x,y,z) = (x+y+z, x-y)$. Suppose $B_1 = \{(1,0,0),(1,-1,0),(1,1,1)\}$, $B_1' = \{(1,1,1),(1,1,0),(1,0,0)\}$, are ordered bases of $\mathbb{R}^3$ and $B_2 = \{(1,1),(-1,0)\}$, $B_2' = \{(1,0),(0,1)\}$ are ordered base of $\mathbb{R}^2$.

Then the matrix representation $[T]_{B_1,B_2}$ is given as

$$T(1,0,0) = (1,1) = 1(1,1) + 0(-1,0),$$
$$T(1,-1,0) = (0,2) = 2(1,1) + 2(-1,0),$$
$$T(1,1,1) = (3,0) = 0(1,1) + (-3)(-1,0),$$

$$[T]_{B_1,B_2} = \begin{bmatrix} 1 & 2 & 0 \\ 0 & 2 & -3 \end{bmatrix}.$$

Now, the transition matrix $[I_{\mathbb{R}^3}]_{B_1',B_1}$ is given as

$$(1,1,1) = 0(1,0,0) + 0(1,-1,0) + 1(1,1,1),$$
$$(1,1,0) = 2(1,0,0) - 1(1,-1,0) + 0(1,1,1),$$
$$(1,0,0) = 1(1,0,0) + 0(1,-1,0) + 0(1,1,1),$$

$$[I_{\mathbb{R}^3}]_{B_1',B_1} = \begin{bmatrix} 0 & 2 & 1 \\ 0 & -1 & 0 \\ 1 & 0 & 0 \end{bmatrix}.$$

Transition matrix $([I_{\mathbb{R}^2}]_{B_2',B_2})^{-1} = [I_{\mathbb{R}^2}]_{B_2,B_2'}$. Hence, from

$$(1,1) = 1(1,0) + 1(0,1),$$
$$(-1,0) = -1(1,0) + 0(0,1),$$

$$[I_{\mathbb{R}^2}]_{B_2,B_2'} = ([I_{\mathbb{R}^2}]_{B_2',B_2})^{-1} = \begin{bmatrix} 0 & -1 \\ 1 & 0 \end{bmatrix}.$$

Then

$$\left([I_{\mathbb{R}^2}]_{B'_2,B_2}\right)^{-1}[T]_{B_1,B_2}[I_{\mathbb{R}^3}]_{B'_1,B_1}$$

$$= \begin{bmatrix} 1 & -1 \\ 1 & -0 \end{bmatrix} \begin{bmatrix} 1 & 2 & 0 \\ 0 & 2 & -3 \end{bmatrix} \begin{bmatrix} 0 & 2 & 1 \\ 0 & -1 & 0 \\ 1 & 0 & 0 \end{bmatrix}$$

$$= \begin{bmatrix} 3 & 2 & 1 \\ 0 & 0 & 1 \end{bmatrix}.$$

Now we show that above matrix is equal to $[T]_{B'_1,B'_2}$.
Since

$$T(1,1,1) = (3,0) = 3(1,0) + 0(0,1),$$
$$T(1,1,0) = (2,0) = 2(1,0) + 0(0,1),$$
$$T(1,0,0) = (1,1) = 1(1,0) + 1(0,1),$$

$$[T]_{B'_1,B'_2} = \begin{bmatrix} 3 & 2 & 1 \\ 0 & 0 & 1 \end{bmatrix}.$$

Example 5.2.6. Let $T : \mathbb{R}^3 \to P_2(x)$ be the linear transformation and let $A = \begin{bmatrix} 1 & 2 & 2 \\ -1 & 0 & 1 \\ 1 & 0 & 0 \end{bmatrix}$ be the matrix of T related to the bases,

$$B_1 = \{(1,0,0),(1,1,0),(1,1,1)\} \text{ of } \mathbb{R}^3 \quad \text{and} \quad B_2 = \{1,1+x,1+x+x^2\} \text{ of } P_2(x),$$

respectively, i. e., $[T]_{B_1,B_2} = A$.

In this example, we compute the matrix of T related to the standard bases $B'_1 = \{(1,0,0),(0,1,0),(0,0,1)\}$ and $B'_2 = \{1,x,x^2\}$ of $\mathbb{R}^3$ and $P_2(x)$ respectively, even we do not know the definition of T.

From Theorem 5.2.3, we have $[T]_{B'_1,B'_2} = ([I_{P_2(x)}]_{B'_2,B_2})^{-1}[T]_{B_1,B_2}[I_{\mathbb{R}^3}]_{B'_1,B_1}$.
The transition matrix $[I_{\mathbb{R}^3}]_{B'_1,B_1}$ is given as

$$(1,0,0) = 1(1,0,0) + 0(1,1,0) + 0(1,1,1),$$
$$(0,1,0) = (-1)(1,0,0) + 1(1,1,0) + 0(1,1,1),$$
$$(0,0,1) = 0(1,0,0) + (-1)(1,1,0) + 1(1,1,1),$$

$$[I_{\mathbb{R}^3}]_{B'_1,B_1} = \begin{bmatrix} 1 & -1 & 0 \\ 0 & 1 & -1 \\ 0 & 0 & 1 \end{bmatrix}.$$

Next, the transition matrix $[I_{P_2(x)}]_{B_2,B'_2}$ is given as

$$1 = 1 + 0.x + 0.x^2,$$
$$1 + x = 1 + 1.x + 0.x^2,$$
$$1 + x + x^2 = 1 + 1.x + 1.x^2,$$

$$[I_{P_2(x)}]_{B_2,B_2'} = \begin{bmatrix} 1 & 1 & 1 \\ 0 & 1 & 1 \\ 0 & 0 & 1 \end{bmatrix}.$$

Since

$$\left([I_{P_2(x)}]_{B_2',B_2}\right)^{-1} = [I_{P_2(x)}]_{B_2,B_2'}, \quad \left([I_{P_2(x)}]_{B_2',B_2}\right)^{-1} = \begin{bmatrix} 1 & 1 & 1 \\ 0 & 1 & 1 \\ 0 & 0 & 1 \end{bmatrix}.$$

Hence,

$$[T]_{B_1',B_2'} = \begin{bmatrix} 1 & 1 & 1 \\ 0 & 1 & 1 \\ 0 & 0 & 1 \end{bmatrix} \begin{bmatrix} 1 & 2 & 2 \\ -1 & 0 & 1 \\ 1 & 0 & 0 \end{bmatrix} \begin{bmatrix} 1 & -1 & 0 \\ 0 & 1 & -1 \\ 0 & 0 & 1 \end{bmatrix}$$

$$= \begin{bmatrix} 1 & 1 & 1 \\ 0 & 0 & 1 \\ 1 & -1 & 0 \end{bmatrix}.$$

To verify above computation, we have $T(a, b, c) = (a - b)x^2 + cx + a + b + c$, which is given in Example 5.1.7 or we can obtain T from the given matrix A and bases B_1 and B_2.
Now, we obtain the matrix $[T]_{B_1',B_2'}$ representing T by its definition.
Since

$$T(1, 0, 0) = 1 + x^2 = 1.1 + 0.x + 1.x^2,$$

$$T(0, 1, 0) = 1 - x^2 = 1.1 + 0.x + (-1)x^2,$$

$$T(0, 0, 1) = 1 + x = 1.1 + 1.x + 0.x^2,$$

$$[T]_{B_1',B_2'} = \begin{bmatrix} 1 & 1 & 1 \\ 0 & 0 & 1 \\ 1 & -1 & 0 \end{bmatrix}.$$

Thus, the above computation is correct.

Exercises

(5.2.1) Verify the result $[T]_{B',B'} = ([I]_{B',B})^{-1}[T]_{B,B}[I]_{B',B}$ for the following linear transformations:
(i) $T(x, y, z) = (2x - y, x + z, y)$, where $T : \mathbb{R}^3 \to \mathbb{R}^3$ is a linear transformation and

$$B = \big\{(1, 0, 0), (1, 1, 0), (1, 1, 1)\big\}, \quad B' = \big\{(-2, 1, 0), (-1, 0, 1), (0, 1, 0)\big\}$$

are bases of $\mathbb{R}^3$.
(ii) $T : P_2(x) \to P_2(x)$ is a linear transformation defined by $T(f(x)) = f(x - 1)$ and $B = \{1, x, x^2\}$, $B' = \{1, 1 + x, 1 + x + x^2\}$ are bases of $P_2(x)$.

(iii) $T : P_3(x) \rightarrow P_3(x)$ is a linear transformation defined by $T(f(x)) = f(2x+1)$ and $B = \{1, x, x^2, x^3\}$, $B' = \{1, 1+x, 1+x+x^2, 1+x+x^2+x^3\}$ are bases of $P_3(x)$.

(5.2.2) If

$$[T]_{B,B} = \begin{bmatrix} 1 & 2 & 3 \\ 3 & 2 & 1 \\ -1 & 1 & 0 \end{bmatrix},$$

where $T : \mathbb{R}^3 \rightarrow \mathbb{R}^3$ is a linear transformation and B is the standard ordered basis of $\mathbb{R}^3$, then find $[T]_{B',B'}$ where $B' = \{(-2, 1, 0), (-1, 0, 1), (0, 1, 0)\}$.

(5.2.3) If

$$[T]_{B,B} = \begin{bmatrix} -2 & -1 & 0 & 1 \\ 1 & -1 & 2 & 0 \\ 1 & 2 & 3 & 4 \\ 4 & 0 & 2 & -1 \end{bmatrix},$$

where $T : P_3(x) \rightarrow P_3(x)$ is a linear transformation and B is the standard ordered basis of $P_3(x)$, then for the basis $B' = \{1, 1+x, 1+x+x^2, 1+x+x^2+x^3\}$ of $P_3(x)$, find $[T]_{B',B'}$.

(5.2.4) Let $T : \mathbb{R}^3 \rightarrow P_2(x)$ be a linear transformation defined by $T(a, b, c) = ax^2 + (b-c)x + a - b + c$. Then for the ordered bases,

$$B_1 = \{(1, 0, 0), (1, 1, 0), (1, 1, 1)\},$$
$$B_1' = \{(1, 2, 2), (1, 1, 2), (2, 2, 3)\} \text{ of } \mathbb{R}^3 \quad \text{and}$$
$$B_2 = \{1, x, x^2\}, \quad B_2' = \{1, 1+x, 1+x+x^2\} \text{ of } P_2(x),$$

verify the result $[T]_{B_1',B_2'} = ([I_{P_2(x)}]_{B_2',B_2})^{-1} [T]_{B_1,B_2} [I_{\mathbb{R}^3}]_{B_1',B_1}$.

(5.2.5) Let $T : P_2(x) \rightarrow P_3(x)$ be the linear transformation defined by $T(f(x)) = f'(x) - \int_0^x f(t)dt$. Then for the ordered bases $B_1 = \{1, x, x^2\}, B_1' = \{1, 1+x, 1+x+x^2\}$ of $P_2(x)$ and ordered bases $B_2 = \{1, x, x^2, x^3\}, B_2' = \{1, 1+x, 1+x+x^2, 1+x+x^2+x^3\}$ of $P_3(x)$, verify the result $[T]_{B_1',B_2'} = ([I_{P_3(x)}]_{B_2',B_2})^{-1} [T]_{B_1,B_2} [I_{P_2(x)}]_{B_1',B_1}$.

5.3 Rank of a matrix

Let $A = [a_{ij}]$ be an $m \times n$ matrix with entries in a field F. Then m rows of A can be considered as m vectors, $(a_{11}, a_{12}, \ldots, a_{1n}); (a_{21}, a_{22}, \ldots, a_{2n}); \ldots, (a_{m1}, a_{m2}, \ldots, a_{mn})$ of the vector space F^n over F. The subspace of F^n generated by above m vectors is called the row space of the matrix. Similarly, from the column vectors of the matrix we can define the column space of the matrix.

Definition 5.3.1. The dimension of the row space of a matrix is called row rank of the matrix. In other words, "the maximum number of linearly independent rows of a matrix is called the row rank of the matrix."

Similarly, the maximum number of linearly independent columns of a matrix is called the column rank of the matrix.

Clearly, row rank of A = column rank of A^t.

Theorem 5.3.2. *Let $L_A : F^n \to F^m$ be a left multiplication linear transformation defined by $L_A(X) = AX$, where $A = [a_{ij}]$ is an $m \times n$ matrix with entries in the field F and $X \in F^n$ is a column vector. Then for the standard bases B_1 and B_2 of F^n and F^m, respectively, $[L_A]_{B_1,B_2} = A$ and rank of L_A = column rank of A.*

Proof. Let $B_1 = \{v_1, v_2, \ldots, v_n\} \subseteq F^n$ and $B_2 = \{w_1, w_2, \ldots, w_m\} \subseteq F^m$, where $v_j = (0, \ldots, 0, 1, 0, \ldots, 0)$, j-th coordinate is 1 for all $1 \le j \le n$ and $w_i = (0, \ldots, 0, 1, 0, \ldots, 0)$, i-th coordinate is 1 $\forall 1 \le i \le m$.

Then

$$
L_A(v_j) = Av_j = \begin{bmatrix} a_{11} & a_{12} & \cdots & a_{1j} & \cdots & a_{1n} \\ a_{21} & a_{22} & \cdots & a_{2j} & \cdots & a_{2n} \\ \vdots & \vdots & \cdots & \vdots & \cdots & \vdots \\ a_{m1} & a_{m2} & \cdots & a_{mj} & \cdots & a_{mn} \end{bmatrix} \begin{bmatrix} 0 \\ 0 \\ \vdots \\ 1 \\ 0 \\ \vdots \\ 0 \end{bmatrix} = \begin{bmatrix} a_{1j} \\ a_{2j} \\ \vdots \\ a_{mj} \end{bmatrix}
$$

is the j-th column of the matrix A. Thus, $L_A(v_j) = \sum_{i=1}^{m} a_{ij} w_i$, $\forall 1 \le j \le n$. Hence, $[L_A]_{B_1,B_2} = [a_{ij}] = A$.

Since $L_A(v_j)$ is the j-th column of the matrix A $\forall j$, the range space of L_A is equal to the column space of the matrix A.

That is, $\dim(L_A(F^n))$ = column rank of A. Hence, $\mathrm{rank}(L_A)$ = column rank of A. □

Theorem 5.3.3. *Let V and W be finite-dimensional vector spaces over the field F with ordered bases $B_1 = \{v_1, v_2, \ldots, v_n\}$ and $B_2 = \{w_1, w_2, \ldots, w_m\}$, respectively. If $T : V \to W$ is a linear transformation, then $\mathrm{rank}(T)$ = column rank of $[T]_{B_1,B_2}$.*

Proof. Since V is an n-dimensional vector space over F and W is an m-dimensional vector space over F, V is isomorphic to F^n and W is isomorphic to F^m. Let T_1 be a linear transformation from F^n to V defined by $T_1(x_1, x_2, \ldots, x_n) = x_1 v_1 + x_2 v_2 + \cdots + x_n v_n$.

Then,

$$T_1(1, 0, \ldots, 0) = 1.v_1 + 0.v_2 + \cdots + 0.v_n = v_1,$$
$$T_1(0, 1, 0, \ldots, 0) = 0.v_1 + 1.v_2 + 0.v_3 + \cdots + 0.v_n = v_2,$$
$$\vdots$$
$$T_1(0, \ldots, 0, 1) = 0.v_1 + 0.v_2 + \cdots + 1.v_n = v_n.$$

This shows that T_1 takes standard basis of F^n to the basis of V_1, hence T_1 is an isomorphism.

Similarly, the linear transformation $T_2 : F^m \to W$ defined by $T_2(y_1, y_2, \ldots, y_m) = y_1 w_1 + y_2 w_2 + \cdots + y_m w_m$, is an isomorphism. From the composition of linear transformations,

$$F^n \xrightarrow{T_1} V \xrightarrow{T} W \xrightarrow{T_2^{-1}} F^m,$$

we shall show that $T_2^{-1} \circ T \circ T_1 = L_A$, where $A = [T]_{B_1, B_2} = [a_{ij}]_{m \times n}$.

Let $e_j = (0, \ldots, 0, 1, 0, \ldots, 0) \in F^n$. Then

$$(T_2^{-1} \circ T \circ T_1)(e_j) = T_2^{-1} \circ T(T_1(e_j)) = T_2^{-1} \circ T(v_j)$$

$$= T_2^{-1}(T(v_j)) = T_2^{-1}\left(\sum_{i=1}^{m} a_{ij} w_i\right) = \sum_{i=1}^{m} a_{ij} T_2^{-1}(w_i)$$

$$= a_{1j}(1, 0, \ldots, 0) + a_{2j}(0, 1, 0, \ldots, 0) + \cdots + a_{mj}(0, 0, \ldots, 0, 1)$$

$$= (a_{1j}, a_{2j}, \ldots, a_{mj}).$$

Also, we have $L_A(e_j) = (a_{1j}, a_{2j}, \ldots, a_{mj})$, which is the j-th column of the matrix A. Since $T_2^{-1} \circ T \circ T_1$ and L_A agree on the basis, $T_2^{-1} \circ T \circ T_1 = L_A$.

Since T_1 and T_2 are nonsingular linear transformations,

$$\mathrm{rank}(T_2^{-1} \circ T \circ T_1) = \mathrm{rank}(T) = \mathrm{rank}(L_A).$$

Hence, $\mathrm{rank}(T) = $ column rank of A. $\qquad\square$

Using above result, we shall show that row rank and column rank of a matrix is same.

Theorem 5.3.4. *The row rank and column rank of a matrix are equal.*

Proof. Let A be any $m \times n$ matrix and let $T : V \to W$ be a linear transformation such that $[T]_{B_1, B_2} = A$, where B_1 and B_2 are ordered bases of V and W, respectively. Then column rank of $A = \mathrm{rank}(T)$.

We have that the matrix representation of T^t with respect to dual bases is $[T^t]_{B_2^*, B_1^*} = A^t$. Then $\mathrm{rank}(T^t) = $ column rank of $A^t = $ row rank of A. Since $\mathrm{rank}(T^t) = \mathrm{rank}(T)$, row rank of $A = $ column rank of A. $\qquad\square$

In view of the above result, we have the following definition.

Definition 5.3.5. The rank of a matrix A is the maximum number of linearly independent rows of A, or equivalently, the maximum number of linearly independent columns of A.

Theorem 5.3.6. *Let A and B be matrices of order $m \times n$ and $n \times p$, respectively. Then $\mathrm{rank}(AB) \leq \min\{\mathrm{rank}(A), \mathrm{rank}(B)\}$.*

Proof. Let $T_2 \circ T_1$ be the linear transformation that represents the matrix AB, where T_1 represents B and T_2 represents A. Then $\text{rank}(T_2 \circ T_1) = \text{rank}(AB)$.

We have that $\text{rank}(T_2 \circ T_1) \leq \min\{\text{rank}(T_2), \text{rank}(T_1)\}$. Hence, $\text{rank}(AB) \leq \min\{\text{rank}(A), \text{rank}(B)\}$. □

Theorem 5.3.7. *Let A be an n-square matrix. Then* $\text{rank}(A) = n$ *if and only if A is invertible.*

Proof. Let $T : V \to V$ be a linear transformation representing the n-square matrix A. Then $\dim(V) = n$. Suppose $\text{rank}(A) = n$. Then $\text{rank}(T) = n$.

From the rank-nullity theorem, we have

$$\text{rank}(T) + \text{null}(T) = n$$
$$\Rightarrow n + \text{null}(T) = n$$
$$\Rightarrow \text{null}(T) = 0.$$

This proves that T is one-one onto. Hence, T is invertible. Therefore, matrix of $T = A$ is invertible.

Conversely, if A is invertible then T is invertible, and hence $\text{rank}(T) = n = \text{rank}(A)$. □

Theorem 5.3.8. *If A is an n-square nonsingular matrix and B is any n-square matrix. Then* $\text{rank}(AB) = \text{rank}(BA) = \text{rank}(B)$.

Proof. Let T_1 be the linear transformation representing the matrix A and T_2 be the linear transformation representing the matrix B. Since A is nonsingular, T_1 is nonsingular.

Using the result, $\text{rank}(T_1 \circ T_2) = \text{rank}(T_2 \circ T_1) = \text{rank}(T_2)$, we have $\text{rank}(AB) = \text{rank}(BA) = \text{rank}(B)$. □

Theorem 5.3.9. *Let A and B are similar matrices. Then* $\text{rank}(A) = \text{rank}(B)$.

Proof. Since A and B are similar matrices, there exist a nonsingular matrix P such that $B = P^{-1}AP$. Hence, from above result we have $\text{rank}(B) = \text{rank}(P^{-1}AP) = \text{rank}(AP) = \text{rank}(A)$. □

Definition 5.3.10. *A* matrix N of order $m \times n$ is said to be in normal form if $N = \begin{bmatrix} I_r & 0_{r\ n-r} \\ 0_{m-r\ r} & 0_{m-r\ n-r} \end{bmatrix}$, where $r \leq \min\{m, n\}$ and 0_{mn} denotes the zero matrix of order $m \times n$.

For example, consider the matrix

$$A = \begin{bmatrix} 1 & 0 & 0 & 0 & 0 \\ 0 & 1 & 0 & 0 & 0 \\ 0 & 0 & 0 & 0 & 0 \\ 0 & 0 & 0 & 0 & 0 \end{bmatrix},$$

of order 4×5. Since

$$A = \begin{bmatrix} 1 & 0 & \cdot & 0 & 0 & 0 \\ 0 & 1 & \cdot & 0 & 0 & 0 \\ \cdot & \cdot & \cdot & \cdot & \cdot & \cdot \\ 0 & 0 & \cdot & 0 & 0 & 0 \\ 0 & 0 & \cdot & 0 & 0 & 0 \end{bmatrix} = \begin{bmatrix} I_2 & 0_{23} \\ 0_{22} & 0_{23} \end{bmatrix},$$

matrix A is in normal form.

Theorem 5.3.11. *Every $m \times n$ matrix is equivalent to a matrix in normal form.*

Proof. Let A be an $m \times n$ matrix with entries in a field F. Let $T : V \rightarrow W$ be a linear transformation such that $[T]_{B_1,B_2} = A$, where $B_1 = \{v_1, v_2, \ldots, v_n\}$ and $B_2 = \{w_1, w_2, \ldots, w_m\}$ are ordered bases of V and W, respectively. If A is a zero matrix, then it is in normal form.

Suppose A is a nonzero matrix and rank of $A = r$. Then rank$(A) = $ rank$(T) = r$. Hence, dimension of Ker $T = \dim(V) - $ rank$(T) = n - r$.

Let $\{x_1, x_2, \ldots, x_{n-r}\}$ be a basis of Ker T. Then $T(x_1) = T(x_2) = \cdots = T(x_n - r) = 0$.

Since Ker T is a subspace of V, basis $\{x_1, x_2, \ldots, x_{n-r}\}$ can be extended to a basis $B_1' = \{e_1, e_2, \ldots, e_r, x_1, x_2, \ldots, x_{n-r}\}$ of V. Similarly, the set $\{T(e_1) = f_1, T(e_2) = f_2, \ldots, T(e_r) = f_r\}$ is a basis of Image(T). Since Image(T) is a subspace of W, basis $\{f_1, f_2, \ldots, f_r\}$ can be extended to a basis $B_2' = \{f_1, f_2, \ldots, f_r, y_1, y_2, \ldots, y_{m-r}\}$ of W.

Then the matrix $[T]_{B_1', B_2'}$ is equivalent to the matrix $[T]_{B_1,B_2} = A$.

Now, we shall show that the matrix $[T]_{B_1',B_2'}$, is in normal form:

$$T(e_1) = f_1 = 1.f_1 + 0.f_2 + \cdots + 0.f_r + 0.y_1 + \cdots + 0.y_{m-r},$$
$$T(e_2) = f_2 = 0.f_1 + 1.f_2 + 0.f_3 + \cdots + 0.f_r + 0.y_1 + \cdots + 0.y_{m-r},$$

$$\vdots$$

$$T(e_r) = f_r = 0.f_1 + 0.f_2 + \cdots + 0.f_r - 1 + 1.f_r + 0y_1 + \cdots + 0.y_{m-r},$$
$$T(x_1) = 0 = 0.f_1 + 0.f_2 + \cdots + 0.f_r + 0.y_1 + \cdots + 0.y_{m-r},$$

$$\vdots$$

$$T(x_{n-r}) = 0 = 0.f_1 + 0.f_2 + \cdots + 0.f_r + 0.y_1 + \cdots + 0.y_{m-r}.$$

Hence, $[T]_{B_1',B_2'} = \begin{bmatrix} I_r & 0_{r\ n-r} \\ 0_{m-r\ r} & 0_{m-r\ n-r} \end{bmatrix}$ is a matrix in normal form.

It is evident from the above result that the rank of a matrix is equal to the order of the identity matrix in its normal form. □

Corollary 5.3.12. *Two $m \times n$ matrices of the same rank are equivalent.*

Proof. Let A and B be $m \times n$ matrices such that rank$(A) = $ rank$(B) = r$. Then A and B will be equivalent to a normal form matrix,

$$N = \begin{bmatrix} I_r & 0_{r\ n-r} \\ 0_{m-r\ r} & 0_{m-r\ n-r} \end{bmatrix}.$$

Since the relation of being "equivalent to" is an equivalence relation on the set of all $m \times n$ matrices $M_{m,n}(F)$, matrix A and B will be equivalent, i. e., $A \cong N \cong B \Rightarrow A \cong B$. □

Theorem 5.3.13. *Let A and B be matrices of order $m \times n$ with entries in a field F. Then* $\operatorname{rank}(A + B) \leq \operatorname{rank}(A) + \operatorname{rank}(B)$.

Proof. Let $A = [a_{ij}]_{m \times n} = [A_1 A_2 \cdots A_n]$ and $B = [b_{ij}]_{m \times n} = [B_1 B_2 \cdots B_n]$, where $A_i \in F^m$ and $B_i \in F^m$ are column vectors of A and B, respectively. Also, $A + B = [A_1 + B_1 \; A_2 + B_2 \cdots A_n + B_n]$.

We know that the rank of a matrix A is the dimension of column the space of A. That is, $\operatorname{rank}(A) = \dim(\operatorname{span}(A_1, A_2, \ldots, A_n))$.

Similarly, we have

$$\operatorname{rank}(B) = \dim(\operatorname{span}(B_1, B_2, \ldots, B_n)) \quad \text{and}$$
$$\operatorname{rank}(A + B) = \dim(\operatorname{span}(A_1 + B_1, A_2 + B_2, \ldots, A_n + B_n)).$$

Clearly, $\operatorname{span}(A_1, A_2, \ldots, A_n)$, $\operatorname{span}(B_1, B_2, \ldots, B_n)$ and $\operatorname{span}(A_1 + B_1, A_2 + B_2, \ldots, A_n + B_n)$ are subspaces of the m-dimensional vector space F^m.

We claim that

$$\operatorname{span}(A_1 + B_1, A_2 + B_2, \ldots, A_n + B_n)$$
$$\subseteq \operatorname{span}(A_1, A_2, \ldots, A_n) + \operatorname{span}(B_1, B_2, \ldots, B_n).$$

Suppose the vector $X \in \operatorname{span}(A_1 + B_1, A_2 + B_2, \ldots, A_n + B_n)$.
Then $X = k_1(A_1 + B_1) + k_2(A_2 + B_2) + \cdots + k_n(A_n + B_n)$ for some scalars $k_1, k_2, \ldots, k_n$. This gives that

$$X = (k_1 A_1 + k_2 A_2 + \cdots + k_n A_n) + (k_1 B_1 + k_2 B_2 + \cdots + k_n B_n)$$
$$\in \operatorname{span}(A_1, A_2, \ldots, A_n) + \operatorname{span}(B_1, B_2, \ldots, B_n).$$

Hence, the claim is proved.

Now, using the result; if U and V are two subspaces of a vector space, $\dim(U + V) \leq \dim(U) + \dim(V)$, we have

$$\operatorname{rank}(A + B) = \dim(\operatorname{span}(A_1 + B_1, A_2 + B_2, \ldots, A_n + B_n))$$
$$\leq \dim(\operatorname{span}(A_1, A_2, \ldots, A_n) + \operatorname{span}(B_1, B_2, \ldots, B_n))$$
$$\leq \dim(\operatorname{span}(A_1, A_2, \ldots, A_n)) + \dim(\operatorname{span}(B_1, B_2, \ldots, B_n))$$
$$= \operatorname{rank}(A) + \operatorname{rank}(B).$$
□

Exercises

(5.3.1) (a) Find the rank of the matrix $[a_{ij}]_{5 \times 7}$, where $a_{ij} = 1 \; \forall ij$.
(b) Find the rank of the matrix $[a_{ij}]_{5 \times 7}$, where $a_{ij} = 10 \; \forall ij$.

(5.3.2)   If the rank of a matrix A is zero, then show that A is a zero matrix.

(5.3.3)   Let A be a 10-square invertible matrix. Then find the rank of A.

(5.3.4)   Let A and B be 10-square matrices. If rank(A) = 10 and rank(B) = 5, then find the rank(AB).

(5.3.5)   Let A and B be matrices such that $A + B$ and AB both are defined. If rank(A) = 5 and rank(B) = 7, then find the upper limits of rank($A + B$) and rank(AB), respectively.

(5.3.6)   If A is a matrix of order $m \times n, m \neq n$, then show that either the row vectors are the column vectors of A are linearly dependent.

(5.3.7)   Let A and B be real matrices of orders $m \times n$ and $n \times m$, respectively. If $m < n$, then show that BA is a singular matrix.

(5.3.8)   Let J denotes the matrix of order 10×10 with all entries 1 and let B be a 50-square matrix given by

$$B = \begin{bmatrix} J & 0 & 0 & 0 & 0 \\ 0 & J & 0 & 0 & 0 \\ 0 & 0 & J & 0 & 0 \\ 0 & 0 & 0 & J & 0 \\ 0 & 0 & 0 & 0 & J \end{bmatrix}.$$

Then find the rank of B.

5.4 Elementary matrix operations

In this section, we introduce the elementary operations that are helpful to obtain the rank of a matrix, solution of system of linear equations, inverse of a matrix, normal form of a matrix and many more properties in coming sections and chapters.

Definition 5.4.1. Let A be a matrix of order $m \times n$. Then the following operations on the rows of A are called elementary row operations:

(1) Interchanging any two rows of A.

We shall use the notation $R_i \leftrightarrow R_j$ to denote the interchange of i-th and j-th, rows of A.

(2) Multiplying a row of A by a nonzero constant.

Multiplication of i-th row of A by a nonzero scalar k will be denoted by $R_i \rightarrow kR_i$.

(3) Adding a constant multiple of a row of A to another row.

We use the notation $R_i \rightarrow R_i + kR_j$ to denote that the i-th row of A is replaced by i-th row $+ k \times j$-th row of A.

Similarly, by replacing rows with columns in the aforementioned operations elementary column operations can be defined. The following lists the three elementary column operations in the form of notation:

- $C_i \leftrightarrow C_j$
- $C_i \rightarrow kC_i$
- $C_i \rightarrow C_i + kC_j$, where C represents the column.

Elementary row or elementary column operations are called elementary operations.

Example 5.4.2. Let

$$A = \begin{bmatrix} 3 & 4 & 2 & -1 \\ 2 & 4 & 6 & 2 \\ -2 & 4 & 5 & 0 \end{bmatrix},$$

be a 3×4 matrix. Then the matrix after applying elementary row operation $R_1 \leftrightarrow R_3$ on A is

$$\begin{bmatrix} -2 & 4 & 5 & 0 \\ 2 & 4 & 6 & 2 \\ 3 & 4 & 2 & -1 \end{bmatrix}.$$

The above matrix is obtained by interchanging first and third rows of A.

The matrix after applying the elementary row operation $R_2 \rightarrow 5R_2$ on A is

$$\begin{bmatrix} 3 & 4 & 2 & -1 \\ 10 & 20 & 30 & 10 \\ -2 & 4 & 5 & 0 \end{bmatrix}.$$

If we apply the elementary row operation $R_1 \rightarrow R_1 + 2R_3$ on A, then we get the following matrix

$$\begin{bmatrix} -1 & 12 & 12 & -1 \\ 2 & 4 & 6 & 2 \\ -2 & 4 & 5 & 0 \end{bmatrix}.$$

Similarly, on the application of elementary column operations $C_1 \leftrightarrow C_3$, $C_2 \rightarrow 5C_2$ and $C_1 \rightarrow C_1 + 2C_3$ on A we get the matrices

$$\begin{bmatrix} 2 & 4 & 3 & -1 \\ 6 & 4 & 2 & 2 \\ 5 & 4 & -2 & 0 \end{bmatrix}, \quad \begin{bmatrix} 3 & 20 & 2 & -1 \\ 2 & 20 & 6 & 2 \\ -2 & 20 & 5 & 0 \end{bmatrix} \quad \text{and} \quad \begin{bmatrix} 7 & 4 & 2 & -1 \\ 14 & 4 & 6 & 2 \\ 8 & 4 & 5 & 0 \end{bmatrix},$$

respectively.

Definition 5.4.3. A matrix obtained by performing an elementary operation on the identity matrix I_n is called an elementary matrix.

For example, if you take

$$I_3 = \begin{bmatrix} 1 & 0 & 0 \\ 0 & 1 & 0 \\ 0 & 0 & 1 \end{bmatrix},$$

then the matrix obtained by performing elementary row operation $R_1 \leftrightarrow R_3$ is an elementary matrix

$$\begin{bmatrix} 0 & 0 & 1 \\ 0 & 1 & 0 \\ 1 & 0 & 0 \end{bmatrix}.$$

This matrix can be obtained by applying the elementary column operation $C_1 \leftrightarrow C_3$ on I_3, also.

Thus, any elementary matrix can be obtained in at least two ways, either by performing an elementary row operation on I_n or by performing an elementary column operation on I_n.

Similarly, more elementary matrices can be obtained with respect to various elementary operations on I_n.

In Example 5.4.2, we have obtained the matrix

$$\begin{bmatrix} -2 & 4 & 5 & 0 \\ 2 & 4 & 6 & 2 \\ 3 & 4 & 2 & -1 \end{bmatrix},$$

after performing elementary row operation $R_1 \leftrightarrow R_3$ on A.

Let

$$I_3 = \begin{bmatrix} 1 & 0 & 0 \\ 0 & 1 & 0 \\ 0 & 0 & 1 \end{bmatrix}.$$

Then the elementary matrix obtained from the operation $R_1 \leftrightarrow R_3$ on I_3 is

$$E_1 = \begin{bmatrix} 0 & 0 & 1 \\ 0 & 1 & 0 \\ 1 & 0 & 0 \end{bmatrix}.$$

Now,

$$E_1 A = \begin{bmatrix} 0 & 0 & 1 \\ 0 & 1 & 0 \\ 1 & 0 & 0 \end{bmatrix} \begin{bmatrix} 3 & 4 & 2 & -1 \\ 2 & 4 & 6 & 2 \\ -2 & 4 & 5 & 0 \end{bmatrix} = \begin{bmatrix} -2 & 4 & 5 & 0 \\ 2 & 4 & 6 & 2 \\ 3 & 4 & 2 & -1 \end{bmatrix}.$$

Thus, the effect of elementary row operation $R_1 \leftrightarrow R_3$ on A is same as the pre-multiplication of elementary matrix obtained from the same elementary row operation $R_1 \leftrightarrow R_3$.

Similarly, the matrix

$$\begin{bmatrix} 2 & 4 & 3 & -1 \\ 6 & 4 & 2 & 2 \\ 5 & 4 & -2 & 0 \end{bmatrix}$$

is obtained from A by performing elementary column operation $C_1 \leftrightarrow C_3$.

Let

$$I_4 = \begin{bmatrix} 1 & 0 & 0 & 0 \\ 0 & 1 & 0 & 0 \\ 0 & 0 & 1 & 0 \\ 0 & 0 & 0 & 1 \end{bmatrix}$$

be the identity matrix of order 4×4.

Consider the elementary matrix obtained by performing the elementary column operation $C_1 \leftrightarrow C_3$ on I_4,

$$E_2 = \begin{bmatrix} 0 & 0 & 1 & 0 \\ 0 & 1 & 0 & 0 \\ 1 & 0 & 0 & 0 \\ 0 & 0 & 0 & 1 \end{bmatrix}.$$

Then

$$AE_2 = \begin{bmatrix} 3 & 4 & 2 & -1 \\ 2 & 4 & 6 & 2 \\ -2 & 4 & 5 & 0 \end{bmatrix} \begin{bmatrix} 0 & 0 & 1 & 0 \\ 0 & 1 & 0 & 0 \\ 1 & 0 & 0 & 0 \\ 0 & 0 & 0 & 1 \end{bmatrix} = \begin{bmatrix} 2 & 4 & 3 & -1 \\ 6 & 4 & 2 & 2 \\ 5 & 4 & -2 & 0 \end{bmatrix}.$$

In this case, we have that the effect of column operation $C_1 \leftrightarrow C_3$ on A is same as the post multiplication of elementary matrix obtained from the same elementary column operation $C_1 \leftrightarrow C_3$.

In this manner, we can prove that for each elementary (row or column) operation on a matrix there exists an elementary matrix whose multiplication gives the equal effect.

Theorem 5.4.4. *Let A be an $m \times n$ matrix and let B be a matrix obtained from A by performing an elementary row (column) operation. Then there exists an $m \times m$ ($n \times n$) elementary matrix E obtained by performing the corresponding row (column) operation on I_m (I_n) such that $B = EA(AE)$. Conversely, if E is an elementary $m \times m$ ($n \times n$) matrix, then $EA(AE)$ is a matrix that can be obtained by performing an elementary row (column) operation on A.*

Remark. An elementary operation on a matrix is nothing but a multiplication with a suitable matrix.

Theorem 5.4.5. *Elementary matrices are invertible.*

Proof. We know that an n-square elementary matrix is obtained by performing an elementary row or elementary column operations on I_n. Hence, we prove the result for the elementary matrices obtained by elementary row operations, and the result will be proved in a similar manner for the elementary matrices obtained by elementary column operations.

Here, we need to consider only three cases (one for each type of operation).

Let the elementary matrix E is obtained by the elementary row operation $R_i \leftrightarrow R_j$ on I_n. Then I_n can be obtained by the same elementary row operation $R_i \leftrightarrow R_j$ on E, i. e., $EE = I_n$. Hence, E is invertible.

Now, suppose the elementary matrix E is obtained by the elementary row operation $R_i \rightarrow kR_i$, $k \neq 0$, on I_n. Let $\bar{E}$ be the elementary matrix obtained from I_n by the elementary row operation $R_i \rightarrow \frac{1}{k}R_i$ on I_n. Then $E\bar{E} = \bar{E}E = I_n$. Hence, E is invertible.

Finally, suppose that the elementary matrix E is obtained by the elementary row operation $R_i \rightarrow R_i + kR_j$, $k \neq 0$, $i \neq j$ on I_n. Observe that I_n can be obtained by elementary row operation $R_i \rightarrow R_i - kR_j$ on E, where $k \neq 0$ and $i \neq j$. If $\bar{E}$ is the elementary matrix obtained from I_n by the elementary row operation $R_i \rightarrow R_i - kR_j$, then $\bar{E}E = I_n$. Hence, E is invertible. □

We use the example below to demonstrate the above result.

Example 5.4.6. Let

$$E = \begin{bmatrix} 0 & 1 & 0 \\ 1 & 0 & 0 \\ 0 & 0 & 1 \end{bmatrix}$$

be the elementary matrix obtained by the elementary row operation $R_1 \leftrightarrow R_2$ on

$$I_3 = \begin{bmatrix} 1 & 0 & 0 \\ 0 & 1 & 0 \\ 0 & 0 & 1 \end{bmatrix}.$$

Then $EE = I_3$.
Let

$$E = \begin{bmatrix} 1 & 0 & 0 \\ 0 & 5 & 0 \\ 0 & 0 & 1 \end{bmatrix}$$

be obtained by the elementary row operation $R_2 \rightarrow 5R_2$ on I_3. If

$$\bar{E} = \begin{bmatrix} 1 & 0 & 0 \\ 0 & \frac{1}{5} & 0 \\ 0 & 0 & 1 \end{bmatrix},$$

which is obtained from I_3 by the elementary row operation $R_2 \rightarrow \frac{1}{5}R_2$ on I_3, then $E\bar{E} = \bar{E}E = I_3$.

Similarly, if we perform the elementary row operation $R_3 \rightarrow R_3 + 3R_1$ on I_3 we get the elementary matrix

$$E = \begin{bmatrix} 1 & 0 & 0 \\ 0 & 1 & 0 \\ 3 & 0 & 1 \end{bmatrix}.$$

Let

$$\bar{E} = \begin{bmatrix} 1 & 0 & 0 \\ 0 & 1 & 0 \\ -3 & 0 & 1 \end{bmatrix}$$

be the elementary matrix, which is obtained by the elementary row operation $R_3 \rightarrow R_3 - 3R_1$ on I_3. Then $E\bar{E} = \bar{E}E = I_3$.

Theorem 5.4.7. *Elementary operations (row or column) on a matrix are rank preserving.*

Proof. Let the matrix B be obtained by performing elementary row and column operations on the matrix A. Since an elementary row operation on A is equivalent to premultiplication of an elementary matrix with A and elementary column operations on A is equivalent to post multiplication of an elementary matrix with A,

$$B = E_p \cdots E_2 E_1 A E_1' E_2' \cdots E_q', \quad \text{where } E_1, E_2, \ldots, E_p, E_1' E_2' \cdots E_q'$$

are elementary matrices.

Since all the elementary matrices are nonsingular,

$$\text{rank}(B) = \text{rank}(E_p \cdots E_2 E_1 A E_1' E_2' \cdots E_q') = \text{rank}(A) \text{ or } B \cong A. \qquad \square$$

Elementary matrices are useful to convert a matrix into a normal form. The following is the illustration to convert a matrix into normal form using elementary matrices.
In general, any matrix A of order $m \times n$ can be written as

$$A = I_m A I_n.$$

Now, suppose $E_1, E_2, \ldots, E_r$ and $E_1', E_2', \ldots, E_s'$ are elementary matrices such that $E_r \cdots E_2 E_1 A E_1' E_2' \cdots E_s'$ is a matrix in normal form N.

Then $N = E_r \cdots E_2 E_1 I_m A I_n E_1' E_2' \cdots E_s' = PAQ$, where $P = E_r \cdots E_2 E_1 I_m$ and $Q = I_n E_1' E_2' \cdots E_s'$.

Since P and Q are the product of elementary matrices and elementary matrices are invertible, P and Q are nonsingular.

Note that P and Q may not be unique because P and Q depend on the sequence of elementary operations.

In view of above illustration, we have the following result.

Theorem 5.4.8. *Every matrix of order $m \times n$ can be converted into a unique normal form matrix.*

The next few examples will demonstrate the algorithms used to transform a matrix A into normal form. Additionally, we identify the nonsingular matrices P and Q such that PAQ is in normal form.

Example 5.4.9. Let

$$A = \begin{bmatrix} 0 & 1 & 2 & -2 \\ 4 & 0 & 2 & 6 \\ 2 & 1 & 3 & 1 \end{bmatrix}$$

be a matrix of order 3×4. In this example, we explain the algorithm to convert A into the normal form using elementary row and column operations.

Step I. Make 1 to the entry located at $(1,1)$-th position. In this case, we apply elementary column operation $C_1 \leftrightarrow C_2$ on A, and we obtain

$$\begin{bmatrix} 1 & 0 & 2 & -2 \\ 0 & 4 & 2 & 6 \\ 1 & 2 & 3 & 1 \end{bmatrix}.$$

Step II. Make the first column's entries zero below 1. In order to do this, we use the basic row operation $R_3 \to R_3 - R_1$ in the matrix above.

$$\begin{bmatrix} 1 & 0 & 2 & -2 \\ 0 & 4 & 2 & 6 \\ 0 & 2 & 1 & 3 \end{bmatrix}.$$

Step III. Make entries zero right side to 1 in the first row. We apply elementary column operations $C_3 \to C_3 - 2C_1$ and $C_4 \to C_4 + 2C_1$ in the above matrix and get the matrix,

$$\begin{bmatrix} 1 & 0 & 0 & 0 \\ 0 & 4 & 2 & 6 \\ 0 & 2 & 1 & 3 \end{bmatrix}.$$

Step IV. Make 1 at $(2,2)$-th place. For this, we apply elementary column operation $\frac{1}{4}C_2$ in the previous matrix and get the matrix,

$$\begin{bmatrix} 1 & 0 & 0 & 0 \\ 0 & 1 & 2 & 6 \\ 0 & \frac{1}{2} & 1 & 3 \end{bmatrix}.$$

Step V. Make entries zero below 1 in column 2. We apply elementary row operation $R_3 \rightarrow R_3 - \frac{1}{2}R_2$ in the above matrix and get the matrix,

$$\begin{bmatrix} 1 & 0 & 0 & 0 \\ 0 & 1 & 2 & 6 \\ 0 & 0 & 0 & 0 \end{bmatrix}.$$

Step VI. Make entries zero right side to 1 in the second row. In order to do this, we apply elementary column operations $C_3 \rightarrow C_3 - 2C_2$ and $C_4 \rightarrow C_4 - 6C_2$ in the above matrix,

$$\begin{bmatrix} 1 & 0 & 0 & 0 \\ 0 & 1 & 0 & 0 \\ 0 & 0 & 0 & 0 \end{bmatrix} \approx \begin{bmatrix} 1 & 0 & \cdot & 0 & 0 \\ 0 & 1 & \cdot & 0 & 0 \\ \cdot & \cdot & & & \cdot \\ 0 & 0 & \cdot & 0 & 0 \end{bmatrix} \approx \begin{bmatrix} I_2 & \cdot & 0 \\ \cdot & \cdot & \cdot \\ 0 & \cdot & 0 \end{bmatrix}.$$

Now, above matrix is in normal form and equivalent to A. Hence, the rank of A is 2.

Example 5.4.10. Consider the matrix

$$A = \begin{bmatrix} 3 & 2 & 1 & 3 \\ 2 & 4 & 3 & 2 \\ 1 & 2 & 3 & 0 \\ 6 & 8 & 7 & 5 \end{bmatrix}.$$

In this example, we apply elementary row and column operations to convert the matrix A into normal form and determine the rank of A.

Applying elementary row operation $R_1 \leftrightarrow R_3$ on A,

$$\begin{bmatrix} 1 & 2 & 3 & 0 \\ 2 & 4 & 3 & 2 \\ 3 & 2 & 1 & 3 \\ 6 & 8 & 7 & 5 \end{bmatrix}.$$

From $R_2 \rightarrow R_2 - 2R_1$, $R_3 \rightarrow R_3 - 3R_1$, $R_4 \rightarrow R_4 - 6R_1$,

$$\begin{bmatrix} 1 & 2 & 3 & 0 \\ 0 & 0 & -3 & 2 \\ 0 & -4 & -8 & 3 \\ 0 & -4 & -11 & 5 \end{bmatrix}.$$

From $C_2 \rightarrow C_2 - 2C_1$, $C_3 \rightarrow C_3 - 3C_1$,

$$\begin{bmatrix} 1 & 0 & 0 & 0 \\ 0 & 0 & -3 & 2 \\ 0 & -4 & -8 & 3 \\ 0 & -4 & -11 & 5 \end{bmatrix}.$$

From $C_2 \rightarrow \frac{-1}{4}C_2$,

$$\begin{bmatrix} 1 & 0 & 0 & 0 \\ 0 & 0 & -3 & 2 \\ 0 & 1 & -8 & 3 \\ 0 & 1 & -11 & 5 \end{bmatrix}.$$

From $R_2 \leftrightarrow R_3$,

$$\begin{bmatrix} 1 & 0 & 0 & 0 \\ 0 & 1 & -8 & 3 \\ 0 & 0 & -3 & 2 \\ 0 & 1 & -11 & 5 \end{bmatrix}.$$

From $R_4 \rightarrow R_4 - R_2$,

$$\begin{bmatrix} 1 & 0 & 0 & 0 \\ 0 & 1 & -8 & 3 \\ 0 & 0 & -3 & 2 \\ 0 & 0 & -3 & 2 \end{bmatrix}.$$

From $C_3 \rightarrow C_3 + 8C_2$, $C_4 \rightarrow C_4 - 3C_2$,

$$\begin{bmatrix} 1 & 0 & 0 & 0 \\ 0 & 1 & 0 & 0 \\ 0 & 0 & -3 & 2 \\ 0 & 0 & -3 & 2 \end{bmatrix}.$$

From $C_3 \to \frac{-1}{3}C_3$,

$$\begin{bmatrix} 1 & 0 & 0 & 0 \\ 0 & 1 & 0 & 0 \\ 0 & 0 & 1 & 2 \\ 0 & 0 & 1 & 2 \end{bmatrix}.$$

From $R_4 \to R_4 - R_3$,

$$\begin{bmatrix} 1 & 0 & 0 & 0 \\ 0 & 1 & 0 & 0 \\ 0 & 0 & 1 & 2 \\ 0 & 0 & 0 & 0 \end{bmatrix}.$$

From $C_4 \to C_4 - 2C_3$,

$$\begin{bmatrix} 1 & 0 & 0 & 0 \\ 0 & 1 & 0 & 0 \\ 0 & 0 & 1 & 0 \\ 0 & 0 & 0 & 0 \end{bmatrix} \approx \begin{bmatrix} 1 & 0 & 0 & \cdot & 0 \\ 0 & 1 & 0 & \cdot & 0 \\ 0 & 0 & 1 & \cdot & 0 \\ \cdot & \cdot & \cdot & \cdot & \cdot \\ 0 & 0 & 0 & \cdot & 0 \end{bmatrix} \approx \begin{bmatrix} I_3 & \cdot & 0 \\ \cdot & \cdot & \cdot \\ 0 & \cdot & 0 \end{bmatrix}.$$

The above matrix is in the normal form, which is equivalent to the matrix A. Hence, rank(A) = 3.

In the following example, we explain the algorithm to find the nonsingular matrices P and Q such that for a given matrix A, PAQ becomes in normal form.

Example 5.4.11. Let

$$A = \begin{bmatrix} 1 & 2 & 3 & 2 \\ 2 & -1 & 4 & 2 \\ 1 & 7 & 5 & 4 \end{bmatrix}$$

be a matrix of order 3×4. Then we can write $A_{3\times 4} = I_3 A_{3\times 4} I_4$, or

$$\begin{bmatrix} 1 & 2 & 3 & 2 \\ 2 & -1 & 4 & 2 \\ 1 & 7 & 5 & 4 \end{bmatrix} = \begin{bmatrix} 1 & 0 & 0 \\ 0 & 1 & 0 \\ 0 & 0 & 1 \end{bmatrix} A \begin{bmatrix} 1 & 0 & 0 & 0 \\ 0 & 1 & 0 & 0 \\ 0 & 0 & 1 & 0 \\ 0 & 0 & 0 & 1 \end{bmatrix}.$$

We proceed with the above equation and apply elementary row and column operations to both sides so that the left side becomes normal form, and the right side changes I_3 to P and I_4 to Q. Note that on the right side, elementary row operations will be applied on I_3 and elementary column operations will be applied on I_4 because elementary row operation is pre-multiplication of an elementary matrix and elementary column operation is post-multiplication of an elementary matrix.

Applying $R_2 \rightarrow R_2 - 2R_1$, $R_3 \rightarrow R_3 - R_1$,

$$\begin{bmatrix} 1 & 2 & 3 & 2 \\ 0 & -5 & -2 & -2 \\ 0 & 5 & 2 & 2 \end{bmatrix} = \begin{bmatrix} 1 & 0 & 0 \\ -2 & 1 & 0 \\ -1 & 0 & 1 \end{bmatrix} A \begin{bmatrix} 1 & 0 & 0 & 0 \\ 0 & 1 & 0 & 0 \\ 0 & 0 & 1 & 0 \\ 0 & 0 & 0 & 1 \end{bmatrix}.$$

From $C_2 \rightarrow C_2 - 2C_1$, $C_3 \rightarrow C_3 - 3C_1$, $C_4 \rightarrow C_4 - 2C_1$,

$$\begin{bmatrix} 1 & 0 & 0 & 0 \\ 0 & -5 & -2 & -2 \\ 0 & 5 & 2 & 2 \end{bmatrix} = \begin{bmatrix} 1 & 0 & 0 \\ -2 & 1 & 0 \\ -1 & 0 & 1 \end{bmatrix} A \begin{bmatrix} 1 & -2 & -3 & -2 \\ 0 & 1 & 0 & 0 \\ 0 & 0 & 1 & 0 \\ 0 & 0 & 0 & 1 \end{bmatrix}.$$

From $C_2 \rightarrow \frac{-1}{5} C_2$,

$$\begin{bmatrix} 1 & 0 & 0 & 0 \\ 0 & 1 & -2 & -2 \\ 0 & -1 & 2 & 2 \end{bmatrix} = \begin{bmatrix} 1 & 0 & 0 \\ -2 & 1 & 0 \\ -1 & 0 & 1 \end{bmatrix} A \begin{bmatrix} 1 & \frac{2}{5} & -3 & -2 \\ 0 & \frac{-1}{5} & 0 & 0 \\ 0 & 0 & 1 & 0 \\ 0 & 0 & 0 & 1 \end{bmatrix}.$$

From $R_3 \rightarrow R_3 + R_2$,

$$\begin{bmatrix} 1 & 0 & 0 & 0 \\ 0 & 1 & -2 & -2 \\ 0 & 0 & 0 & 0 \end{bmatrix} = \begin{bmatrix} 1 & 0 & 0 \\ -2 & 1 & 0 \\ -3 & 1 & 1 \end{bmatrix} A \begin{bmatrix} 1 & \frac{2}{5} & -3 & -2 \\ 0 & \frac{-1}{5} & 0 & 0 \\ 0 & 0 & 1 & 0 \\ 0 & 0 & 0 & 1 \end{bmatrix}.$$

From $C_3 \rightarrow C_3 + 2C_2$, $C_4 \rightarrow C_4 + 2C_2$,

$$\begin{bmatrix} 1 & 0 & . & 0 & 0 \\ 0 & 1 & . & 0 & 0 \\ . & . & . & . & . \\ 0 & 0 & . & 0 & 0 \end{bmatrix} = \begin{bmatrix} 1 & 0 & 0 \\ -2 & 1 & 0 \\ -3 & 1 & 1 \end{bmatrix} A \begin{bmatrix} 1 & \frac{2}{5} & \frac{-11}{5} & \frac{-6}{5} \\ 0 & \frac{-1}{5} & \frac{-2}{5} & \frac{-2}{5} \\ 0 & 0 & 1 & 0 \\ 0 & 0 & 0 & 1 \end{bmatrix}.$$

Thus,

$$P = \begin{bmatrix} 1 & 0 & 0 \\ -2 & 1 & 0 \\ -3 & 1 & 1 \end{bmatrix} \quad \text{and} \quad Q = \begin{bmatrix} 1 & \frac{2}{5} & \frac{-11}{5} & \frac{-6}{5} \\ 0 & \frac{-1}{5} & \frac{-2}{5} & \frac{-2}{5} \\ 0 & 0 & 1 & 0 \\ 0 & 0 & 0 & 1 \end{bmatrix}.$$

In the following definition, we introduce a special form of a matrix called row re-duced echelon form, which is highly helpful from an application point of view.

Definition 5.4.12. A matrix of order $m \times n$ is called in reduced row echelon form if the following conditions are satisfied:
(i)   All the zero rows appear at the bottom of the matrix if any exists.
(ii) The first nonzero entry of each nonzero row is one. This entry is called the pivot and the corresponding column is called the pivot column.
(iii) The pivot entry in a row appears to the left of the pivot entry in any lower row.
(iv) If a column contains a pivot, then all other lower entries in that column are zero.

Similarly, by interchanging the roles of rows and columns in the definition of reduced row echelon form we can define the reduced column echelon form of a matrix.

The following matrices are in row reduced echelon form:

$$
A = \begin{bmatrix} 1 & 2 & 3 & 0 & -5 \\ 0 & 1 & 0 & 3 & 4 \\ 0 & 0 & 1 & 2 & -3 \\ 0 & 0 & 0 & 0 & 0 \end{bmatrix}, \quad B = \begin{bmatrix} 1 & -2 & 3 & 0 & 4 \\ 0 & 0 & 1 & 2 & 0 \\ 0 & 0 & 0 & 1 & 2 \\ 0 & 0 & 0 & 0 & 0 \\ 0 & 0 & 0 & 0 & 0 \end{bmatrix}.
$$

In the matrix A, columns 1, 2 and 3 are pivot columns and the columns 1, 3, 4 are pivot columns in the matrix B.

Columns other than pivot columns of a reduced row echelon form matrix are called free columns.

Theorem 5.4.13. *An $m \times n$ matrix can be reduced to a matrix in reduced row echelon form using elementary row operations.*

Proof. Let $A = [a_{ij}]$ be a nonzero $m \times n$ matrix. Then A can be reduced to reduced row echelon form with the following algorithm:
1.   Start by identifying the leftmost nonzero column.
2.   If the first nonzero column is the j_1-th column, use row operations (interchanging rows) to make $a_{1j_1} \neq 0$. By multiplying the first row with $a_{1j_1}^{-1}$ make a_{1j_1} a pivot, i. e., $a_{1j_1} = 1$.
3.   Using elementary row operations $R_i \rightarrow R_i - a_{ij_1} R_1$, $i > 1$, make all the entries below the pivot equal to zero, i. e., make $a_{2j_1} = a_{3j_1} = \cdots = a_{mj_1} = 0$.

If $a_{ij} = 0$ for all $i \geq 2$, then it is in reduced row echelon form. If not, then we have the following.
4.   Find j_2 the smallest number such that $a_{ij_2} \neq 0$, for some $i \geq 2$ and in this case $j_2 > j_1$. Use row operations as in step 2 and make $a_{2j_2} = 1$.

5. Again repeating step 3 make $a_{3j_2} = a_{4j_2} = \cdots = a_{mj_2} = 0.$

Repeat the above process until the matrix A is reduced in echelon form. □

We use the aforementioned algorithm in the following example to reduce the matrix into a reduced row echelon form.

Example 5.4.14. Consider the matrix

$$A = \begin{bmatrix} 0 & 0 & 3 & 4 \\ 0 & 0 & 2 & 2 \\ 0 & 3 & 6 & 9 \\ 0 & 2 & 1 & 4 \end{bmatrix}.$$

The second column in the given matrix A is the nonzero column. Use the elementary row operation $R_1 \leftrightarrow R_3$ to make a_{12} nonzero. Next, the matrix is obtained:

$$\begin{bmatrix} 0 & 3 & 6 & 9 \\ 0 & 0 & 2 & 2 \\ 0 & 0 & 3 & 4 \\ 0 & 2 & 1 & 4 \end{bmatrix}.$$

Now, we apply $R_1 \rightarrow \frac{1}{3}R_1$ in the above matrix to make $a_{12} = 1$ and then the resulting matrix is

$$\begin{bmatrix} 0 & 1 & 2 & 3 \\ 0 & 0 & 2 & 2 \\ 0 & 0 & 3 & 4 \\ 0 & 2 & 1 & 4 \end{bmatrix}.$$

To make the entries zero below, the pivot a_{12} we apply $R_4 \rightarrow R_4 - 2R_1$ in the above matrix and get the matrix

$$\begin{bmatrix} 0 & 1 & 2 & 3 \\ 0 & 0 & 2 & 2 \\ 0 & 0 & 3 & 4 \\ 0 & 0 & -3 & -2 \end{bmatrix}.$$

Now, $a_{23} = 2 \neq 0.$ From $R_2 \rightarrow \frac{1}{2}R_2,$ a_{23} becomes the pivot,

$$\begin{bmatrix} 0 & 1 & 2 & 3 \\ 0 & 0 & 1 & 1 \\ 0 & 0 & 3 & 4 \\ 0 & 0 & -3 & -2 \end{bmatrix}.$$

Applying $R_3 \to R_3 - 3R_2$, $R_4 \to R_4 + 3R_2$, we get

$$\begin{bmatrix} 0 & 1 & 2 & 3 \\ 0 & 0 & 1 & 1 \\ 0 & 0 & 0 & 1 \\ 0 & 0 & 0 & 1 \end{bmatrix}.$$

Next, from $R_4 \to R_4 - R_3$,

$$\begin{bmatrix} 0 & 1 & 2 & 3 \\ 0 & 0 & 1 & 1 \\ 0 & 0 & 0 & 1 \\ 0 & 0 & 0 & 0 \end{bmatrix}.$$

This is the reduced row echelon form of the matrix A.

Example 5.4.15. Consider the following 7×6 matrix A, which is in the reduced row echelon form:

$$A = \begin{bmatrix} 1 & 2 & 3 & -1 & 4 & 5 \\ 0 & 0 & 1 & -3 & 2 & 0 \\ 0 & 0 & 0 & 1 & 2 & -2 \\ 0 & 0 & 0 & 0 & 1 & 6 \\ 0 & 0 & 0 & 0 & 0 & 1 \\ 0 & 0 & 0 & 0 & 0 & 0 \\ 0 & 0 & 0 & 0 & 0 & 0 \end{bmatrix}.$$

Now, we shall show that the set of nonzero rows $\{R_1, R_2, R_3, R_4, R_5\}$ of A are linearly independent, where R_i denotes the i-th row from the top of the matrix. If possible, suppose that the set $\{R_5, R_4, R_3, R_2, R_1\}$ is linearly dependent. Then one of the rows, let us say R_2, is a linear combination of the preceding rows R_5, R_4 and R_3. In that case, $R_2 = a_3 R_3 + a_4 R_4 + a_5 R_5$, for some constants a_3, a_4, a_5.
That is,

$$(0,0,1,-3,2,0) = a_3(0,0,0,1,2,-2) + a_4(0,0,0,0,1,6) + a_5(0,0,0,0,0,1).$$

From the above equation, we have that the third component $1 = 0a_3 + 0a_4 + 0a_5 = 0$, which is a contradiction. Hence, the set of nonzero rows of matrix A are linearly independent.

Using the aforementioned idea, we prove the following result for any matrix in reduced row echelon form.

Theorem 5.4.16. *Let A be an $m \times n$ matrix in reduced row echelon form. Then the rank of A is equal to the number of its nonzero rows.*

Proof. Let $S = \{R_r, R_{r-1}, \ldots, R_2, R_1\}$ be the set of nonzero rows of the matrix A, where $R_i, 1 \le i \le r$ denotes the i-th row from the top of the matrix A.

Since the rank of a matrix is the maximum number of linearly independent rows, we shall show that S is linearly independent.

Suppose, if possible the set S is linearly dependent. Then there exist a row $R_k \in S$, which is a linear combination of preceding rows, $R_{k+1}, R_{k+2}, \ldots, R_r$. That is, for the scalars $a_{k+1}, a_{k+2}, \ldots, a_r$,

$$R_k = a_{k+1}R_{k+1} + a_{k+2}R_{k+2} + \cdots + a_r R_r.$$

Since A is in the reduced row echelon form, the above equation gives that the pivot of the row R_k is $1 = 0a_{k+1} + 0a_{k+2} + \cdots + 0a_r = 0$, which is a contradiction. Hence, the set S is linearly independent.

This proves that rank(A) = maximum number of linearly independent rows of A = the number of nonzero rows of A.

Since elementary row operations on a matrix A are rank preserving, rank A = rank of A in its reduced row echelon form. □

In the following example, we obtain the rank of a matrix by converting it into the reduced row echelon form.

Example 5.4.17. Consider the matrix

$$A = \begin{bmatrix} 3 & 2 & 1 & 3 \\ 2 & 4 & 3 & 2 \\ 1 & 2 & 3 & 0 \\ 6 & 8 & 7 & 5 \end{bmatrix},$$

given in Example 5.4.9. Now, we find the rank of A by reducing in row echelon form. From the elementary row operation $R_1 \leftrightarrow R_3$ on A, we get

$$\begin{bmatrix} 1 & 2 & 3 & 0 \\ 2 & 4 & 3 & 2 \\ 3 & 2 & 1 & 3 \\ 6 & 8 & 7 & 5 \end{bmatrix}.$$

From the elementary row operations $R_2 \rightarrow R_2 - 2R_1$, $R_3 \rightarrow R_3 - 3R_1$, $R_4 \rightarrow R_4 - 6R_1$, we get

$$\begin{bmatrix} 1 & 2 & 3 & 0 \\ 0 & 0 & -3 & 2 \\ 0 & -4 & -8 & 3 \\ 0 & -4 & -11 & 5 \end{bmatrix}.$$

From the elementary row operation $R_2 \leftrightarrow R_3$, we have

$$\begin{bmatrix} 1 & 2 & 3 & 0 \\ 0 & -4 & -8 & 3 \\ 0 & 0 & -3 & 2 \\ 0 & -4 & -11 & 5 \end{bmatrix}.$$

Applying $R_2 \rightarrow \frac{-1}{4}R_2$, we get

$$\begin{bmatrix} 1 & 2 & 3 & 0 \\ 0 & 1 & 2 & \frac{-3}{4} \\ 0 & 0 & -3 & 2 \\ 0 & -4 & -11 & 5 \end{bmatrix}.$$

From the elementary row operation $R_4 \rightarrow R_4 + 4R_2$, we have

$$\begin{bmatrix} 1 & 2 & 3 & 0 \\ 0 & 1 & 2 & \frac{-3}{4} \\ 0 & 0 & -3 & 2 \\ 0 & 0 & -3 & 2 \end{bmatrix}.$$

Applying $R_3 \rightarrow \frac{-1}{3}R_3$, we get

$$\begin{bmatrix} 1 & 2 & 3 & 0 \\ 0 & 1 & 2 & \frac{-3}{4} \\ 0 & 0 & 1 & \frac{-2}{3} \\ 0 & 0 & -3 & 2 \end{bmatrix}.$$

From the elementary row operation $R_4 \rightarrow R_4 + 3R_3$, we get

$$\begin{bmatrix} 1 & 2 & 3 & 0 \\ 0 & 1 & 2 & \frac{-3}{4} \\ 0 & 0 & 1 & \frac{-2}{3} \\ 0 & 0 & 0 & 0 \end{bmatrix}.$$

This matrix is the reduced row echelon form of A. In this matrix, the number of nonzero rows are three. Hence, rank$(A) = 3$.

Note that we only use elementary row operations to convert a matrix into its reduced row echelon form.

Theorem 5.4.18. *An n-square matrix A is invertible if and only if it can be reduced to the identity matrix I_n by elementary row operations.*

Proof. Suppose $A = [a_{ij}]$ is invertible. Then rank$(A) = n$. In this case, the reduced row echelon form of A will be an upper triangular matrix U with principal diagonal entries 1.

Further, by elementary row operations $R_1 \to R_1 - a_{12}R_2$, $R_2 \to R_2 - a_{23}R_3, \ldots, R_{n-1} \to R_{n-1} - a_{(n-1)n}R_n$, U can be reduced into the identity matrix I_n.

Conversely, suppose that the reduced row echelon form of A is I_n. Then $\text{rank}(A) = n$. Hence, A is invertible.

Any n-square matrix A can be written as $A = I_n A$. If A is invertible, then A can be converted into identity matrix I_n by elementary row operations.

Suppose $E_1, E_2, \ldots, E_r$ are elementary matrices such that

$$E_r \cdots E_2 E_1 A = I_n.$$

Now, from the equation $A = I_n A$ we have the following:

$$E_1 A = E_1 I_n A,$$
$$E_2 E_1 A = E_2 E_1 I_n A,$$

$$\vdots$$

$$E_r E_{r-1} \cdots E_2 E_1 A = E_r E_{r-1} \cdots E_2 E_1 I_n A,$$
$$I_n = BA,$$

where $B = E_r E_{r-1} \cdots E_2 E_1 I_n$.
Then $B = A^{-1}$. □

Since premultiplication of an elementary matrix with a matrix is equivalent to an elementary row operation on the matrix, we can compute the inverse of a matrix A by performing elementary row operations on the equations $A = I_n A$. This method is called the Gauss–Jordan method.

Example 5.4.19. In this example, we compute the inverse of the 4-square matrix

$$A = \begin{bmatrix} 1 & 2 & 0 & 1 \\ 1 & 1 & -1 & -2 \\ 0 & 2 & 1 & 3 \\ -1 & 1 & 2 & 6 \end{bmatrix},$$

using Gauss–Jordan method.

Consider the equation, $A = I_4 A$.

That is,

$$\begin{bmatrix} 1 & 2 & 0 & 1 \\ 1 & 1 & -1 & -2 \\ 0 & 2 & 1 & 3 \\ -1 & 1 & 2 & 6 \end{bmatrix} = \begin{bmatrix} 1 & 0 & 0 & 0 \\ 0 & 1 & 0 & 0 \\ 0 & 0 & 1 & 0 \\ 0 & 0 & 0 & 1 \end{bmatrix} A.$$

Now, we choose elementary row operations in such a way that the left side matrix reduces into the identity matrix I_4. Elementary row operations will be applied on the left side matrix as well as on I_4 in the right side of the equation.

Applying $R_2 \rightarrow R_2 - R_1$, and $R_4 \rightarrow R_4 + R_1$, in the above equation, we get

$$\begin{bmatrix} 1 & 2 & 0 & 1 \\ 0 & -1 & -1 & -3 \\ 0 & 2 & 1 & 3 \\ 0 & 3 & 2 & 7 \end{bmatrix} = \begin{bmatrix} 1 & 0 & 0 & 0 \\ -1 & 1 & 0 & 0 \\ 0 & 0 & 1 & 0 \\ 1 & 0 & 0 & 1 \end{bmatrix} A.$$

From $R_2 \rightarrow (-1)R_2$,

$$\begin{bmatrix} 1 & 2 & 0 & 1 \\ 0 & 1 & 1 & 3 \\ 0 & 2 & 1 & 3 \\ 0 & 3 & 2 & 7 \end{bmatrix} = \begin{bmatrix} 1 & 0 & 0 & 0 \\ 1 & -1 & 0 & 0 \\ 0 & 0 & 1 & 0 \\ 1 & 0 & 0 & 1 \end{bmatrix} A.$$

From $R_3 \rightarrow R_3 - 2R_2, R_4 \rightarrow R_4 - 3R_2$,

$$\begin{bmatrix} 1 & 2 & 0 & 1 \\ 0 & 1 & 1 & 3 \\ 0 & 0 & -1 & -3 \\ 0 & 0 & -1 & -2 \end{bmatrix} = \begin{bmatrix} 1 & 0 & 0 & 0 \\ 1 & -1 & 0 & 0 \\ -2 & 2 & 1 & 0 \\ -2 & 3 & 0 & 1 \end{bmatrix} A.$$

From $R_3 \rightarrow (-1)R_3$,

$$\begin{bmatrix} 1 & 2 & 0 & 1 \\ 0 & 1 & 1 & 3 \\ 0 & 0 & 1 & 3 \\ 0 & 0 & -1 & -2 \end{bmatrix} = \begin{bmatrix} 1 & 0 & 0 & 0 \\ 1 & -1 & 0 & 0 \\ 2 & -2 & -1 & 0 \\ -2 & 3 & 0 & 1 \end{bmatrix} A.$$

From $R_4 \rightarrow R_4 + R_3$,

$$\begin{bmatrix} 1 & 2 & 0 & 1 \\ 0 & 1 & 1 & 3 \\ 0 & 0 & 1 & 3 \\ 0 & 0 & 0 & 1 \end{bmatrix} = \begin{bmatrix} 1 & 0 & 0 & 0 \\ 1 & -1 & 0 & 0 \\ 2 & -2 & -1 & 0 \\ 0 & 1 & -1 & 1 \end{bmatrix} A.$$

From $R_1 \rightarrow R_1 - 2R_2$,

$$\begin{bmatrix} 1 & 0 & -2 & -5 \\ 0 & 1 & 1 & 3 \\ 0 & 0 & 1 & 3 \\ 0 & 0 & 0 & 1 \end{bmatrix} = \begin{bmatrix} -1 & 2 & 0 & 0 \\ 1 & -1 & 0 & 0 \\ 2 & -2 & -1 & 0 \\ 0 & 1 & -1 & 1 \end{bmatrix} A.$$

From $R_1 \rightarrow R_1 + 2R_3, R_2 \rightarrow R_2 - R_3$,

$$\begin{bmatrix} 1 & 0 & 0 & 1 \\ 0 & 1 & 0 & 0 \\ 0 & 0 & 1 & 3 \\ 0 & 0 & 0 & 1 \end{bmatrix} = \begin{bmatrix} 3 & -2 & -2 & 0 \\ -1 & 1 & 1 & 0 \\ 2 & -2 & -1 & 0 \\ 0 & 1 & -1 & 1 \end{bmatrix} A.$$

From $R_1 \rightarrow R_1 - R_4, R_3 \rightarrow R_3 - 3R_4$,

$$\begin{bmatrix} 1 & 0 & 0 & 0 \\ 0 & 1 & 0 & 0 \\ 0 & 0 & 1 & 0 \\ 0 & 0 & 0 & 1 \end{bmatrix} = \begin{bmatrix} 3 & -3 & -1 & -1 \\ -1 & 1 & 1 & 0 \\ 2 & -5 & 2 & -3 \\ 0 & 1 & -1 & 1 \end{bmatrix} A.$$

Thus,

$$A^{-1} = \begin{bmatrix} 3 & -3 & -1 & -1 \\ -1 & 1 & 1 & 0 \\ 2 & -5 & 2 & -3 \\ 0 & 1 & -1 & 1 \end{bmatrix}.$$

Alternatively, to compute the inverse of a matrix A, some authors construct the augmented matrix $[A : I_4]$ in place of equation $A = I_4 A$. That is, to find the inverse of the matrix

$$A = \begin{bmatrix} 1 & 2 & 0 & 1 \\ 1 & 1 & -1 & -2 \\ 0 & 2 & 1 & 3 \\ -1 & 1 & 2 & 6 \end{bmatrix},$$

write the augmented matrix

$$\begin{bmatrix} 1 & 2 & 0 & 1 & \cdot & 1 & 0 & 0 & 0 \\ 1 & 1 & -1 & -2 & \cdot & 0 & 1 & 0 & 0 \\ 0 & 2 & 1 & 3 & \cdot & 0 & 0 & 1 & 0 \\ -1 & 1 & 2 & 6 & \cdot & 0 & 0 & 0 & 1 \end{bmatrix}.$$

Then by elementary row operations $R_2 \rightarrow R_2 - R_1, R_4 \rightarrow R_4 + R_1$, we get

$$\begin{bmatrix} 1 & 2 & 0 & 1 & \cdot & 1 & 0 & 0 & 0 \\ 0 & -1 & -1 & -3 & \cdot & -1 & 1 & 0 & 0 \\ 0 & 2 & 1 & 3 & \cdot & 0 & 0 & 1 & 0 \\ 0 & 3 & 2 & 7 & \cdot & 1 & 0 & 0 & 1 \end{bmatrix}.$$

Similarly, by applying elementary row operations: $R_2 \rightarrow (-1)R_2$, $R_3 \rightarrow R_3 - 2R_2$, $R_4 \rightarrow R_4 - 3R_2$, $R_3 \rightarrow (-1)R_3$, $R_4 \rightarrow R_4 + R_3$, $R_1 \rightarrow R_1 - 2R_2$, $R_1 \rightarrow R_1 + 2R_3$, $R_2 \rightarrow R_2 - R_3$, $R_1 \rightarrow R_1 - R_4$, $R_3 \rightarrow R_3 - 3R_4$ we get

$$
\begin{bmatrix}
1 & 0 & 0 & 0 \\
0 & 1 & 0 & 0 \\
0 & 0 & 1 & 0 \\
0 & 0 & 0 & 1
\end{bmatrix}
\cdot
\begin{matrix}
3 & -3 & -1 & -1 \\
-1 & 1 & 1 & 0 \\
2 & -5 & 2 & -3 \\
0 & 1 & -1 & 1
\end{matrix}
$$

Then

$$
A^{-1} =
\begin{bmatrix}
3 & -3 & -1 & -1 \\
-1 & 1 & 1 & 0 \\
2 & -5 & 2 & -3 \\
0 & 1 & -1 & 1
\end{bmatrix}.
$$

Exercises

(5.4.1) Find the rank of the following matrices by reducing them into normal form:

$$
A =
\begin{bmatrix}
1 & -1 & -2 & -4 \\
-2 & -3 & 1 & 1 \\
3 & 1 & 3 & -2 \\
6 & 3 & 0 & -7
\end{bmatrix},
\quad
B =
\begin{bmatrix}
1 & 4 & 3 & -2 & 1 \\
2 & 3 & 1 & -4 & -3 \\
-1 & 6 & 7 & 2 & 9 \\
3 & -3 & -6 & -6 & -12
\end{bmatrix},
$$

$$
C =
\begin{bmatrix}
1 & 1 & 1 & 1 \\
1 & 2 & 3 & 1 \\
1 & 3 & 3 & 2 \\
2 & 4 & 3 & 3
\end{bmatrix},
\quad
D =
\begin{bmatrix}
1 & 2 & 3 & 4 \\
5 & 6 & 7 & 8 \\
9 & 10 & 11 & 12 \\
13 & 14 & 15 & 16
\end{bmatrix}.
$$

(5.4.2) Find the nonsingular matrices $P_1, P_2, P_3, P_4, Q_1, Q_2, Q_3, Q_4$ such that $P_1AQ_1, P_2BQ_2, P_3CQ_3, P_4DQ_4$ are in normal form where A, B, C, D are the matrices given in the question (5.4.1).

(5.4.3) Find the rank of the matrices A, B, C, D given in the question (5.4.1) by reducing them into echelon form.

(5.4.4) Using elementary operations find the inverse of the following matrices:

$$
A =
\begin{bmatrix}
2 & 3 & -2 & 4 \\
3 & -2 & 1 & 2 \\
3 & 2 & 3 & 4 \\
-2 & 4 & 0 & 5
\end{bmatrix},
\quad
B =
\begin{bmatrix}
1 & 2 & 0 & 1 \\
-1 & 1 & 2 & 6 \\
0 & 2 & 1 & 3 \\
1 & 1 & -1 & -2
\end{bmatrix}.
$$

5.5 System of linear equations

An equation of the form $a_1x_1 + a_2x_2 + \cdots + a_nx_n = b$ is called linear equation in n variables $x_1, x_2, \ldots, x_n$, where $a_1, a_2, \ldots, a_n, b$ are members of a field F.

A system of m linear equations in n variables is a collection of linear equations:

$$a_{11}x_1 + a_{12}x_2 + \cdots + a_{1n}x_n = b_1,$$
$$a_{21}x_1 + a_{22}x_2 + \cdots + a_{2n}x_n = b_2,$$
$$\vdots$$
$$a_{m1}x_1 + a_{m2}x_2 + \cdots + a_{mn}x_n = b_m,$$

where $a_{ij}, b_i, (1 \le i \le m, 1 \le j \le n)$ are elements of a field F.
If

$$A = \begin{bmatrix} a_{11} & a_{12} & \cdots & a_{1n} \\ a_{21} & a_{22} & \cdots & a_{2n} \\ \vdots & & & \\ a_{m1} & a_{m2} & \cdots & a_{mn} \end{bmatrix}, \quad X = \begin{bmatrix} x_1 \\ x_2 \\ \vdots \\ x_n \end{bmatrix} \quad \text{and} \quad B = \begin{bmatrix} b_1 \\ b_2 \\ \vdots \\ b_m \end{bmatrix},$$

then the above system of linear equations can be written in the matrix form as $AX = B$. The matrix A is called the coefficient matrix of the system of linear equations.

A column vector

$$L = \begin{bmatrix} l_1 \\ l_2 \\ \vdots \\ l_n \end{bmatrix} \in F^n$$

is called a solution of $AX = B$, if $AL = B$. The set of all solutions of a system of linear equations is called the solution set of the system.

Definition 5.5.1. A system of m linear equations in n variables, $AX = B$ is called homogeneous if $B = 0$, otherwise the system is called nonhomogeneous.

For example, the system of linear equations

$$x + y + 3z = 2,$$
$$2x - 3y - 4z = 5,$$

written in matrix form

$$\begin{bmatrix} 1 & 1 & 3 \\ 2 & -3 & -4 \end{bmatrix} \begin{bmatrix} x \\ y \\ z \end{bmatrix} = \begin{bmatrix} 2 \\ 5 \end{bmatrix}$$

is nonhomogeneous, whereas the system

$$-x + 2y + 3z = 0,$$
$$6x + 4y - 2z = 0,$$

written in matrix form

$$\begin{bmatrix} -1 & 2 & 3 \\ 6 & 4 & -2 \end{bmatrix} \begin{bmatrix} x \\ y \\ z \end{bmatrix} = \begin{bmatrix} 0 \\ 0 \end{bmatrix}$$

is homogeneous.

Let us examine the following example containing three different systems of linear equations.

Example 5.5.2.
(a) The system of linear equations

$$x + 2y = 5,$$
$$-3x + y = -1,$$

has unique solution

$$\begin{bmatrix} x \\ y \end{bmatrix} = \begin{bmatrix} 1 \\ 2 \end{bmatrix}.$$

(b) The system of linear equations

$$x + 2y = 3,$$
$$3x + 6y = 9,$$

has infinite solutions as

$$\begin{bmatrix} x \\ y \end{bmatrix} = \begin{bmatrix} 1 \\ 1 \end{bmatrix}, \quad \begin{bmatrix} x \\ y \end{bmatrix} = \begin{bmatrix} 3 \\ 0 \end{bmatrix}, \quad \begin{bmatrix} x \\ y \end{bmatrix} = \begin{bmatrix} -197 \\ 100 \end{bmatrix}, \quad \text{etc.}$$

(c) The system of linear equations

$$x - 2y = 1,$$
$$3x - 6y = 5,$$

has no solution.

It is evident from the aforementioned examples that a system of linear equations may have an infinite solution, a unique solution or none at all.

Definition 5.5.3. A system of nonhomogeneous linear equations $AX = B$ is called consistent if it has at least one solution. The system is considered inconsistent if $AX = B$ has no solution.

Since the zero vector $X = 0$ satisfies the equation $AX = 0$, the system of homogeneous equations $AX = 0$ is always consistent. The solution $X = 0$ is called the trivial solution, while a nonzero column vector Y satisfying $AY = 0$ is known as the nontrivial solution.

Definition 5.5.4. Let $AX = B$ be the system of m linear equations in n variables, where

$$A = \begin{bmatrix} a_{11} & a_{12} & \cdots & a_{1n} \\ a_{21} & a_{22} & \cdots & a_{2n} \\ \vdots & & & \\ a_{m1} & a_{m2} & \cdots & a_{mn} \end{bmatrix}, \quad X = \begin{bmatrix} x_1 \\ x_2 \\ \vdots \\ x_n \end{bmatrix} \quad \text{and} \quad B = \begin{bmatrix} b_1 \\ b_2 \\ \vdots \\ b_m \end{bmatrix}.$$

Then the block matrix $[A : B]$ is called the augmented matrix and will be denoted by aug A.
That is,

$$\text{aug } A = \begin{bmatrix} a_{11} & a_{12} & \cdots & a_{1n} & : b_1 \\ a_{21} & a_{22} & \cdots & a_{2n} & : b_2 \\ \vdots & & & & \\ a_{m1} & a_{m2} & \cdots & a_{mn} & : b_m \end{bmatrix}.$$

If order of the coefficient matrix A is $m \times n$, then the order of aug A is $m \times (n + 1)$.
For the system of linear equations,

$$x - 2y + 3z = 2,$$
$$-3x + y + 2z = 1,$$
$$2x + 2y + 2z = 6,$$

the coefficient matrix

$$A = \begin{bmatrix} 1 & -2 & 3 \\ -3 & 1 & 2 \\ 2 & 2 & 2 \end{bmatrix} \quad \text{and} \quad \text{aug } A = \begin{bmatrix} 1 & -2 & 3 & : 2 \\ -3 & 1 & 2 & : 1 \\ 2 & 2 & 2 & : 6 \end{bmatrix}.$$

Theorem 5.5.5. *A system of nonhomogeneous linear equations $AX = B$ is consistent if and only if* rank$(A) = $ rank(aug A).

Proof. Let the order of the coefficient matrix $A = [a_{ij}]$ be $m \times n$. Then from Theorem 5.3.2, the matrix of the left multiplication linear transformation $L_A : F^n \rightarrow F^m$ related to the standard ordered bases of F^n and F^m is A, where $L_A(X) = AX$. Therefore, $L_A(v_j) = j$-th column of A if $B_1 = \{v_1, v_2, \ldots, v_n\}$ is the standard basis of F^n, where the column vector $v_j = (0, 0, \ldots, 0, 1, 0, \ldots, 0) \in F^n$.

As a result, the equation $AX = B$ can be written as $L_A(X) = B$, and hence finding the solution to the system of linear equations $AX = B$ is equivalent to finding $X \in F^n$ such that $L_A(X) = B$.

Now, suppose that the system of linear equations $AX = B$ is consistent. Then there exists an $X \in F^n$ such that $L_A(X) = B$. That is, $B \in \text{Image}(L_A)$. $\text{Image}(L_A)$ is generated by $\{L_A(v_1), L_A(v_2), \ldots, L_A(v_n)\}$, it means $\text{Image}(L_A)$ is generated by the columns of the matrix A. Hence, B is the linear combination of n columns of A. This gives that the maximum number of linearly independent columns of A = maximum number of linearly independent columns of aug A. Hence, $\text{rank}(A) = \text{rank}(\text{aug } A)$.

Conversely, suppose that $\text{rank}(A) = \text{rank}(\text{aug } A)$. Then the column vector B of the matrix aug A can be written as the linear combination of the columns of A. Since the set $\{L_A(v_1), L_A(v_2), \ldots, L_A(v_n)\}$ represents the columns of A, B can be written as

$$\begin{aligned} B &= k_1 L_A(v_1) + k_2 L_A(v_2) + \cdots + k_n L_A(v_n) \\ &= L_A(k_1 v_1 + k_2 v_2 + \cdots + k_n v_n) \\ &= L_A(X), \end{aligned}$$

where $X = k_1 v_1 + k_2 v_2 + \cdots + k_n v_n \in F^n$ and $k_1, k_2, \ldots, k_n$ are scalars. Thus, there exists $X \in F^n$ such that $L_A(X) = B \Rightarrow AX = B$. This proves that X is a solution of the linear equation $AX = B$, and hence it is consistent. □

Recall Example 5.5.2(c),

$$x - 2y = 1,$$
$$3x - 6y = 5.$$

In this case,

$$\text{aug } A = \begin{bmatrix} 1 & -2 & : 1 \\ 3 & -6 & : 5 \end{bmatrix}.$$

From the reduced row echelon form

$$\begin{bmatrix} 1 & -2 & : 1 \\ 0 & 0 & : 2 \end{bmatrix}$$

of aug A, we have that $\text{rank}(A) = 1 \neq \text{rank}(\text{aug } A) = 2$. Hence, the system of linear equations has no solution.

Theorem 5.5.6. *Let $AX = 0$ be a homogeneous system of m linear equations in n variables. Then the solution space denoted by $N(A)$ is a subspace of F^n of dimension $n - \text{rank}(A)$.*

Proof. Since the homogeneous system $AX = 0$ has m equations in n variables, the coefficient matrix is of order $m \times n$.

Consider the linear transformation $L_A : F^n \to F^m$ defined as $L_A(X) = AX$, where $X \in F^n$. If $X' \in F^n$ is a solution of $AX = 0$, then $AX' = 0 \Leftrightarrow L_A(X') = 0 \Leftrightarrow X' \in \text{Ker}(L_A)$. Hence, the solution set $N(A)$ of the homogeneous equation $AX = 0$ is the $\text{Ker}(L_A)$, which is a subspace of F^n. From the rank-nullity theorem, $\dim(\text{Ker}(L_A)) = n - \text{rank}(L_A) = n - \text{rank}(A)$. Hence, the dimension of the solution space $= n - \text{rank}(A)$. Thus, all the solutions of $AX = 0$ are linear combination of $n - \text{rank}(A)$ linearly independent solutions. □

Remark. Since the solution space $N(A)$ of $AX = 0$ is equal to $\text{Ker}(L_A)$, $N(A)$ is a vector space. Hence, if X_1 and X_2 are two solutions and $a_1, a_2 \in F$, then $a_1 X_1 + a_2 X_2 \in N(A)$. That is, $A(a_1 X_1 + a_2 X_2) = a_1 A X_1 + a_2 A X_2 = a_1 0 + a_2 0 = 0$.

Corollary 5.5.7. *Let $AX = 0$ be the homogeneous system of m linear equations in n variables. Then:*
(i) *the system has nontrivial (infinite) solution if $\text{rank}(A) < n$;*
(ii) *the system has a nontrivial (infinite) solution if $m < n$;*
(iii) *the system has only trivial solutions if $\text{rank}(A) = n$.*

Proof. (i) The dimension of the solution space $N(A) = n - \text{rank}(A)$. If $\text{rank}(A) < n$, then $\dim N(A) > 0$. Hence, the system has a nontrivial solution.
(ii) Since $\text{rank}(A) \le \min\{m, n\} = m$, $\text{rank}(A) \le m < n$. Hence, from (i) part, system has a nontrivial solution.
(iii) Given that $\text{rank}(A) = n$. Hence, dimension of $N(A) = n - \text{rank}(A) = n - n = 0$. Hence, the system $AX = 0$ has only trivial solution. □

Theorem 5.5.8. *If $AX = B$ is a consistent system of m linear equations in n variables, and $AX = 0$ is its associated homogeneous system, then the set $X_B + N(A) = \{X_B + X_N : X_N \in N(A)\}$ constitutes a complete set of solutions for $AX = B$, where X_B is any particular solution of $AX = B$ and $N(A)$ is the solution space of $AX = 0$.*

Proof. Given X_B as a solution of $AX = B$ and $X_N \in N(A)$, it follows that $AX_B = B$ and $AX_N = 0$. Thus, $A(X_B + X_N) = AX_B + AX_N = B + 0 = B$. This implies that $X_B + X_N$ is a solution of $AX = B$.

Conversely, if Y_B is any solution of $AX = B$, then $AY_B = B = AX_B \Rightarrow A(Y_B - X_B) = 0$. Thus, $(Y_B - X_B) \in N(A)$. Therefore, there exists $X_N \in N(A)$ such that $Y_B - X_B = X_N$, leading to $Y_B = X_N + X_B$. This proves that $X_B + N(A)$ constitutes the set of all solutions of the linear equation $AX = B$. □

Remarks. (1) The solution set $X_B + N(A)$ for $AX = B$ is not a subspace of F^n, as 0 is not in $X_B + N(A)$.
(2) The comparison between linear equations and their solutions with linear differential equations and their solutions, particularly the complementary function (CF) and particular integral (PI), is noteworthy.

Corollary 5.5.9. *Let $AX = B$ be a consistent nonhomogeneous system of m linear equations in n variables. Then:*
(i)   *$AX = B$ has infinite solutions if rank$(A) < n$;*
(ii)   *$AX = B$ has infinite solutions if $m < n$;*
(iii) *the system has unique solution if rank$(A) = n$.*

Proof. Using Theorem 5.5.8, the complete solution set of $AX = B$ is $X_B + N(A)$. From Corollary 5.5.7, we deduce:
(i)   If rank$(A) < n$, then $N(A)$ has an infinite number of elements, hence $X_B + N(A)$ has an infinite number of elements.
(ii)   If $m < n$, then rank$(A) < n$, and thus the result follows from part (i).
(iii) If rank$(A) = n$, then $N(A) = \{0\}$; hence $X_B + N(A) = X_B$, resulting in a unique solution.   □

Corollary 5.5.10. *Let $AX = B$ be a system of nonhomogeneous n linear equations in n variables. Then the system has exactly one solution $X = A^{-1}B$ if and only if A is invertible.*

Proof. If the system $AX = B$ has only one solution X_B, then $X_B + N(A) = X_B$ implies $N(A) = \{0\}$. That is, the associated homogeneous linear equation $AX = 0$ has only a trivial solution. Thus, from Corollary 5.5.7(iii), rank$(A) = n$. This proves A is invertible.

Conversely, if A is invertible, then rank$(A) = n$. From Corollary 5.5.9(iii), $AX = B$ has a unique solution. As A is invertible, $AX = B \Rightarrow A^{-1}AX = A^{-1}B \Rightarrow X = A^{-1}B$.   □

The subsequent theorem asserts that a given system of linear equations can be altered into an equivalent system with identical solutions but is simpler to solve.

Theorem 5.5.11. *Consider a system of m linear equations in n unknowns denoted by $AX = B$. If aug $\acute{A} = [\acute{A} : \acute{B}]$ represents the reduced row echelon form of aug $A = [A : B]$, then the associated system of equations $\acute{A}X = \acute{B}$ possesses the same set of solutions.*

Proof. The conversion of aug $A = [A : B]$ into its reduced row echelon form is accomplished through elementary row operations. As these operations are equivalent to the premultiplication of invertible elementary matrices, it suffices to demonstrate that the systems $AX = B$ and $EAX = EB$ yield equivalent solutions, where E denotes an invertible square matrix of order m.

Suppose Y is a solution of $AX = B$, implying $AY = B$. Then $EAY = E(AY) = EB$, indicating that Y also serves as a solution for $EAX = EB$.

Conversely, if Y is a solution of $EAX = EB$, then $EAY = EB$. Consequently, $AY = E^{-1}EB = B$, demonstrating that Y also satisfies $AX = B$.   □

Thus, to solve the system of linear equations $AX = B$, we transform aug $A = [A : B]$ into its reduced row echelon form aug $\acute{A} = [\acute{A} : \acute{B}]$, and then address the corresponding system of linear equations $\acute{A}X = \acute{B}$.

Example 5.5.12. Consider the following system of linear equations:

$$x - 4y - z + t = 3,$$
$$2x - 8y + z - 4t = 9,$$
$$-x + 4y - 2z + 5t = -6.$$

In matrix form, this system can be expressed as

$$\begin{bmatrix} 1 & -4 & -1 & 1 \\ 2 & -8 & 1 & -4 \\ -1 & 4 & -2 & 5 \end{bmatrix} \begin{bmatrix} x \\ y \\ z \\ t \end{bmatrix} = \begin{bmatrix} 3 \\ 9 \\ -6 \end{bmatrix}.$$

If

$$A = \begin{bmatrix} 1 & -4 & -1 & 1 \\ 2 & -8 & 1 & -4 \\ -1 & 4 & -2 & 5 \end{bmatrix}, \quad X = \begin{bmatrix} x \\ y \\ z \\ t \end{bmatrix} \quad \text{and} \quad B = \begin{bmatrix} 3 \\ 9 \\ -6 \end{bmatrix},$$

then the given system is $AX = B$.

To test the system's consistency, we form the augmented matrix:

$$\text{aug } A = \begin{bmatrix} 1 & -4 & -1 & 1 & \cdot & 3 \\ 2 & -8 & 1 & -4 & \cdot & 9 \\ -1 & 4 & -2 & 5 & \cdot & -6 \end{bmatrix}.$$

We reduce this matrix to its reduced row echelon form using elementary row operations.

From $R_2 \to R_2 - 2R_1$, $R_3 \to R_3 + R_1$,

$$\begin{bmatrix} 1 & -4 & -1 & 1 & \cdot & 3 \\ 0 & 0 & 3 & -6 & \cdot & 3 \\ 0 & 0 & -3 & 6 & \cdot & -3 \end{bmatrix}.$$

From $R_2 \to \frac{1}{3}R_2$,

$$\begin{bmatrix} 1 & -4 & -1 & 1 & \cdot & 3 \\ 0 & 0 & 1 & -2 & \cdot & 1 \\ 0 & 0 & -3 & 6 & \cdot & -3 \end{bmatrix}.$$

From $R_3 \rightarrow R_3 + 3R_2$,

$$\begin{bmatrix} 1 & -4 & -1 & 1 & \cdot & 3 \\ 0 & 0 & 1 & -2 & \cdot & 1 \\ 0 & 0 & 0 & 0 & \cdot & 0 \end{bmatrix}.$$

From the above matrix, it is evident that $\text{rank}(A) = 2 = \text{rank}(\text{aug}(A))$.

Hence, from Theorem 5.5.5, the system is consistent. In the given system, the number of unknowns $= 4$ and $\text{rank}(A) = 2 < 4$. Hence, from Corollary 5.5.9, the system has infinite $(4 - 2 + 1 = 3$, linearly independent) solutions.

From the reduced row echelon form of $\text{aug}\, A$, we get the following system having the same solutions:

$$x - 4y - z + t = 3,$$
$$z - 2t = 1.$$

To solve this system, we divide the variables x, y, z, t into two sets. The first set is $\{x, z\}$, which consists of leftmost variables of each equations of the system. The second set consists of all the remaining variables, i. e., $\{y, t\}$. We assign parameter values to each variable in the second set and then solve for the variables in the first set in terms of these parameters.

Let $y = r_1$ and $t = r_2$. Then:

$$z = 1 + 2t = 1 + 2r_2,$$
$$x = 3 + 4y + z - t = 4 + 4r_1 + r_2.$$

Thus,

$$\begin{bmatrix} x \\ y \\ z \\ t \end{bmatrix} = \begin{bmatrix} 4 + 4r_1 + r_2 \\ r_1 \\ 1 + 2r_2 \\ r_2 \end{bmatrix} = \begin{bmatrix} 4 \\ 0 \\ 1 \\ 0 \end{bmatrix} + r_1 \begin{bmatrix} 4 \\ 1 \\ 0 \\ 0 \end{bmatrix} + r_2 \begin{bmatrix} 1 \\ 0 \\ 2 \\ 1 \end{bmatrix},$$

which is of the form $X_B + N(A)$. Hence, according to Theorems 5.5.8 and 5.5.11, all solutions of the given system will be of the form:

$$\left\{ \begin{bmatrix} 4 \\ 0 \\ 1 \\ 0 \end{bmatrix} + r_1 \begin{bmatrix} 4 \\ 1 \\ 0 \\ 0 \end{bmatrix} + r_2 \begin{bmatrix} 1 \\ 0 \\ 2 \\ 1 \end{bmatrix} \right\} : r_1, r_2 \in \mathbb{R}.$$

Note that

$$\begin{bmatrix} 4 \\ 0 \\ 1 \\ 0 \end{bmatrix}$$

is a particular solution of $AX = B$ and

$$\left\{ \begin{bmatrix} 4 \\ 1 \\ 0 \\ 0 \end{bmatrix}, \begin{bmatrix} 1 \\ 0 \\ 2 \\ 1 \end{bmatrix} \right\}$$

generates the solution space $N(A)$ of the homogeneous part $AX = 0$ of the system. Also, we can compare the solution as CF + PI, where

$$\text{PI} = \begin{bmatrix} 4 \\ 0 \\ 1 \\ 0 \end{bmatrix} \quad \text{and} \quad \text{CF} = \left\{ r_1 \begin{bmatrix} 4 \\ 1 \\ 0 \\ 0 \end{bmatrix} + r_2 \begin{bmatrix} 1 \\ 0 \\ 2 \\ 1 \end{bmatrix} \right\} : r_1, r_2 \in \mathbb{R}.$$

Example 5.5.13. Consider the following system of linear equations:

$$-x_1 + 2x_2 + 3x_3 = -2,$$
$$2x_1 - 5x_2 + x_3 = 2,$$
$$3x_1 - 8x_2 + 5x_3 = 2,$$
$$5x_1 - 12x_2 - x_3 = 6.$$

In the matrix form, this system can be written as

$$\begin{bmatrix} -1 & 2 & 3 \\ 2 & -5 & 1 \\ 3 & -8 & 5 \\ 5 & -12 & -1 \end{bmatrix} \begin{bmatrix} x_1 \\ x_2 \\ x_3 \end{bmatrix} = \begin{bmatrix} -2 \\ 2 \\ 2 \\ 6 \end{bmatrix}.$$

Let

$$\text{aug } A = \left[\begin{array}{ccc|c} -1 & 2 & 3 & -2 \\ 2 & -5 & 1 & 2 \\ 3 & -8 & 5 & 2 \\ 5 & -12 & -1 & 6 \end{array} \right].$$

Now, we convert the aug A in the reduced row echelon form.

From $R_1 \to (-1)R_1$,

$$\begin{bmatrix} 1 & -2 & -3 & \cdot & 2 \\ 2 & -5 & 1 & \cdot & 2 \\ 3 & -8 & 5 & \cdot & 2 \\ 5 & -12 & -1 & \cdot & 6 \end{bmatrix}.$$

From $R_2 \to R_2 - 2R_1$, $R_3 \to R_3 - 3R_1$ and $R_4 \to R_4 - 5R_1$,

$$\begin{bmatrix} 1 & -2 & -3 & \cdot & 2 \\ 0 & -1 & 7 & \cdot & -2 \\ 0 & -2 & 14 & \cdot & -4 \\ 0 & -2 & 14 & \cdot & -4 \end{bmatrix}.$$

From $R_2 \to (-1)R_2$,

$$\begin{bmatrix} 1 & -2 & -3 & \cdot & 2 \\ 0 & 1 & -7 & \cdot & 2 \\ 0 & -2 & 14 & \cdot & -4 \\ 0 & -2 & 14 & \cdot & -4 \end{bmatrix}.$$

From $R_3 \to R_3 + 2R_2$, $R_4 \to R_4 + 2R_2$,

$$\begin{bmatrix} 1 & -2 & -3 & \cdot & 2 \\ 0 & 1 & -7 & \cdot & 2 \\ 0 & 0 & 0 & \cdot & 0 \\ 0 & 0 & 0 & \cdot & 0 \end{bmatrix}.$$

This gives that $\text{rank}(A) = \text{rank}(\text{aug } A) = 2 < 3$, the number of unknowns. Hence, the given system of linear equations is consistent and has infinite number of solutions.
From the reduced row echelon form of the matrix aug A, the equivalent system is

$$x_1 - 2x_2 - 3x_3 = 2,$$
$$x_2 - 7x_3 = 2.$$

Put $x_3 = r$. Then $x_2 = 2 + 7r$ and $x_1 = 2 + 2x_2 + 3x_3 = 6 + 17r$.
Hence, the solution of the given system is

$$\begin{bmatrix} x_1 \\ x_2 \\ x_3 \end{bmatrix} = \begin{bmatrix} 6 + 17r \\ 2 + 7r \\ r \end{bmatrix} = \begin{bmatrix} 6 \\ 2 \\ 0 \end{bmatrix} + r \begin{bmatrix} 17 \\ 7 \\ 1 \end{bmatrix}.$$

Thus, all solutions of the given system will be of the form:

$$\left\{ \begin{bmatrix} 6 \\ 2 \\ 0 \end{bmatrix} + r \begin{bmatrix} 17 \\ 7 \\ 1 \end{bmatrix} : r \in \mathbb{R} \right\}.$$

Example 5.5.14. In this example, we solve the following homogeneous system of linear equations:

$$x + 2y + 3z = 0,$$
$$2x + 4y + 3z + 2t = 0,$$
$$3x + 2y + z + 3t = 0,$$
$$6x + 8y + 7z + 5t = 0.$$

In the matrix form, the system is

$$\begin{bmatrix} 1 & 2 & 3 & 0 \\ 2 & 4 & 3 & 2 \\ 3 & 2 & 1 & 3 \\ 6 & 8 & 7 & 5 \end{bmatrix} \begin{bmatrix} x \\ y \\ z \\ t \end{bmatrix} = \begin{bmatrix} 0 \\ 0 \\ 0 \\ 0 \end{bmatrix}.$$

Let

$$A = \begin{bmatrix} 1 & 2 & 3 & 0 \\ 2 & 4 & 3 & 2 \\ 3 & 2 & 1 & 3 \\ 6 & 8 & 7 & 5 \end{bmatrix}.$$

Now we find the rank of A by reducing into reduced row echelon form.

Applying elementary row operations $R_2 \rightarrow R_2 - 2R_1$, $R_3 \rightarrow R_3 - 3R_1$, $R_4 \rightarrow R_4 - 6R_1$, we have

$$\begin{bmatrix} 1 & 2 & 3 & 0 \\ 0 & 0 & -3 & 2 \\ 0 & -4 & -8 & 3 \\ 0 & -4 & -11 & 5 \end{bmatrix}.$$

From $R_2 \leftrightarrow R_3$,

$$\begin{bmatrix} 1 & 2 & 3 & 0 \\ 0 & -4 & -8 & 3 \\ 0 & 0 & -3 & 2 \\ 0 & -4 & -11 & 5 \end{bmatrix}.$$

From $R_2 \rightarrow \frac{-1}{4}R_2$,

$$\begin{bmatrix} 1 & 2 & 3 & 0 \\ 0 & 1 & 2 & \frac{-3}{4} \\ 0 & 0 & -3 & 2 \\ 0 & -4 & -11 & 5 \end{bmatrix}.$$

From $R_4 \rightarrow R_4 + 4R_2$,

$$\begin{bmatrix} 1 & 2 & 3 & 0 \\ 0 & 1 & 2 & \frac{-3}{4} \\ 0 & 0 & -3 & 2 \\ 0 & 0 & -3 & 2 \end{bmatrix}.$$

From $R_3 \rightarrow \frac{-1}{3}R_3$,

$$\begin{bmatrix} 1 & 2 & 3 & 0 \\ 0 & 1 & 2 & \frac{-3}{4} \\ 0 & 0 & 1 & \frac{-2}{3} \\ 0 & 0 & -3 & 2 \end{bmatrix}.$$

From $R_4 \rightarrow R_4 - 3R_3$,

$$\begin{bmatrix} 1 & 2 & 3 & 0 \\ 0 & 1 & 2 & \frac{-3}{4} \\ 0 & 0 & 1 & \frac{-2}{3} \\ 0 & 0 & 0 & 0 \end{bmatrix}.$$

From the above matrix, it is evident that rank$(A) = 3 < 4$, the number of variables. Hence, the given system of equations has infinite solutions in the form of $4 - 3 = 1$ parametric variable.

From the echelon form matrix, the equivalent system is

$$x + 2y + 3z = 0,$$

$$y + 2z - \frac{3}{4}t = 0,$$

$$z - \frac{2}{3}t = 0.$$

Put $t = r$. Then

$$z = \frac{2}{3}r,$$

$$y = -2z + \frac{3}{4}t = -2\left(\frac{2}{3}r\right) + \frac{3}{4}r = -\frac{7}{12}r,$$

$$x = -2y - 3z = -2\left(-\frac{7}{12}r\right) - 3\left(\frac{2}{3}r\right) = -\frac{5}{6}r.$$

Hence, the solution is of the form

$$\left\{ r \begin{bmatrix} \frac{-5}{6} \\ \frac{-7}{12} \\ \frac{2}{3} \\ 1 \end{bmatrix} : r \in \mathbb{R} \right\}.$$

Example 5.5.15. Find the values of a and b for which the system of linear equations

$$x + y + z = 2,$$
$$x + 2y + 5z = 10,$$
$$x + y + (a - 5)z = b - 5$$

has (i) no solution, (ii) unique solution and (iii) infinite solutions.

Rewrite the given system in matrix form $AX = B$:

$$\begin{bmatrix} 1 & 1 & 1 \\ 1 & 2 & 5 \\ 1 & 1 & a-5 \end{bmatrix} \begin{bmatrix} x \\ y \\ z \end{bmatrix} = \begin{bmatrix} 2 \\ 10 \\ b-10 \end{bmatrix}.$$

Now, we convert the augmented matrix

$$\text{aug } A = \begin{bmatrix} 1 & 1 & 1 & \cdot & 2 \\ 1 & 2 & 5 & \cdot & 10 \\ 1 & 1 & a-5 & \cdot & b-10 \end{bmatrix}$$

into the reduced row echelon form.

Through elementary row operations $R_2 \rightarrow R_2 - R_1$ and $R_3 \rightarrow R_3 - R_1$, the resulting matrix becomes

$$\begin{bmatrix} 1 & 1 & 1 & \cdot & 2 \\ 0 & 1 & 4 & \cdot & 8 \\ 0 & 0 & a-6 & \cdot & b-12 \end{bmatrix}.$$

(i) The given system has no solution if rank$(A) \neq$ rank$(\text{aug }A)$. This discrepancy arises only when rank rank$(A) = 2$ and rank$(\text{aug }A) = 3$. Thus, from the above matrix, we deduce $a - 6 = 0 \Rightarrow a = 6$ and $b - 12 \neq 0 \Rightarrow b \neq 12$. Consequently, the system has no solution when $a = 6$ and $b \neq 12$.

(ii) The given system has unique solution if rank$(A) =$ rank$(\text{aug }A) = 3$. Hence, from the above matrix we have $a - 6 \neq 0 \Rightarrow a \neq 6$.

(iii) The given system has infinite number of solutions if rank(A) = rank(aug A) < 3. This happens only if rank(A) = rank(aug A) = 2. Hence, from the above matrix, we conclude $a - 6 = 0 \Rightarrow a = 6$ and $b - 12 = 0 \Rightarrow b = 12$. Consequently, the system has infinite solutions when $a = 6$ and $b = 12$.

Example 5.5.16. Find the values of a for which the system of linear equations:

$$x + y + z = 1,$$
$$3x + 2y + 5z = a + 1,$$
$$3x + 9z = a^2 - 1$$

is consistent.

Rewrite the given system in matrix form $AX = B$:

$$\begin{bmatrix} 1 & 1 & 1 \\ 3 & 2 & 5 \\ 3 & 0 & 9 \end{bmatrix} \begin{bmatrix} x \\ y \\ z \end{bmatrix} = \begin{bmatrix} 1 \\ a + 1 \\ a^2 - 1 \end{bmatrix}.$$

Now, we convert the augmented matrix

$$\text{aug } A = \begin{bmatrix} 1 & 1 & 1 & \cdot & 1 \\ 3 & 2 & 5 & \cdot & a + 1 \\ 3 & 0 & 9 & \cdot & a^2 - 1 \end{bmatrix}$$

into the reduced row echelon form.

From $R_2 \to R_2 - 3R_1$ and $R_3 \to R_3 - 3R_1$,

$$\begin{bmatrix} 1 & 1 & 1 & \cdot & 1 \\ 0 & -1 & 2 & \cdot & a - 2 \\ 0 & -3 & 6 & \cdot & a^2 - 4 \end{bmatrix}.$$

From $R_2 \to (-1)R_2$,

$$\begin{bmatrix} 1 & 1 & 1 & \cdot & 1 \\ 0 & 1 & -2 & \cdot & -a + 2 \\ 0 & -3 & 6 & \cdot & a^2 - 4 \end{bmatrix}.$$

From $R_3 \to R_3 + 3R_2$,

$$\begin{bmatrix} 1 & 1 & 1 & \cdot & 1 \\ 0 & 1 & -2 & \cdot & -a + 2 \\ 0 & 0 & 0 & \cdot & a^2 - 3a + 2 \end{bmatrix}.$$

A linear system represented by $AX = B$ is consistent when the rank of the coefficient matrix A equals the rank of the augmented matrix aug A. Hence, from the above matrix it is clear that $\text{rank}(A) = \text{rank}(\text{aug } A)$ if and only if $\alpha^2 - 3\alpha + 2 = 0 \Rightarrow (\alpha - 1)(\alpha - 2) = 0$. Consequently, the system is consistent when α equals either 1 or 2.

ⓘ Exercises

(5.5.1) Solve the following system of linear equations:
(a)

$$x - 2y + 2z - t = -14,$$
$$8y - 7z + 5t = 59,$$
$$7y - 5z + t = 46,$$
$$-2x + 5y - 3z - 3t = 26.$$

(b)
$$x - y + z - t = 2,$$
$$x - y + z + t = 0,$$
$$2x - 2y + 2z + t = 1,$$
$$-2x + 2y - 2z + t = -3.$$

(5.5.2) Find the dimension and a basis for the solution space of the following system of linear equations:
(a)

$$x + 2y + 3z + t = 0,$$
$$-x - 2y + 5z - 2t = 0,$$
$$7x + 4y + 3z + 7t = 0.$$

(b)
$$2x + 4y - 3z + 2t - 5u = 0,$$
$$5x - 5y + 7z - 3t + 5u = 0,$$
$$-x + 5y - 3z + 4t + 7u = 0.$$

(5.5.3) Solve the following system of linear equations:
(a)

$$x_1 - x_2 + 2x_3 + x_4 = 3,$$
$$2x_1 - x_2 + 4x_3 - x_4 = 4,$$
$$2x_1 + x_2 - x_3 - x_4 = 1,$$
$$5x_1 + 2x_2 - 4x_4 = 3.$$

(b)
$$x_1 + 3x_2 - 5x_3 = -4,$$
$$2x_1 - x_2 + 3x_3 = 5,$$
$$7x_1 + 4x_3 = 11.$$

(5.5.4) Find the solution set for the following nonhomogeneous system of linear equations:

$$-x + y - z - 3u - v = 1,$$
$$-2x + 2y - z - 4u - 3v = 3,$$
$$2z + 4u - 2v = 2,$$
$$-x + y + z + u - 3v = 3.$$

(5.5.5) For what values of a and β the following system of linear equations have (i) no solution (ii) unique solution and (iii) infinite no of solutions:

(a) $x_1 + 2x_2 - x_3 = 1, x_1 + ax_2 + 3x_3 = 2, 2x_1 + 3x_2 + ax_3 = 3.$
(b) $x_1 + x_2 + x_3 = 3, 2x_1 + 2x_2 + (1 + 2a)x_3 = 10, x_1 + 2x_2 + 3ax_3 = \beta.$
(c) $x_1 + x_2 + 2x_3 = 3, x_1 + 2x_2 + ax_3 = 5, x_2 + 2x_3 = \beta - 3.$
(d) $x_1 + x_2 + ax_3 = 1, x_1 + ax_2 + x_3 = 1, ax_1 + x_2 + x_3 = 1.$

5.6 Miscellaneous

Example 5.6.1. Let A be a matrix of order $m \times n$. Then prove that $\mathrm{rank}(A) = \mathrm{rank}(A^t A) = \mathrm{rank}(AA^t).$

Solution. Consider the linear transformation $L_A : \mathbb{R}^n \to \mathbb{R}^m$ defined by $L_A(X) = AX$. Then we have $\mathrm{rank}(L_A) + \mathrm{null}(L_A) = n.$

Let $X \in \ker(L_A)$. Then $AX = 0 \Rightarrow A^t(AX) = 0 \Rightarrow X \in \ker(L_{A^t A}) \Rightarrow \ker(L_A) \subseteq \ker(L_{A^t A}).$

Conversely, suppose that $X \in \ker(L_{A^t A})$. Then

$$A^t AX = 0 \Rightarrow X^t A^t AX = 0 \Rightarrow (AX)^t (AX) = 0 \Rightarrow \|AX\|^2 = 0 \Rightarrow AX = 0$$
$$\Rightarrow X \in \ker(L_A) \Rightarrow \ker(L_{A^t A}) \subseteq \ker(L_A).$$

Thus, $\ker(L_{A^t A}) = \ker(L_A) \Rightarrow \dim(\ker(L_{A^t A})) = \dim(\ker(L_A))$. That is, $\mathrm{null}(L_A) = \mathrm{null}(L_{A^t A}).$

Additionally, we have

$$\mathrm{rank}(L_A) + \mathrm{null}(L_A) = n \quad \text{and} \quad \mathrm{rank}(L_{A^t A}) + \mathrm{null}(L_{A^t A}) = n.$$

From these equations, we deduce that $\mathrm{rank}(L_A) = \mathrm{rank}(L_{A^t A})$. Since $\mathrm{rank}(L_A) = \mathrm{rank}(A)$, it follows that $\mathrm{rank}(A) = \mathrm{rank}(A^t A)$. Replacing A with A^t in the above equation yields

$$\mathrm{rank}(A^t) = \mathrm{rank}(AA^t).$$

Thus, from the result that $\mathrm{rank}(A) = \mathrm{rank}(A^t)$, we conclude

$$\mathrm{rank}(A) = \mathrm{rank}(A^t A) = \mathrm{rank}(AA^t).$$

Example 5.6.2. Let A be a real matrix of order 6×5 such that $AX = 0$ if and only if $X = 0$, where X is a 5×1 vector and 0 denotes the null vector. Determine the rank of A.

Solution. Let $L_A : \mathbb{R}^5 \rightarrow \mathbb{R}^6$ be the linear transformation defined by $L_A(X) = AX$. Then we know that $\text{rank}(L_A) + \text{null}(L_A) = 5$. Since $AX = 0$, implying $L_A(X) = 0$, we have $\text{null}(L_A) = 0$. Therefore, $\text{rank}(L_A) = 5$. As A corresponds to the matrix representing the linear transformation L_A with respect to the standard bases of $\mathbb{R}^5$ and $\mathbb{R}^6$, it follows that $\text{rank}(A) = 5$.

Example 5.6.3. Let A be a matrix of order 30×70. If $\text{rank}(A) = 15$, then find the dimension of the solution space of $AX = 0$.

Solution. Let $L_A : \mathbb{R}^{70} \rightarrow \mathbb{R}^{30}$ be the linear transformation defined by $L_A(X) = AX$. Since L_A represents the matrix A, $\text{rank}(A) = \text{rank}(L_A) = 15$.

We have $\text{rank}(L_A) + \text{null}(L_A) = 70 \Rightarrow 15 + \text{null}(L_A) = 70 \Rightarrow \text{null}(L_A) = 55$. Hence, the dimension of the solution space of $AX = 0$ is 55.

Sylvester rank inequality

Let A and B be matrices of order $m \times p$ and $p \times n$, respectively. Then the Sylvester rank inequality states

$$\text{rank}(AB) \geq \text{rank}(A) + \text{rank}(B) - p.$$

If A and B are n-square matrices, then the inequality becomes

$$\text{rank}(AB) \geq \text{rank}(A) + \text{rank}(B) - n.$$

For an n-square matrix A, we have

$$\text{rank}(A^2) \geq \text{rank}(A) + \text{rank}(A) - n = 2\ \text{rank}(A) - n,$$

which simplifies to

$$\text{rank}(A^2) \geq 2\ \text{rank}(A) - n.$$

Similarly, for A^3:

$$\text{rank}(A^3) \geq \text{rank}(A^2) + \text{rank}(A) - n$$
$$\geq 2\,\text{rank}(A) - n + \text{rank}(A) - n$$
$$= 3\,\text{rank}(A) - 2n,$$

i. e., $\text{rank}(A^3) \geq 3\ \text{rank}(A) - 2n$.

This can be generalized to

$$\text{rank}(A^k) \geq k\ \text{rank}(A) - (k-1)n.$$

Example 5.6.4. Let A and B be n-square real matrices such that $AB = BA = 0$, and $A + B$ is invertible. Then show that (a) $\text{rank}(A) + \text{rank}(B) = n$ and (b) $\text{null}(A) + \text{null}(B) = n$.

Solution.

(a) Given $AB = 0$, it implies $\text{rank}(AB) = 0$. According to the Sylvester rank inequality, which states $\text{rank}(A) + \text{rank}(B) - n \leq \text{rank}(AB)$, we get

$$\text{rank}(A) + \text{rank}(B) - n \leq 0 \Rightarrow \text{rank}(A) + \text{rank}(B) \leq n.$$

Additionally, since $A + B$ is invertible, $\text{rank}(A + B) = n$. By the property that $\text{rank}(A + B) \leq \text{rank}(A) + \text{rank}(B)$, we obtain

$$n \leq \text{rank}(A) + \text{rank}(B).$$

Thus, combining these inequalities, we conclude

$$n \leq \text{rank}(A) + \text{rank}(B) \leq n \Rightarrow \text{rank}(A) + \text{rank}(B) = n.$$

(b) Using the results $\text{null}(A) = n - \text{rank}(A)$ and $\text{null}(B) = n - \text{rank}(B)$, we have

$$\text{null}(A) + \text{null}(B) = 2n - (\text{rank}(A) + \text{rank}(B)) = 2n - n = n.$$

Example 5.6.5. Consider real matrices A of order 3×5 and B of order 5×3. If $AB = I_3$, where I_3 is the 3-square identity matrix, determine the ranks of A and B.

Solution. Define the corresponding linear transformations $L_A : \mathbb{R}^5 \to \mathbb{R}^3$ and $L_B : \mathbb{R}^3 \to \mathbb{R}^5$.

Given $AB = I_3$, it implies $L_A \circ L_B = I$, where I denotes the identity linear transformation.

Since $L_A \circ L_B$ is bijective, L_B is injective (one-to-one) and L_A is surjective (onto). Consequently, the nullity of L_B is zero, and $\text{rank}(L_A) = 3$.

As $\text{null}(L_B) = 0$, we deduce that $\text{rank}(L_B) = 3 - 0 = 3$.

Therefore, $\text{rank}(A) = 3 = \text{rank}(B)$.

Example 5.6.6. Consider a 7-square real matrix A, and suppose the dimension of the solution space of $AX = 0$ is at least 3. Show that $\text{rank}(A^2) \leq 4$.

Solution. Let $L_A : \mathbb{R}^7 \to \mathbb{R}^7$ be the corresponding linear transformation defined as $L_A X = AX$.

Given that the dimension of the solution space of $AX = 0$ is at least 3, we have $\text{nullity}(L_A) \geq 3$. Hence, $\text{rank}(L_A) \leq 4$, and consequently, $\text{rank}(A) \leq 4$.

Using the result $\text{rank}(AB) \leq \min\{\text{rank}(A), \text{rank}(B)\}$, we have $\text{rank}(A^2) \leq \text{rank}(A) \leq 4$.

Example 5.6.7. In this example, we introduce another method using the rank of a matrix to determine whether a set of vectors is linearly dependent or linearly independent.

Consider the set of vectors $(2, 1, 1), (1, -1, 1), (3, 0, 2)$ in $\mathbb{R}^3$, as given in Example 2.3.5.

Let $a_1(2, 1, 1) + a_2(1, -1, 1) + a_3(3, 0, 2) = (0, 0, 0)$, for some scalars $a_1, a_2, a_3 \in \mathbb{R}$. Then

$$(2a_1 + a_2 + 3a_3, a_1 - a_2, a_1 + a_2 + 2a_3) = (0, 0, 0)$$
$$\Rightarrow 2a_1 + a_2 + 3a_3 = 0, a_1 - a_2 = 0, \text{and } a_1 + a_2 + 2a_3 = 0.$$

From the definition of linearly independent vectors, the given set of vectors will be linearly independent if and only if $a_1 = a_2 = a_3 = 0$. That is, if the above system of linear equations has a trivial solution, then the set of vectors will be linearly independent. Similarly, if the system of equations has a nontrivial solution, then the given set of vectors will be linearly dependent.

Let

$$A = \begin{bmatrix} 2 & 1 & 3 \\ 1 & -1 & 0 \\ 1 & 1 & 2 \end{bmatrix}$$

be the coefficient matrix of the system of linear equations.

If rank(A) = 3 (the number of unknowns or the number of vectors), then the system of linear equations has trivial solutions, and hence the given set of vectors will be linearly independent.

If rank(A) < 3 (the number of unknowns or the number of vectors), then the system of linear equations has nontrivial solutions, and hence the given set of vectors will be linearly dependent.

Reducing A into row echelon form, we find that rank(A) = 2 < 3. Hence, the given set of vectors is linearly dependent.

We observe that the column vectors of the coefficient matrix A are the same as the given vectors.

In view of the above illustrations, we have the following method to test the linear dependence and linear independence of a set of vectors.

Method:

(1) Construct a matrix A using given vectors as column vectors of A.
(2) Find rank(A).
(3) If rank(A) = the numbers of vectors, then the given set of vectors is linearly independent.
(4) If rank(A) < the numbers of vectors, then the given set of vectors is linearly dependent.

Example 5.6.8. Determine whether the set of vectors

$$\{(1, -1, 0, 2), (0, 1, 1, -1), (2, 1, 2, 1), (3, -2, 1, 6)\}$$

is linearly dependent or linearly independent.

Solution. Consider the matrix of vectors

$$A = \begin{bmatrix} 1 & 0 & 2 & 3 \\ -1 & 1 & 1 & -2 \\ 0 & 1 & 2 & 1 \\ 2 & -1 & 1 & 6 \end{bmatrix}.$$

Reducing matrix A into row echelon form yields rank$(A) = 4$, which is equal to the number of vectors. Thus, the given set of vectors is linearly independent.

Example 5.6.9. Test whether the set of vectors

$$\{(1, -1, 2, -1), (3, 1, 0, 1), (1, 1, -1, 1)\}$$

is linearly dependent or linearly independent.

Solution. Consider the matrix of vectors:

$$A = \begin{bmatrix} 1 & 3 & 1 \\ -1 & 1 & 1 \\ 2 & 0 & -1 \\ -1 & 1 & 1 \end{bmatrix}.$$

Reducing A into row echelon form we get that rank$(A) = 2$, which is less than 3, the number of vectors. Hence, the given set of vectors is linearly dependent.

Example 5.6.10. Let $U = \{(x, y, z, u) \in \mathbb{R}^4 : x + y + z = 0, 2x + 2z - u = 0, 2y + u = 0\}$ and $V = \{(x, y, z, u) \in \mathbb{R}^4 : x + z - 2u = 0, x + z + u = 0, y - u = 0\}$. Determine dim$(U)$, dim$(V)$ and dim$(U + V)$.

Solution. For U, the dimension dim(U) is equivalent to the number of linearly independent solutions of the system of equations $x + y + z = 0, 2x + 2z - u = 0, 2y + u = 0$. The coefficient matrix A for this system is

$$\begin{bmatrix} 1 & 0 & 1 & 0 \\ 2 & 0 & 2 & -1 \\ 0 & 2 & 0 & 1 \end{bmatrix}.$$

As rank$(A) = 2$, we have dim$(U) = 4 -$ rank$(A) = 4 - 2 = 2$.

Similarly, for V, the dimension dim(V) is calculated using the coefficient matrix B:

$$B = \begin{bmatrix} 1 & 0 & 1 & -2 \\ 1 & 0 & 1 & 1 \\ 0 & 1 & 0 & -1 \end{bmatrix}.$$

With rank$(B) = 3$, we find dim$(V) = 4 -$ rank$(B) = 4 - 3 = 1$.

To find $\dim(U+V)$, we use the formula $\dim(U+V) = \dim(U) + \dim(V) - \dim(U \cap V)$. The intersection $U \cap V$ consists of solutions common to both sets of equations, i. e., $U \cap V = \{(x,y,z,u) \in \mathbb{R}^4 : x+y+z = 0, 2x+2z-u = 0, 2y+u = 0, x+z-2u = 0, x+z+u = 0, y-u = 0\}$. The coefficient matrix C for this system is

$$C = \begin{bmatrix} 1 & 0 & 1 & 0 \\ 2 & 0 & 2 & -1 \\ 0 & 2 & 0 & 1 \\ 1 & 0 & 1 & -2 \\ 1 & 0 & 1 & 1 \\ 0 & 1 & 0 & -1 \end{bmatrix}.$$

The rank of C is 3, implying $\dim(U \cap V) = 4 - \operatorname{rank}(C) = 4 - 3 = 1$. Therefore, $\dim(U+V) = 2 + 1 - 1 = 2$.

6 Inner product spaces

We have observed that the notion of a vector space extends the linear characteristics of $\mathbb{R}^n$. Throughout this exploration, we employed the operations of addition and scalar multiplication from $\mathbb{R}^n$, yet we overlooked other significant aspects such as vector length and the angle between vectors. To address this, an inner product is introduced on a vector space, drawing from the properties of the dot product in $\mathbb{R}^n$. Throughout this chapter, all vector spaces are defined over either the real numbers or the complex numbers.

6.1 Inner products and norms

Definition 6.1.1. An inner product on a vector space V over the field F ($F = \mathbb{R}$ or $F = \mathbb{C}$) is a mapping $\langle\,\rangle : V \times V \to F$ that satisfies the following properties:

(i) $\langle a_1 v_1 + a_2 v_2, v_3 \rangle = a_1 \langle v_1, v_3 \rangle + a_2 \langle v_2, v_3 \rangle$, where $a_1, a_2 \in F$ and $v_1, v_2, v_3 \in V$.

(ii) $\langle v_1, v_2 \rangle = \overline{\langle v_2, v_1 \rangle}$, the complex conjugate of $\langle v_1, v_2 \rangle$, for all $v_1, v_2, v_3 \in V$.

(iii) $\langle v_1, v_1 \rangle \geq 0$ and $\langle v_1, v_1 \rangle = 0 \Leftrightarrow v_1 = 0$.

A vector space V equipped with an inner product is called an inner product space. If $F = \mathbb{R}$, V is referred to as a real inner product space, and if $F = \mathbb{C}$, it is a complex inner product space.

Note.

(1) The symbol $\langle u, v \rangle$ represents the image of $(u, v) \in V \times V$ under the mapping $\langle\,\rangle$.

(2) The property $\langle v_1, v_2 \rangle = \overline{\langle v_2, v_1 \rangle}$ implies that $\langle v_1, v_1 \rangle = \overline{\langle v_1, v_1 \rangle}$. Therefore, $\langle v_1, v_1 \rangle$ yields a real number for all $v_1 \in V$, allowing for meaningful interpretation of the inequality in property (iii).

(3) Property (i) implies that if $a_1 = a_2 = 1$, then $\langle v_1 + v_2, v_3 \rangle = \langle v_1, v_3 \rangle + \langle v_2, v_3 \rangle$, and if $a_1 = 1$ and $a_2 = 0$, then $\langle a_1 v_1, v_3 \rangle = a_1 \langle v_1, v_3 \rangle$ for all $v_1, v_2, v_3 \in V$.

(4) For $v, v_1, v_2 \in V$ and $a_1, a_2 \in F$,

$$\langle v, a_1 v_1 + a_2 v_2 \rangle = \overline{\langle a_1 v_1 + a_2 v_2, v \rangle} = \overline{a_1 \langle v_1, v \rangle + a_2 \langle v_2, v \rangle} = \overline{a_1 \langle v_1, v \rangle} + \overline{a_2 \langle v_2, v \rangle}$$

$$= \bar{a}_1 \overline{\langle v_1, v \rangle} + \bar{a}_2 \overline{\langle v_2, v \rangle} = \bar{a}_1 \langle v, v_1 \rangle + \bar{a}_2 \langle v, v_2 \rangle.$$

In particular, $\langle v, v_1 + v_2 \rangle = \langle v, v_1 \rangle + \langle v, v_2 \rangle$ and $\langle v, a_1 v_1 \rangle = \bar{a}_1 \langle v, v_1 \rangle$.

Further, $\langle 0, v \rangle = \langle 0.0, v \rangle = 0 \langle 0, v \rangle = 0$ and similarly, $\langle v, 0 \rangle = \overline{\langle 0, v \rangle} = 0, \forall v \subset V$.

(5) Every subspace of an inner product space is itself an inner product space.

The inner products described in the subsequent example are known as standard inner products.

https://doi.org/10.1515/9783111516035-006

Example 6.1.2.

(a) In $\mathbb{R}^n$, the inner product $\langle x, y \rangle = x_1 y_1 + x_2 y_2 + \cdots + x_n y_n$ is defined for $x = (x_1, x_2, \ldots, x_n)$ and $y = (y_1, y_2, \ldots, y_n)$. This inner product is known as the Euclidean inner product and the pair $(\mathbb{R}^n, \langle \rangle)$ is called the Euclidean space.

Similarly, in $\mathbb{C}^n$, the inner product is given by $\langle z, v \rangle = z_1 \bar{v}_1 + z_2 \bar{v}_2 + \cdots + z_n \bar{v}_n$, where $z = (z_1, z_2, \ldots, z_n)$ and $v = (v_1, v_2, \ldots, v_n)$ belong to $\mathbb{C}^n$. Here, $(\mathbb{C}^n, \langle \rangle)$ forms the standard unitary space.

It is worth noting that $\langle z, v \rangle = v^* z$, where $z, v \in \mathbb{C}^n$ represent column vectors ($n \times 1$ matrices). In case $z, v \in \mathbb{R}^n$, the inner product $\langle z, v \rangle = v^* z = v^t z$. Both of these inner products are commonly referred to as dot products.

(b) In the vector space $C(\mathbb{R}^{[a,b]})$, consisting of all continuous real valued functions on the closed interval $[a, b]$, the inner product $\langle f, g \rangle = \int_a^b f(x)g(x)dx$ is defined for $f, g \in C(\mathbb{R}^{[a,b]})$. Similarly, in the vector space $C(\mathbb{C}^{[a,b]})$ comprising continuous complex-valued functions on the interval $[a, b] \subset \mathbb{R}$, the inner product is given by $\langle f, g \rangle = \int_a^b f(x)\overline{g(x)}dx$, where $f, g \in C(\mathbb{C}^{[a,b]})$.

The vector space of polynomials $P(x)$ over $\mathbb{R}$ is a subspace of $C(\mathbb{R}^{[a,b]})$, while the vector space of polynomials $P(x)$ over $\mathbb{C}$ is a subspace of $C(\mathbb{C}^{[a,b]})$. Thus, both are considered inner product spaces.

(c) For the vector space $M_{m,n}(\mathbb{R})$, which consists of all $m \times n$ matrices over $\mathbb{R}$, the inner product $\langle A, B \rangle = \text{tr}(B^t A)$ is defined for $A, B \in M_{m,n}(\mathbb{R})$. Similarly, in the vector space $M_{m,n}(\mathbb{C})$, the inner product is given by $\langle A, B \rangle = \text{tr}(B^* A)$ for $A, B \in M_{m,n}(\mathbb{C})$.

It is important to note that on a vector space V, more than one inner product can be defined. The inner product spaces mentioned above are considered standard inner product spaces unless otherwise specified.

Remark.

(1) Consider the result: $\mathbb{R}^{m \times n} \cong M_{m,n}(\mathbb{R})$ implies $\mathbb{R}^m \cong M_{m,1}(\mathbb{R})$.

Let

$$X = \begin{bmatrix} x_1 \\ x_2 \\ \vdots \\ x_m \end{bmatrix} \quad \text{and} \quad Y = \begin{bmatrix} y_1 \\ y_2 \\ \vdots \\ y_m \end{bmatrix}$$

be column vectors in $\mathbb{R}^m$ or equivalently matrices of order $m \times 1$ in $M_{m,1}(\mathbb{R})$. Then $\langle X, Y \rangle = Y^t X = \text{tr}(Y^t X)$. It follows from below:

$$\text{tr}(Y^t X) = \text{tr}\left(\begin{bmatrix} y_1 & y_2 & \cdots & y_m \end{bmatrix} \begin{bmatrix} x_1 \\ x_2 \\ \vdots \\ x_m \end{bmatrix} \right)$$

$$= \text{tr}([y_1 x_1 + y_2 x_2 + \cdots + y_m x_m])$$

$$= y_1 x_1 + y_2 x_2 + \cdots + y_m x_m$$
$$= x_1 y_1 + x_2 y_2 + \cdots + x_m y_m$$
$$= Y^t X.$$

Similarly, if X and Y are in $\mathbb{C}^m$, then the standard inner product $\langle X, Y \rangle = x_1 \bar{y}_1 + x_2 \bar{y}_2 + \cdots + x_m \bar{y}_m = Y^* X = \text{tr}(Y^* X)$.

Let

$$A = \begin{bmatrix} a_{11} & a_{12} & a_{13} \\ a_{21} & a_{22} & a_{23} \end{bmatrix} \quad \text{and} \quad B = \begin{bmatrix} b_{11} & b_{12} & b_{13} \\ b_{21} & b_{22} & b_{23} \end{bmatrix},$$

be two real matrices of order 2×3. Then

$$B^t A = \begin{bmatrix} b_{11} & b_{21} \\ b_{12} & b_{22} \\ b_{13} & b_{23} \end{bmatrix} \begin{bmatrix} a_{11} & a_{12} & a_{13} \\ a_{21} & a_{22} & a_{23} \end{bmatrix}$$

$$= \begin{bmatrix} b_{11}a_{11} + b_{21}a_{21} & - & - \\ - & b_{12}a_{12} + b_{22}a_{22} & - \\ - & - & b_{13}a_{13} + b_{23}a_{23} \end{bmatrix}.$$

Hence,

$$\text{tr}(B^t A) = b_{11}a_{11} + b_{21}a_{21} + b_{12}a_{12} + b_{22}a_{22} + b_{13}a_{13} + b_{23}a_{23}$$
$$= a_{11}b_{11} + a_{12}b_{12} + a_{13}b_{13} + a_{21}b_{21} + a_{22}b_{22} + a_{23}b_{23}.$$

It is shown that $M_{m,n}(\mathbb{R}) \cong \mathbb{R}^{m \times n}$. Hence, $M_{2,3}(\mathbb{R}) \cong \mathbb{R}^{2 \times 3} = \mathbb{R}^6$ with the isomorphism $A \leftrightarrow (a_{11}, a_{12}, a_{13}, a_{21}, a_{22}, a_{23})$. Suppose $X = (a_{11}, a_{12}, a_{13}, a_{21}, a_{22}, a_{23})$ and $Y = (b_{11}, b_{12}, b_{13}, b_{21}, b_{22}, b_{23})$ are column vectors in $\mathbb{R}^6$. Then $\langle X, Y \rangle = Y^t X = a_{11}b_{11} + a_{12}b_{12} + a_{13}b_{13} + a_{21}b_{21} + a_{22}b_{22} + a_{23}b_{23} = \langle A, B \rangle = \text{tr}(B^t A)$.

Thus, the inner product defined in $M_{m,n}(\mathbb{R})$ is nothing but the dot product defined in $\mathbb{R}^{m \times n}$. Similarly, the inner product defined in $M_{m,n}(\mathbb{C})$ is same as dot product defined in $\mathbb{C}^{m \times n}$.

(2) Throughout, the vectors in $\mathbb{R}^m$ or $\mathbb{C}^m$ we consider as column matrices of order $m \times 1$ and call them as column vectors.

Example 6.1.3. Let $V = \{\{x_n\} \in \mathbb{R}^N : \sum_{n=1}^{\infty} x_n^2 < \infty\}$, that is, V is a vector space of squares summable sequences. Then V is an inner product space with the inner product defined by $\langle \{a_n\}, \{b_n\} \rangle = \sum a_n b_n$, where $\{a_n\}, \{b_n\} \in V$. This inner product space is called the l^2-space.

Definition 6.1.4. Let V be an inner product space. Then for any $v \in V$, the norm or length of v denoted by $\|v\|$ is defined $\|v\| = +\sqrt{\langle v, v \rangle}$.

Note that $\langle v, v \rangle$ is a nonnegative real number for all $v \in V$. Hence, $\|v\|$ is a nonnegative real number. Also, the function $v \to \|v\|$ is called norm function from V to $\mathbb{R}$.

Example 6.1.5.

(a) Let $x = (x_1, x_2, \ldots, x_n) \in \mathbb{R}^n$. Then

$$\|x\| = +\sqrt{\langle x, x \rangle} = \sqrt{x_1^2 + x_2^2 + \cdots + x_n^2}.$$

For $z = (z_1, z_2, \ldots, z_n) \in \mathbb{C}^n$,

$$\|z\| = +\sqrt{\langle z, z \rangle} = \sqrt{z_1 \bar{z}_1 + z_2 \bar{z}_2 + \cdots + z_n \bar{z}_n}$$
$$= \sqrt{|z_1|^2 + |z_2|^2 + \cdots + |z_n|^2}.$$

(b) Let $f \in C(\mathbb{R}^{[a,b]})$. Then

$$\|f\| = \sqrt{\langle f, f \rangle} = \sqrt{\int_a^b f(x)f(x)dx} = \sqrt{\int_a^b (f(x))^2 dx},$$

and for $f \in C(\mathbb{C}^{[a,b]})$,

$$\|f\| = \sqrt{\langle f, f \rangle} = \sqrt{\int_a^b f(x)\overline{f(x)}dx} = \sqrt{\int_a^b |f(x)|^2 dx}.$$

(c) Let $A = [a_{ij}] \in M_{m,n}(\mathbb{R})$. Then

$$\|A\| = \sqrt{\langle A, A \rangle} = \sqrt{\operatorname{tr}(A^t A)} = \sqrt{\sum_{i,j} a_{ij}^2}.$$

Similarly, for $A = [a_{ij}] \in M_{m,n}(\mathbb{C})$,

$$\|A\| = \sqrt{\operatorname{tr}(A^* A)} = \sqrt{\sum_{i,j} |a_{ij}|^2}.$$

Theorem 6.1.6 (Cauchy–Schwarz inequality). *Let V be an inner product space. Then for any $u, v \in V$, $|\langle u, v \rangle| \le \|u\|\|v\|$. Further, equality holds if and only if u and v are linearly dependent.*

Proof. If $u = 0$, then the result is true for all $v \in V$. Now, we prove the inequality for $u \ne 0$. From the definition of inner product, we have $\langle au + v, au + v \rangle \ge 0$, for all $a \in F$ and $v \in V$. Then,

$$a\langle u, au + v \rangle + \langle v, au + v \rangle \ge 0,$$

$$a\bar{a}\langle u, u\rangle + a\langle u, v\rangle + \bar{a}\langle v, u\rangle + \langle v, v\rangle \geq 0.$$

The above inequality is true for all $a \in F$. Take $a = -\frac{\langle v, u\rangle}{\|u\|^2}$. Then

$$\frac{\langle v, u\rangle}{\|u\|^2}\frac{\overline{\langle v, u\rangle}}{\|u\|^2}\|u\|^2 - \frac{\langle v, u\rangle}{\|u\|^2}\langle u, v\rangle - \frac{\overline{\langle v, u\rangle}}{\|u\|^2}\langle v, u\rangle + \|v\|^2 \geq 0, \quad \text{or}$$

$$\langle v, u\rangle\frac{\overline{\langle v, u\rangle}}{\|u\|^2} - \frac{\langle v, u\rangle}{\|u\|^2}\langle u, v\rangle - \frac{\overline{\langle v, u\rangle}}{\|u\|^2}\langle v, u\rangle + \|v\|^2 \geq 0, \quad \text{or}$$

$$-\frac{\langle v, u\rangle}{\|u\|^2}\langle u, v\rangle + \|v\|^2 \geq 0, \quad \text{or}$$

$$-\frac{\overline{\langle u, v\rangle}}{\|u\|^2}\langle u, v\rangle + \|v\|^2 \geq 0, \quad \text{or}$$

$$-\frac{|\langle u, v\rangle|^2}{\|u\|^2} + \|v\|^2 \geq 0, \quad \text{or}$$

$$\|v\|^2 \geq \frac{|\langle u, v\rangle|^2}{\|u\|^2}, \quad \text{or}$$

$$\|u\|^2\|v\|^2 \geq |\langle u, v\rangle|^2.$$

Hence, $|\langle u, v\rangle| \leq \|u\|\|v\|$.

To prove the remaining part, suppose $|\langle u, v\rangle| = \|u\|\|v\|$. Now, we shall show that $\{u, v\}$ are linearly dependent. That is, $u = av$, $a \in F$. Suppose if possible, $\{u, v\}$ are linearly independent. Then $u + av \neq 0$, $\forall a \in F$. Hence, from the steps of the proof we have $\langle u + av, u + av\rangle \geq 0$, $\forall a \in F$. This implies $|\langle u, v\rangle| \leq \|u\|\|v\|$, which is a contradiction to our assumption that $|\langle u, v\rangle| = \|u\|\|v\|$. Hence, $\{u, v\}$ are linearly dependent.

Conversely, suppose that $\{u, v\}$ are linearly dependent. Then $u = av$, for some $a \in F$ and so

$$|\langle u, v\rangle| = |\langle av, v\rangle| = |a\langle v, v\rangle| = |a\|v\|^2| = |a|\|v\|^2$$
$$= |a|\|v\|\|v\| = \|av\|\|v\| = \|u\|\|v\|. \qquad \square$$

Theorem 6.1.7. *Let V be an inner product space. Then the norm function $\|\| : V \to \mathbb{R}$ satisfies the following conditions:*
(i) $\|v\| \geq 0$ *and* $\|v\| = 0$ *if and only if* $v = 0$;
(ii) $\|av\| = |a|\|v\|$ $\forall a \in F$;
(iii) $\|u + v\| \leq \|u\| + \|v\|$, *called the triangle inequality.*

Proof. (i) $\|v\| = +\sqrt{\langle v, v\rangle}$. From the definition of inner product, we have $\langle v, v\rangle \geq 0$ and $\langle v, v\rangle = 0 \Leftrightarrow v = 0$. Hence, the result is proved.
(ii) We have $\|av\|^2 = \langle av, av\rangle = a\bar{a}\langle v, v\rangle = |a|^2\|v\|^2$, which implies $\|av\| = |a|\|v\|$.

(iii)

$$\|u + v\|^2 = \langle u + v, u + v \rangle$$
$$= \langle u, u \rangle + \langle u, v \rangle + \langle v, u \rangle + \langle v, v \rangle$$
$$= \|u\|^2 + \langle u, v \rangle + \overline{\langle u, v \rangle} + \|v\|^2$$
$$= \|u\|^2 + 2R\langle u, v \rangle + \|v\|^2.$$

Since, if $z \in \mathbb{C}$, $z + \bar{z} = 2\times$ real part of z. $R(\langle u, v \rangle)$ denotes the real part of $\langle u, v \rangle$. Now, from the property $R(z) \le |z|$ we have

$$\|u + v\|^2 \le \|u\|^2 + 2|\langle u, v \rangle| + \|v\|^2$$
$$\le \|u\|^2 + 2\|u\|\|v\| + \|v\|^2 = (\|u\| + \|v\|)^2.$$

Hence, $\|u + v\| \le \|u\| + \|v\|$.

$\square$

Corollary 6.1.8. *If v is an inner product space, then for $u, v \in V$, $\|\|u\| - \|v\|\| \le \|u - v\|$.*

Proof. $\|u\| = \|(u - v) + v\| \le \|u - v\| + \|v\|$, by triangle inequality.
Hence, $\|u\| - \|v\| \le \|u - v\|$.
Interchanging u and v in above inequality, we get

$$\|v\| - \|u\| \le \|v - u\| = \|(-1)(u - v)\| = |-1|\|(u - v)\| = \|u - v\|.$$

This shows that $\pm(\|u\| - \|v\|) \le \|u - v\|$. Hence, $\|\|u\| - \|v\|\| \le \|u - v\|$.

$\square$

Theorem 6.1.9 (Parallelogram law). *Let V be an inner product space. Then $\|u - v\|^2 + \|u + v\|^2 = 2(\|u\|^2 + \|v\|^2)$ for all $u, v \in V$.*

Proof.

$$\|u - v\|^2 = \langle u - v, u - v \rangle = \langle u, u \rangle - \langle v, u \rangle - \langle u, v \rangle + \langle v, v \rangle \quad \text{and}$$
$$\|u + v\|^2 = \langle u + v, u + v \rangle = \langle u, u \rangle + \langle v, u \rangle + \langle u, v \rangle + \langle v, v \rangle.$$

Adding above two equations, we get

$$\|u - v\|^2 + \|u + v\|^2 = 2\langle u, u \rangle + 2\langle v, v \rangle \quad \text{or,}$$
$$\|u - v\|^2 + \|u + v\|^2 = 2(\|u\|^2 + \|v\|^2).$$

Thus, the parallelogram law states that the sum of areas formed by lengths of the four sides of a parallelogram equals the sum of areas formed by diagonals.

$\square$

Example 6.1.10. Let $x = (a_1, a_2, \ldots, a_n)$ and $y = (\frac{1}{a_1}, \frac{1}{a_2}, \ldots, \frac{1}{a_n})$, where $a_i \ne 0$ $\forall 1 \le i \le n$ be in $\mathbb{R}^n$. Then by the Cauchy–Schwarz inequality,

$$|\langle x,y\rangle| \le \|x\|\|y\|,$$

$$a_1\frac{1}{a_1} + a_2\frac{1}{a_2} + \cdots + a_n\frac{1}{a_n} \le \sqrt{a_1^2 + a_2^2 + \cdots + a_n^2}\sqrt{\frac{1}{a_1^2} + \frac{1}{a_2^2} + \cdots + \frac{1}{a_n^2}},$$

$$|n| \le \sqrt{\sum_{i=1}^{n} a_i^2}\sqrt{\sum_{i=1}^{n} \frac{1}{a_i^2}},$$

$$n^2 \le \sum_{i=1}^{n} a_i^2 \sum_{i=1}^{n} \frac{1}{a_i^2}.$$

In particular, for $a_1, a_2 \ne 0$, $(a_1^2 + a_2^2)(\frac{1}{a_1^2} + \frac{1}{a_2^2}) \ge 4$.

Further, if $x = (\sqrt{a_1}, \sqrt{a_2}, \ldots, \sqrt{a_n})$ and $y = (\frac{1}{\sqrt{a_1}}, \frac{1}{\sqrt{a_2}}, \ldots, \frac{1}{\sqrt{a_n}})$, where $a_i \rangle 0 \ \forall i$, then $n^2 \le \sum_{i=1}^{n} a_i \sum_{i=1}^{n} \frac{1}{a_i}$. In particular, for $a_1, a_2 > 0$, $4 \le (a_1 + a_2)(\frac{1}{a_1} + \frac{1}{a_2})$.

Below, we recall the definition of a metric space and then we shall show that every inner product space is a metric space.

Definition 6.1.11. Let X be a set. Then a function $d : X \times X \to \mathbb{R}$ is called metric or distance on X if it satisfies the following conditions:
(i) $d(x,y) \ge 0$ and $d(x,y) = 0 \Leftrightarrow x = y \ \forall x, y \in X$.
(ii) $d(x,y) = d(y,x) \ \forall x, y \in X$.
(iii) $d(x,z) \le d(x,y) + d(y,z) \ \forall x, y, z \in X$.

The set X with the metric d denoted by (X, d) is called a metric space.

Theorem 6.1.12. *Let V be an inner product space. Then (V, d) is a metric space with the metric d defined by $d(x,y) = \|x - y\|$, where $(x,y) \in V \times V$.*

Proof. (i) $d(x,y) = \|x - y\| \ge 0$, by the property of norm.
       Also, $d(x,y) = 0 \Leftrightarrow \|x - y\| = 0 \Leftrightarrow x - y = 0 \Leftrightarrow x = y$.
(ii) $d(x,y) = \|x - y\| = \|(-1)(y - x)\| = |-1|\|y - x\| = \|y - x\| = d(y,x)$.
(iii) $d(x,z) = \|x - z\| = \|(x - y) + (y - z)\| \le \|x - y\| + \|y - z\| = d(x,y) + d(y,z)$.

Hence, $d(x,z) \le d(x,y) + d(y,z) \ \forall x, y, z \in X$.   □

Thus, an inner product space is a metric space but a metric space may not be a vector space, which is clear from the following example.

Example 6.1.13.
(a) Let $X = \{1, 2, 3, 4, 5, 6\}$. Then the map $d : X \times X \to \mathbb{R}$, defined by

$$d(x,y) = \begin{cases} 0, & \text{if } x = y, \\ 1, & \text{if } x \ne y \end{cases}$$

is a metric space but X is not a vector space, and hence not a normed space.

(b) The set $\{x \in \mathbb{R}^n : \|x\| = 1\}$ being a subset of a metric space $\mathbb{R}^n$ is a metric space but not a vector space, hence not a normed space.

Thus, the concept of a metric space is a generalization of the normed vector space.

Example 6.1.14. Let $v = (1+i, 1-i, 2i) \in \mathbb{C}^3$. If $\|av\| = 1$ for $a \in \mathbb{C}$, then $|a|\|v\| = 1 \Rightarrow |a| = \frac{1}{\|v\|} = \frac{1}{\sqrt{(1+i,1-i,2i,1+i,1-i,2i)}} = \frac{1}{\sqrt{(1+i)(1-i)+(1-i)(1+i)+2i(-2i)}} = \frac{1}{\sqrt{2+2+4}} = \frac{1}{2\sqrt{2}}$. Hence, a is in a circle of radius $\frac{1}{2\sqrt{2}}$ in $\mathbb{C}$.

Exercises

(6.1.1)   Let $v_1 = (1, -2, 3, -1)$ and $v_2 = (4, -3, 2, -1)$ be two vectors in $\mathbb{R}^4$. Then find all a_1 and a_2 in $\mathbb{R}$ such that $\|a_1 v_1\| = 1 = \|a_2 v_2\|$.

(6.1.2)   Let $v_1 = (1+i, 2i, 1-2i, 2)$ and $v_2 = (1-i, -i, i, 1+3i)$ be two vectors in $\mathbb{C}^4$. Then find all a_1 and a_2 in $\mathbb{C}$ such that $\|a_1 v_1\| = \|a_2 v_2\| = 1$.

(6.1.3)   Let $V = \mathbb{C}^4$ be the vector space with the standard inner product. If $x = (1+i, 1-i, 2, i)$ and $y = (1-i, 1, 1+2i, -1+i)$ are in V, then compute $\langle x, y \rangle$, $\|x\|$, $\|y\|$ and $\|x+y\|^2$. Also, verify Cauchy's and triangle inequality.

(6.1.4)   Let $f(x) = 1+x$ and $g(x) = e^x$ be the vectors in $C(R^{[0,1]})$. Then using the inner product $\langle f, g \rangle = \int_0^1 f(x)g(x)dx$, find $\|f\|$, $\|g\|$, $\langle f, g \rangle$ and $\|f+g\|^2$. Also, verify Cauchy's and triangle inequality.

(6.1.5)   Let $V = M_{2,2}(\mathbb{C})$ with the inner product $\langle A, B \rangle = \text{tr}(B^* A)$. If

$$A = \begin{bmatrix} 1-i & 1+i \\ -2+i & -i \end{bmatrix} \quad \text{and} \quad B = \begin{bmatrix} i & -3+2i \\ 2 & 4i \end{bmatrix},$$

then find $\langle A, B \rangle$, $\|A\|$, $\|B\|$ and $\|A+B\|^2$. Also, verify Cauchy's and triangle inequality.

(6.1.6)   Let V be an inner product space over $\mathbb{R}$. Then show that $\langle x, y \rangle = \frac{1}{4}(\|x+y\|^2 - \|x-y\|^2) \ \forall x, y \in V$. Also, verify for $x = (1, 2)$ and $y = (-1, 3)$ in $\mathbb{R}^2$.

(6.1.7)   For the inner product space V over $\mathbb{C}$, show that

$$\langle x, y \rangle = \frac{1}{4}\left(\|x+y\|^2 - \|x-y\|^2 + i\|x+iy\|^2 - i\|x-iy\|^2\right) \quad \forall x, y \in V.$$

Also, verify for $x = (1+i, -i)$ and $y = (1-i, i)$ in $\mathbb{C}^2$.

(6.1.8)   For $(x, y) \in \mathbb{R}^2$, show that $\|(x, y)\| = \max\{|x|, |y|\}$ is a norm on $\mathbb{R}^2$.

(6.1.9)   Prove that the following are norms on the given vector space V:
(i) $V = M_{m,n}(F)$, $\|A\| = \max |a_{ij}| \ \forall A = [a_{ij}] \in V$.
(ii) $V = C(R^{[0,1]})$, $\|f\| = \max_{x \in [0,1]} |f(x)|$, and $\|f\| = \int_0^1 |f(x)| dx \ \forall f \in V$.

(6.1.10) Let $x = (x_1, x_2)$ and $y = (y_1, y_2)$. Then show that:
(i) $\langle x, y \rangle = x_1 y_1 + 2x_2 y_2$,
(ii) $\langle x, y \rangle = 2x_1 y_1 + 5x_2 y_2$, and
(iii) $\langle x, y \rangle = 4x_1 y_1 - x_1 y_2 - x_2 y_1 + 2x_2 y_2$,
are inner products on $\mathbb{R}^2$. If $x = (1, 0)$ and $y = (1, 1)$, then compute $\|x\|$, $\|y\|$, $\|x+y\|^2$ and verify Cauchy's and triangle inequality with respect to the above norms.

(6.1.11) Show that $\langle x, y \rangle = \sum_{i=1}^n c_i x_i y_i$ is an inner product on $\mathbb{R}^n$, where $x = (x_1, x_2, \ldots, x_n)$, $y = (y_1, y_2, \ldots, y_n) \in \mathbb{R}^n$ and $c_1, c_2, \ldots, c_n$ are positive real numbers.

(6.1.12) For the real numbers $x_1, x_2, \ldots, x_n$, show that $(x_1 + x_2 + \cdots + x_n)^2 \leq n(x_1^2 + x_2^2 + \cdots + x_n^2)$.

(6.1.13) Let $(V_1, \langle\ \rangle_1)$ and $(V_2, \langle\ \rangle_2)$ be inner product spaces. Then prove that the vector space $V_1 \times V_2$ is an inner product space with the inner product defined as $\langle (x_1, x_2), (y_1, y_2) \rangle = \langle x_1, y_1 \rangle_1 + \langle x_2, y_2 \rangle_2$, where $(x_1, x_2), (y_1, y_2) \in V_1 \times V_2$.

6.2 Orthogonality and the Gram–Schmidt process

We begin this section by introducing the notion of angle between two vectors of a real inner product space. Let V be a real inner product space. Then by the Cauchy–Schwarz inequality, we have

$$|\langle u, v \rangle| \leq \|u\|\|v\|, \quad \text{for all } u, v \in V.$$

If u and v are nonzero vectors, then $\frac{|\langle u,v \rangle|}{\|u\|\|v\|} \leq 1 \Rightarrow -1 \leq \frac{\langle u,v \rangle}{\|u\|\|v\|} \leq 1$.

In view of the bijective map $\cos\theta : [0, \pi] \rightarrow [-1, 1]$, we have that far each real number $\frac{\langle u,v \rangle}{\|u\|\|v\|}$, there exists a unique $\theta \in [0, \pi]$ such that $\cos\theta = \frac{\langle u,v \rangle}{\|u\|\|v\|}$. Thus, we have the following definition.

Definition 6.2.1. If u and v are two nonzero vectors in a real inner product space V, then the angle θ between u and v is defined as $\cos\theta = \frac{\langle u,v \rangle}{\|u\|\|v\|}$.

If $\langle u, v \rangle = 0$, then $\cos\theta = 0 \Rightarrow \theta = \frac{\pi}{2}$. In this case, we say that u and v are perpendicular to each other. Generalizing this concept for real or complex inner product spaces, we have the following definition.

Definition 6.2.2. Let V be an inner product space. Then two vectors u and v are said to be orthogonal if $\langle u, v \rangle = 0$, and we denote $u \perp v$.

Note that, since $\langle 0, v \rangle = 0 \ \forall v \in V$, $0 \perp v \ \forall v \in V$. Conversely, if $u \perp v \ \forall v \in V$, then $\langle u, u \rangle = 0 \Rightarrow u = 0$. This shows that $u \in V$ is orthogonal to every vector in V if and only if $u = 0$.

Theorem 6.2.3 (Pythagoras theorem). *In an inner product space V if $u, v \in V$ such that $u \perp v$, then $\|u + v\|^2 = \|u - v\|^2 = \|u\|^2 + \|v\|^2$.*

Proof.

$$\begin{aligned}
\|u + v\|^2 &= \langle u + v, u + v \rangle \\
&= \langle u, u \rangle + \langle u, v \rangle + \langle v, u \rangle + \langle v, v \rangle \\
&= \langle u, u \rangle + \langle v, v \rangle \\
&= \|u\|^2 + \|v\|^2.
\end{aligned}$$

Hence, $\|u + v\|^2 = \|u\|^2 + \|v\|^2$.

If we replace $v = -v$ in the above equality, we get $\|u - v\|^2 = \|u\|^2 + \| - v\|^2 = \|u\|^2 + \|v\|^2$. $\qquad\square$

In the following remark, we discuss about the converse of the above result.

Remark. Suppose $\|u+v\|^2 = \|u\|^2 + \|v\|^2$. We have that $\|u+v\|^2 = \langle u+v, u+v \rangle = \|u\|^2 + \|v\|^2 + \langle u, v \rangle + \langle v, u \rangle$. Then $\|u+v\|^2 = \|u+v\|^2 + \langle u, v \rangle + \overline{\langle u, v \rangle} \Rightarrow \langle u, v \rangle + \overline{\langle u, v \rangle} = 0 \Rightarrow 2R\langle u, v \rangle = 0$. If V is a real inner product space, then $2R\langle u, v \rangle = 0 \Rightarrow \langle u, v \rangle = 0 \Rightarrow u \perp v$, but in a complex inner product space this is not true. Thus, in a real inner product space V, $u \perp v$ if and only if $\|u + v\|^2 = \|u\|^2 + \|v\|^2$.

Definition 6.2.4. Let S be a nonempty subset of an inner product space V. Then S is called an orthogonal set if all vectors in S are mutually orthogonal. That is, $\langle u, v \rangle = 0$ $\forall u \neq v \in S$.

An orthogonal set S is called orthonormal if $\|u\| = 1$, for all $u \in S$. Thus, the set $S = \{v_1, v_2, \ldots, v_n\} \subset V$ is an orthonormal set if

$$\langle v_i, v_j \rangle = \begin{cases} 1, & \text{if } i = j, \\ 0, & \text{if } i \neq j. \end{cases}$$

Theorem 6.2.5. *An orthogonal set of nonzero vectors is linearly independent.*

Proof. Let S be an orthogonal set of nonzero vectors in an inner product space V. Then for distinct elements $v_1, v_2, \ldots, v_n$ in S and scalars $a_1, a_2, \ldots, a_n$, suppose $a_1 v_1 + a_2 v_2 + \cdots + a_n v_n = 0$. Then for $1 \leq r \leq n$,

$$\langle a_1 v_1 + a_2 v_2 + \cdots + a_n v_n, v_r \rangle = \langle 0, v_r \rangle$$
$$\Rightarrow a_r \langle v_r, v_r \rangle = 0$$
$$\Rightarrow a_r \|v_r\|^2 = 0.$$

Since $v_r \neq 0$ $\forall 1 \leq r \leq n$, $a_r = 0$ $\forall r$.

Thus, every finite subset $\{v_1, v_2, \ldots, v_n\}$ of S is linearly independent. Hence, S is linearly independent. $\square$

Definition 6.2.6. A basis of a vector space, which is also an orthonormal set is called an orthonormal basis.

Example 6.2.7.
(a) Let $S = \{e_1, e_2, \ldots, e_n\}$ be the standard basis of $\mathbb{R}^n$. Since $\|e_i\| = 1$ and $\langle e_i, e_j \rangle = 0$ $\forall 1 \leq i, j \leq n$, S is an orthonormal set, and hence an orthonormal basis of $\mathbb{R}^n$.
(b) The set $S = \{\frac{1}{\sqrt{3}}(1,1,1), \frac{1}{\sqrt{6}}(-2,1,1), \frac{1}{\sqrt{2}}(0,-1,1)\}$ is an orthonormal basis of $\mathbb{R}^3$. Every proper subset of S is an orthonormal set in $\mathbb{R}^3$ but not an orthonormal basis of $\mathbb{R}^3$.
(c) Let $V = C(\mathbb{R}^{[-\pi,\pi]})$ be the vector space of continuous functions with the inner product defined by $\langle f, g \rangle = \int_{\pi}^{-\pi} f(x)g(x)dx$. Then the set $\{1, \cos x, \cos 2x, \cos 3x, \ldots, \sin x, \sin 2x, \sin 3x, \ldots\}$ is an orthogonal set in V.

Theorem 6.2.8 (Bessel's inequality). *Let $S = \{v_1, v_2, \ldots, v_n\}$ be an orthonormal set in an inner product space V. Then for $v \in V$, $\sum_{i=1}^{n} |\langle v, v_i \rangle|^2 \leq \|v\|^2$ and equality holds if and only if S is an orthonormal basis.*

Proof. Since $\langle v, v_i \rangle \in F \; \forall v_i \in S$, the linear combination

$$\sum_{i=1}^{n} \langle v, v_i \rangle v_i \in V.$$

Then by the definition of inner product on V,

$$\left\langle v - \sum_{i=1}^{n} \langle v, v_i \rangle v_i, \; v - \sum_{i=1}^{n} \langle v, v_i \rangle v_i \right\rangle \geq 0, \quad \text{or}$$

$$\left\langle v, v - \sum_{i=1}^{n} \langle v, v_i \rangle v_i \right\rangle + \left\langle -\sum_{i=1}^{n} \langle v, v_i \rangle v_i, \; v - \sum_{i=1}^{n} \langle v, v_i \rangle v_i \right\rangle \geq 0, \quad \text{or}$$

$$\langle v, v \rangle - \sum_{i=1}^{n} \overline{\langle v, v_i \rangle} \langle v, v_i \rangle - \sum_{i=1}^{n} \langle v, v_i \rangle \langle v_i, v \rangle + \left\langle \sum_{i=1}^{n} \langle v, v_i \rangle v_i, \; \sum_{i=1}^{n} \langle v, v_i \rangle v_i \right\rangle \geq 0.$$

Since the set S is orthonormal, $\langle v_i, v_i \rangle = 1$ and $\langle v_i, v_j \rangle = 0 \; \forall i \neq j$.
Then

$$\|v\|^2 - \sum_{i=1}^{n} |\langle v, v_i \rangle|^2 - \sum_{i=1}^{n} \langle v, v_i \rangle \overline{\langle v, v_i \rangle} + \sum_{i=1}^{n} \langle v, v_i \rangle \overline{\langle v, v_i \rangle} \geq 0, \quad \text{or}$$

$$\|v\|^2 - \sum_{i=1}^{n} |\langle v, v_i \rangle|^2 \geq 0, \quad \text{or}$$

$$\|v\|^2 \geq \sum_{i=1}^{n} |\langle v, v_i \rangle|^2.$$

Now suppose S is an orthonormal basis of V. Then $v \in V$ can be written as $v = a_1 v_1 + a_2 v_2 + \cdots + a_n v_n$ for some scalars $a_1, a_2, \ldots, a_n$.
Now $\langle v, v_i \rangle = \langle a_1 v_1 + a_2 v_2 + \cdots + a_n v_n, v_i \rangle = a_i \; \forall i$ gives that $v = \sum_{i=1}^{n} a_i v_i = \sum_{i=1}^{n} \langle v, v_i \rangle v_i$.
Hence, $\langle v - \sum_{i=1}^{n} \langle v, v_i \rangle v_i, \; v - \sum_{i=1}^{n} \langle v, v_i \rangle v_i \rangle = 0$.
On expansion given as above, we have

$$\|v\|^2 = \sum_{i=1}^{n} |\langle v, v_i \rangle|^2.$$

Conversely, suppose $\|v\|^2 = \sum_{i=1}^{n} |\langle v, v_i \rangle|^2 \; \forall v$. Since S being an orthonormal set is linearly independent, it is sufficient to show that every vector $v \in V$ can be written as linear combination of elements of V:

$$\left\| v - \sum_{i=1}^{n} \langle v, v_i \rangle v_i \right\|^2$$

$$= \left\langle v - \sum_{i=1}^{n} \langle v, v_i \rangle v_i, v - \sum_{i=1}^{n} \langle v, v_i \rangle v_i \right\rangle$$

$$= \|v\|^2 - \sum_{i=1}^{n} |\langle v, v_i \rangle|^2$$

$$= \sum_{i=1}^{n} |\langle v, v_i \rangle|^2 - \sum_{i=1}^{n} |\langle v, v_i \rangle|^2$$

$$= 0.$$

This shows that $v - \sum_{i=1}^{n} \langle v, v_i \rangle v_i = 0$. That is, $v = \sum_{i=1}^{n} \langle v, v_i \rangle v_i \Rightarrow v = \sum_{i=1}^{n} a_i v_i$, where $a_i = \langle v, v_i \rangle \ \forall i$. □

Corollary 6.2.9. *Let V be an n-dimensional inner product space and let $S = \{v_1, v_2, \ldots, v_n\}$ be its orthonormal subset. Then $\langle v, v_i \rangle = 0 \ \forall 1 \le i \le n$ if and only if $v = 0$.*

Proof. Since $\dim(V) = n$ and S is linearly independent, S is an orthonormal basis of V. From Bessel's inequality, we have

$$\|v\|^2 = \sum_{i=1}^{n} |\langle v, v_i \rangle|^2 \quad \forall v \in V.$$

Then

$$\langle v, v_i \rangle = 0 \ \forall i \Rightarrow \|v\|^2 = 0 \Rightarrow v = 0.$$

If $v = 0$, then $\sum_{i=1}^{n} |\langle v, v_i \rangle|^2 = 0 \Rightarrow |\langle v, v_i \rangle|^2 = 0 \Rightarrow \langle v, v_i \rangle = 0 \ \forall v_i$. □

Definition 6.2.10. Let S be a subset of an inner product space V. Then $S^{\perp} = \{v \in V : \langle v, u \rangle = 0 \ \forall u \in S\}$ is called an orthogonal complement of S in V.

Example 6.2.11.
(a) Let V be an inner product space. Then $\{0\}^{\perp} = V$ and $\{V\}^{\perp} = \{0\}$.
(b) Let $V = \mathbb{R}^3$. Then for $S = \{(1, 1, 0), (0, 1, -1)\}$,

$$S^{\perp} = \{(x, y, z) \in \mathbb{R}^3 : \langle (x, y, z), (1, 1, 0) \rangle = 0 \text{ and } \langle (x, y, z), (0, 1, -1) \rangle = 0\}$$

$$= \{(x, y, z) \in \mathbb{R}^3 : x + y = 0 \text{ and } y - z = 0\}$$

$$= \{x(1, -1, -1) : x \in \mathbb{R}\}$$

$$= \langle \{(1, -1, -1)\} \rangle.$$

That is, $S^{\perp}$ is a vector space generated by $(1, -1, -1) \in \mathbb{R}^3$.

Theorem 6.2.12. *Let S be a subset of an inner product space V. Then $S^\perp$ is a subspace of V.*

Proof. Since $0 \in S^\perp$, $S^\perp \neq \phi$. Let $v_1, v_2 \in S^\perp$. Then for $a_1, a_2 \in F$,

$$\langle a_1 v_1 + a_2 v_2, u \rangle = a_1 \langle v_1, u \rangle + a_2 \langle v_2, u \rangle = 0 \quad \forall u \in S.$$

This shows that $a_1 v_1 + a_2 v_2 \in S^\perp$. Hence, $S^\perp$ is a subspace of V. □

The following theorem and example illustrate the importance of an orthonormal basis in an inner product space.

Theorem 6.2.13. *Let V be an inner product space and let $B = \{v_1, v_2, \ldots, v_n\}$ be its orthonormal basis. Then for any $v = \sum_{i=1}^n a_i v_i$, $a_i = \langle v, v_i \rangle$ $\forall i$ and hence $v = \sum_{i=1}^n \langle v, v_i \rangle v_i$.*

Proof. Given that $v = \sum_{i=1}^n a_i v_i$. Then $\langle v, v_i \rangle = \langle \sum_{i=1}^n a_i v_i, v_i \rangle = a_i \langle v_i, v_i \rangle = a_i \|v_i\|^2 = a_i$ $\forall 1 \leq i \leq n$. Hence, $v = \sum_{i=1}^n a_i v_i = \sum_{i=1}^n \langle v, v_i \rangle v_i$. □

Example 6.2.14. Let $V = \mathbb{R}^3$ and $B = \{v_1 = \frac{1}{\sqrt{3}}(1,1,1), v_2 = \frac{1}{\sqrt{6}}(-2,1,1), v_3 = \frac{1}{\sqrt{2}}(0,-1,1)\}$ be an orthonormal basis of V. If $v = (3,2,-4) \in \mathbb{R}^3$, then v can be expressed as $v = a_1 v_1 + a_2 v_2 + a_3 v_3$. From the above theorem,

$$a_1 = \langle v, v_1 \rangle = \left\langle (3,2,-4), \frac{1}{\sqrt{3}}(1,1,1) \right\rangle = \frac{1}{\sqrt{3}},$$

$$a_2 = \langle v, v_2 \rangle = \left\langle (3,2,-4), \frac{1}{\sqrt{6}}(-2,1,1) \right\rangle = \frac{-8}{\sqrt{6}},$$

$$a_3 = \langle v, v_3 \rangle = \left\langle (3,2,-4), \frac{1}{\sqrt{2}}(0,-1,1) \right\rangle = \frac{-6}{\sqrt{2}}.$$

Hence,

$$(3,2,-4) = \frac{1}{\sqrt{3}}\frac{1}{\sqrt{3}}(1,1,1) + \frac{-8}{\sqrt{6}}\frac{1}{\sqrt{6}}(-2,1,1) + \frac{-6}{\sqrt{2}}\frac{1}{\sqrt{2}}(0,-1,1)$$
$$= \frac{1}{3}(1,1,1) + \frac{-8}{6}(-2,1,1) + \frac{-6}{2}(0,-1,1).$$

Thus, in case of an orthonormal basis, obtaining coefficients of a vector to express as a linear combination of the basis vectors is easier than the expression in the form of basis.

Now we provide an algorithm to obtain an orthonormal basis from a given basis of a finite-dimensional inner product space called Gram–Schmidt process.

Theorem 6.2.15 (Gram–Schmidt). *Let $\{v_1, v_2, \ldots, v_n\}$ be a basis of an inner product space V. Then there exists an orthonormal basis $\{u_1, u_2, \ldots, u_n\}$ of V.*

Proof. To prove the theorem, we use the following construction process.

Let $u_1 = \frac{v_1}{\|v_1\|}$. Then u_1 is a unit vector.
Define $w_2 = v_2 - \langle v_2, u_1 \rangle u_1$. Then

$$\langle w_2, u_1 \rangle = \langle v_2 - \langle v_2, u_1 \rangle u_1, u_1 \rangle = \langle v_2, u_1 \rangle - \langle v_2, u_1 \rangle \langle u_1, u_1 \rangle$$
$$= \langle v_2, u_1 \rangle - \langle v_2, u_1 \rangle$$
$$= 0.$$

Hence, $w_2 \perp u_1$. Also, $w_2 \neq 0$. For if $w_2 = 0$, then $v_2 = \langle v_2, u_1 \rangle u_1 = \langle v_2, u_1 \rangle \frac{v_1}{\|v_1\|} = \alpha v_1$,
where $\alpha = \frac{\langle v_2, u_1 \rangle}{\|v_1\|} \in F$. This implies that the set $\{v_1, v_2\}$ is linearly dependent, and hence
the set $\{v_1, v_2, \ldots, v_n\}$ is linearly dependent, which is a contradiction. So, let $u_2 = \frac{w_2}{\|w_2\|}$.
Then $\{u_1, u_2\}$ is an orthonormal set.
Now, let $w_3 = v_3 - \langle v_3, u_1 \rangle u_1 - \langle v_3, u_2 \rangle u_2$. Then

$$\langle w_3, u_1 \rangle = \langle w_3, u_2 \rangle = 0 \quad \text{and} \quad w_3 \neq 0.$$

For otherwise, $\{v_1, v_2, v_3\}$ is linearly dependent, a contradiction. Letting $u_3 = \frac{w_3}{\|w_3\|}$,
$\{u_1, u_2, u_3\}$ is an orthonormal set.
Proceeding as above by induction, define $w_k = v_k - \sum_{k=1}^{k-1} \langle v_k, u_k \rangle u_k$. Then $u_k = \frac{w_k}{\|w_k\|}$
and $\langle u_k, u_i \rangle = 0 \ \forall 1 \leq i \leq k - 1$. Hence, $\{u_1, u_2, \ldots, u_k\}$ is an orthonormal set. Thus, by
induction we get orthonormal set $\{u_1, u_2, \ldots, u_n\}$ which is a basis of V. $\qquad\square$

Remark. Using the Gram–Schmidt process, every linearly independent subset $\{v_1, v_2, \ldots, v_r\}$ of an inner product space V can be converted into an orthonormal set $\{u_1, u_2, \ldots, u_r\}$ such that $\text{span}\{v_1, v_2, \ldots, v_r\} = \text{span}\{u_1, u_2, \ldots, u_r\}$.

Example 6.2.16. Let $S = \{v_1 = (1, 0, 1), v_2 = (1, 2, -2), v_3 = (2, -1, 1)\}$ be a basis of the inner
product space $\mathbb{R}^3$. We apply the Gram–Schmidt process to convert S to an orthonormal
basis $\{u_1, u_2, u_3\}$.
Then $u_1 = \frac{v_1}{\|v_1\|} = \frac{(1,0,1)}{\sqrt{1^2+0^2+1^2}} = \frac{1}{\sqrt{2}}(1, 0, 1)$.
Let

$$w_2 = v_2 - \langle v_2, u_1 \rangle u_1$$
$$= (1, 2, -2) - \left\langle (1, 2, -2), \frac{1}{\sqrt{2}}(1, 0, 1) \right\rangle \frac{1}{\sqrt{2}}(1, 0, 1)$$
$$= (1, 2, -2) - \left(\frac{1}{\sqrt{2}} - \frac{2}{\sqrt{2}} \right) \frac{1}{\sqrt{2}}(1, 0, 1)$$
$$= (1, 2, -2) + \frac{1}{2}(1, 0, 1)$$
$$= \left(\frac{3}{2}, 2, \frac{-3}{2} \right).$$

Then

$$u_2 = \frac{w_2}{\|w_2\|} = \frac{(\frac{3}{2}, 2, \frac{-3}{2})}{\sqrt{(\frac{3}{2})^2 + 2^2 + (\frac{-3}{2})^2}}$$

$$= \frac{2}{\sqrt{34}}\left(\frac{3}{2}, 2, \frac{-3}{2}\right)$$

$$= \frac{1}{\sqrt{34}}(3, 4, -3).$$

Let

$$w_3 = v_3 - \langle v_3, u_1 \rangle u_1 - \langle v_3, u_2 \rangle u_2$$

$$= (2, -1, 1) - \left\langle (2, -1, 1), \frac{1}{\sqrt{2}}(1, 0, 1) \right\rangle \frac{1}{\sqrt{2}}(1, 0, 1)$$

$$- \left\langle (2, -1, 1), \frac{1}{\sqrt{34}}(3, 4, -3) \right\rangle \frac{1}{\sqrt{34}}(3, 4, -3)$$

$$= (2, -1, 1) - \left(\frac{3}{\sqrt{2}}\right)\frac{1}{\sqrt{2}}(1, 0, 1) - \left(\frac{-1}{\sqrt{34}}\right)\frac{1}{\sqrt{34}}(3, 4, -3)$$

$$= (2, -1, 1) - \frac{3}{2}(1, 0, 1) + \frac{1}{34}(3, 4, -3)$$

$$= \frac{1}{34}(20, -30, -20)$$

$$= \frac{5}{17}(2, -3, -2).$$

Then

$$u_3 = \frac{w_3}{\|w_3\|} = \frac{\frac{5}{17}(2, -3, -2)}{\sqrt{(\frac{10}{17})^2 + (\frac{-15}{17})^2 + (\frac{-10}{17})^2}}$$

$$= \frac{\frac{5}{17}(2, -3, -2)}{\sqrt{\frac{25}{17}}}$$

$$= \frac{1}{\sqrt{17}}(2, -3, -2).$$

Hence, the set $\{\frac{1}{\sqrt{2}}(1, 0, 1), \frac{1}{\sqrt{34}}(3, 4, -3), \frac{1}{\sqrt{17}}(2, \ 3, -2)\}$ is an orthonormal basis of $\mathbb{R}^3$.

Example 6.2.17. Let $V = C(\mathbb{R}^{[0,1]})$ be the inner product space with the inner product defined by

$$\langle f, g \rangle = \int_0^1 f(x)g(x)dx.$$

Then we find the orthonormal basis of the subspace W generated by the basis $\{f_1(x) = x, f_2(x) = x^2, f_3(x) = x^3\}$.

Let $u_1(x) = \frac{f_1(x)}{\|f_1(x)\|}$, where

$$\|f_1(x)\| = \sqrt{\langle f_1(x), f_1(x)\rangle} = \sqrt{\int_0^1 f_1(x)f_1(x)dx} = \sqrt{\int_0^1 x^2 dx} = \frac{1}{\sqrt{3}}.$$

Then $u_1(x) = \frac{x}{\frac{1}{\sqrt{3}}} = \sqrt{3}x$.

Let

$$w_2(x) = f_2(x) - \langle f_2(x), u_1(x)\rangle u_1(x)$$
$$= x^2 - \langle x^2, \sqrt{3}x\rangle \sqrt{3}x$$
$$= x^2 - \left(\int_0^1 x^2\sqrt{3}x dx\right)\sqrt{3}x$$
$$= x^2 - \left(\int_0^1 \sqrt{3}x^3 dx\right)\sqrt{3}x$$
$$= x^2 - \frac{3}{4}x.$$

Then $u_2(x) = \frac{w_2(x)}{\|w_2(x)\|} = \frac{x^2 - \frac{3}{4}x}{\sqrt{\int_0^1 (x^2 - \frac{3}{4}x)^2 dx}} = \frac{x^2 - \frac{3}{4}x}{\sqrt{\int_0^1 (x^4 - \frac{3}{2}x^3 + \frac{9}{16}x^2)dx}} = \frac{x^2 - \frac{3}{4}x}{\frac{1}{80}} = 80x^2 - 60x.$

Let

$$w_3(x) = f_3(x) - \langle f_3(x), u_1(x)\rangle u_1(x) - \langle f_3(x), u_2(x)\rangle u_2(x)$$
$$= x^3 - \langle x^3, \sqrt{3}x\rangle \sqrt{3}x - \langle x^3, 80x^2 - 60x\rangle 80x^2 - 60x$$
$$= x^3 - \left(\int_0^1 \sqrt{3}x^4 dx\right)\sqrt{3}x - \left(\int_0^1 x^3(80x^2 - 60x)dx\right)(80x^2 - 60x)$$
$$= x^3 - \frac{3}{5}x - \frac{4}{3}(80x^2 - 60x)$$
$$= x^3 - \frac{320}{3}x^2 + \frac{3}{5}x + 80x$$
$$= x^3 - \frac{320}{3}x^2 + \frac{403}{5}x.$$

Put $u_3(x) = \frac{w_3(x)}{\|w_3(x)\|} = \frac{x^3 - \frac{320}{3}x^2 + \frac{403}{5}x}{\sqrt{\int_0^1 (x^3 - \frac{320}{3}x^2 + \frac{403}{5}x)^2 dx}} = \frac{x^3 - \frac{320}{3}x^2 + \frac{403}{5}x}{\frac{2}{5}\sqrt{\frac{18266}{21}}} = \frac{\sqrt{21}}{\sqrt{18266}}(\frac{5}{2}x^3 - 800x^2 + \frac{403}{2}x).$

Thus, $\{\sqrt{3}x, x^2 - \frac{3}{4}x, \frac{\sqrt{21}}{\sqrt{18266}}(\frac{5}{2}x^3 - 800x^2 + \frac{403}{2}x)\}$ is an orthonormal basis of W.

Example 6.2.18. Let $V = \mathbb{C}^3$ be the inner product space and let W be the subspace of V generated by the basis $\{v_1 = (1 - i, 0, i), v_2 = (1, 1 + i, 0)\}$. Now we use the Gram–Schmidt process to find the orthonormal basis of W.

Let $u_1 = \frac{v_1}{\|v_1\|} = \frac{(1-i,0,i)}{\sqrt{\langle(1-i,0,i),(1-i,0,i)\rangle}} = \frac{(1-i,0,i)}{\sqrt{(1-i)(1+i)+0\cdot0+i(-i)}} = \frac{(1-i,0,i)}{\sqrt{3}} = (\frac{1}{\sqrt{3}} - \frac{i}{\sqrt{3}}, 0, \frac{i}{\sqrt{3}}).$

Let

$$w_2 = v_2 - \langle v_2, u_1 \rangle u_1$$

$$= (1, 1 + i, 0) - \left\langle (1, 1 + i, 0), \left(\frac{1}{\sqrt{3}} - \frac{i}{\sqrt{3}}, 0, \frac{i}{\sqrt{3}}\right) \right\rangle \left(\frac{1}{\sqrt{3}} - \frac{i}{\sqrt{3}}, 0, \frac{i}{\sqrt{3}}\right)$$

$$= (1, 1 + i, 0) - \left(1 \cdot \left(\frac{1}{\sqrt{3}} + \frac{i}{\sqrt{3}}\right) + (1 + i) \cdot 0 + 0\right) \left(\frac{1}{\sqrt{3}} - \frac{i}{\sqrt{3}}, 0, \frac{i}{\sqrt{3}}\right)$$

$$= (1, 1 + i, 0) - \left(\frac{1}{\sqrt{3}} + \frac{i}{\sqrt{3}}\right) \left(\frac{1}{\sqrt{3}} - \frac{i}{\sqrt{3}}, 0, \frac{i}{\sqrt{3}}\right)$$

$$= (1, 1 + i, 0) - \left(\frac{2}{3}, 0, \frac{-1 + i}{3}\right)$$

$$= \left(\frac{1}{3}, 1 + i, \frac{1 - i}{3}\right).$$

Then put

$$u_2 = \frac{w_2}{\|w_2\|} = \frac{(\frac{1}{3}, 1 + i, \frac{1-i}{3})}{\sqrt{\langle(\frac{1}{3}, 1 + i, \frac{1-i}{3}), (\frac{1}{3}, 1 + i, \frac{1-i}{3})\rangle}} = \frac{(\frac{1}{3}, 1 + i, \frac{1-i}{3})}{\sqrt{\frac{1}{3} \cdot \frac{1}{3} + (1 + i)(1 - i) + \frac{1-i}{3}\frac{1+i}{3}}}$$

$$= \frac{3}{\sqrt{21}}\left(\frac{1}{3}, 1 + i, \frac{1 - i}{3}\right) = \left(\frac{1}{\sqrt{21}}, \frac{3 + 3i}{\sqrt{21}}, \frac{1 - i}{\sqrt{21}}\right).$$

Thus, the set $\{(\frac{1}{\sqrt{3}} - \frac{i}{\sqrt{3}}, 0, \frac{i}{\sqrt{3}}), (\frac{1}{\sqrt{21}}, \frac{3+3i}{\sqrt{21}}, \frac{1-i}{\sqrt{21}})\}$ is an orthonormal basis of W.

Theorem 6.2.19. *If U is a subspace of a finite-dimensional inner product space V, then $V = U \oplus U^\perp$. That is, a finite-dimensional inner product space can be written as the direct sum of a subspace and its orthogonal complement.*

Proof. Since U is finite-dimensional, there exists an orthonormal basis $\{u_1, u_2, \ldots, u_r\}$ of U. Let $v \in V$. Then

$$\left\langle v - \sum_{i=1}^{r} \langle v, u_i \rangle u_i, u_i \right\rangle$$

$$= \langle v, u_i \rangle - \left\langle \sum_{i=1}^{r} \langle v, u_i \rangle u_i, u_i \right\rangle$$

$$= \langle v, u_i \rangle - \langle v, u_i \rangle$$

$$= 0,$$

by the property

$$\langle u_i, u_j \rangle = \begin{cases} 1, & \text{if } i = j, \\ 0, & \text{if } i \neq j. \end{cases}$$

This shows that $v' = v - \sum_{i=1}^{r} \langle v, u_i \rangle u_i$ is orthogonal to each u_i, $1 \leq i \leq r$.
For $u = a_1 u_1 + a_2 u_2 + \cdots + a_r u_r \in U$,

$$\langle v', u \rangle = \langle v', a_1 u_1 + a_2 u_2 + \cdots + a_r u_r \rangle$$
$$= \bar{a}_1 \langle v', u_1 \rangle + \bar{a}_2 \langle v', u_2 \rangle + \cdots + \bar{a}_r \langle v', u_r \rangle$$
$$= 0.$$

Thus, $v' \perp u \ \forall u \in U$. Hence, $v' \in U^\perp$.
Now, $v' = v - \sum_{i=1}^{r} \langle v, u_i \rangle u_i \Rightarrow v' + \sum_{i=1}^{r} \langle v, u_i \rangle u_i = v$, where $v' \in U^\perp$ and $\sum_{i=1}^{r} \langle v, u_i \rangle u_i \in U$. Hence, $V = U + U^\perp$.
Now, let $v \in U \cap U^\perp$. Then $\langle v, v \rangle = 0 \Rightarrow v = 0$. That is, $U \cap U^\perp = \{0\}$. This proves that $U \oplus U^\perp = V$. $\qquad\square$

The above result is not true for infinite-dimensional inner product spaces. For more detail and examples, one can refer any good book of functional.

From the above theorem, it is clear that if U is a subspace of a finite- dimensional inner product space V, then every vector $v \in V$ can be expressed uniquely as $v = u_1 + u_2$, where $u_1 \in U$ and $u_2 \in U^\perp$.

In this case, u_1 is called the orthogonal projection of v on U and u_2 is called the orthogonal projection of v on $U^\perp$ denoted by $\text{Proj}_U(v)$ and $\text{Proj}_{U^\perp}(v)$, respectively. Then $v = u_1 + u_2 = \text{Proj}_U(v) + \text{Proj}_{U^\perp}(v)$. Thus, from $v = \text{Proj}_U(v) + (v - \text{Proj}_U(v))$, $v - \text{Proj}_U(v)$ is called the complement of v orthogonal to U.

Theorem 6.2.20. *If U is a subspace of a finite-dimensional inner product space V and $\{u_1, u_2, \ldots, u_m\}$ is an orthonormal basis of U, then for any vector $v \in V$, $\text{Proj}_U(v) = \langle v, u_1 \rangle u_1 + \langle v, u_2 \rangle u_2 + \cdots + \langle v, u_m \rangle u_m$.*

Proof. Given that $v \in V$. Then v can be written as $v = \text{Proj}_U(v) + \text{Proj}_{U^\perp}(v)$, where $\text{Proj}_U(v) \in U$. Since $\{u_1, u_2, \ldots, u_m\}$ is an orthonormal basis of U, $\text{Proj}_U(v)$ can be expressed as:

$$\text{Proj}_U(v) = \sum_{i=1}^{m} \langle \text{Proj}_U(v), u_i \rangle u_i.$$

Since $\text{Proj}_{U^\perp}(v)$ is orthogonal to U, $\langle \text{Proj}_{U^\perp}(v), u_i \rangle = 0$, $\forall 1 \leq i \leq m$.
Then

$$\text{Proj}_U(v) = \sum_{i=1}^{m} \langle \text{Proj}_U(v), u_i \rangle u_i$$

$$= \sum_{i=1}^{m} (\langle \text{Proj}_U(v), u_i \rangle + \langle \text{Proj}_{U^\perp}(v), u_i \rangle) u_i$$

$$= \sum_{i=1}^{m} \langle \text{Proj}_U(v) + \text{Proj}_{U^\perp}(v), u_i \rangle u_i$$

$$= \sum_{i=1}^{m} \langle v, u_i \rangle u_i.$$

$\square$

Corollary 6.2.21. *If U is a subspace of a finite-dimensional inner product space V and $\{u_1, u_2, \ldots, u_m\}$ is an orthogonal basis of U, then for any vector $v \in V$, $\text{Proj}_U(v) = \frac{\langle v, u_1 \rangle u_1}{\|u_1\|^2} + \frac{\langle v, u_2 \rangle u_2}{\|u_2\|^2} + \cdots + \frac{\langle v, u_m \rangle u_m}{\|u_m\|^2}$.*

Proof. Given that $\{u_1, u_2, \ldots, u_m\}$ is an orthogonal basis of U. Then $\{\frac{u_1}{\|u_1\|} + \frac{u_2}{\|u_2\|} + \cdots + \frac{u_m}{\|u_m\|}\}$ is an orthonormal basis of U. Hence, from above theorem we have

$$\text{Proj}_U(v) = \left\langle v, \frac{u_1}{\|u_1\|} \right\rangle \frac{u_1}{\|u_1\|} + \left\langle v, \frac{u_2}{\|u_2\|} \right\rangle \frac{u_2}{\|u_2\|} + \cdots + \left\langle v, \frac{u_m}{\|u_m\|} \right\rangle \frac{u_m}{\|u_m\|}$$

$$= \frac{\langle v, u_1 \rangle u_1}{\|u_1\|^2} + \frac{\langle v, u_2 \rangle u_2}{\|u_2\|^2} + \cdots + \frac{\langle v, u_m \rangle u_m}{\|u_m\|^2}.$$

$\square$

Example 6.2.22. Let U be a subspace of the vector space $\mathbb{R}^3$ with the orthonormal basis $\{(\frac{1}{\sqrt{3}}, \frac{1}{\sqrt{3}}, \frac{1}{\sqrt{3}}), (0, \frac{-1}{\sqrt{2}}, \frac{-1}{\sqrt{2}})\}$. Then the orthogonal projection of the vector $(1, 2, 3) \in \mathbb{R}^3$ on U is

$$\text{Proj}_U(1, 2, 3) = \left\langle (1, 2, 3), \left(\frac{1}{\sqrt{3}}, \frac{1}{\sqrt{3}}, \frac{1}{\sqrt{3}} \right) \right\rangle \left(\frac{1}{\sqrt{3}}, \frac{1}{\sqrt{3}}, \frac{1}{\sqrt{3}} \right)$$

$$+ \left\langle (1, 2, 3), \left(0, \frac{-1}{\sqrt{2}}, \frac{-1}{\sqrt{2}} \right) \right\rangle \left(0, \frac{-1}{\sqrt{2}}, \frac{-1}{\sqrt{2}} \right)$$

$$= \frac{6}{\sqrt{3}} \left(\frac{1}{\sqrt{3}}, \frac{1}{\sqrt{3}}, \frac{1}{\sqrt{3}} \right) + \frac{1}{\sqrt{2}} \left(0, \frac{-1}{\sqrt{2}}, \frac{-1}{\sqrt{2}} \right)$$

$$= \left(\frac{6}{3}, \frac{6}{3}, \frac{6}{3} \right) + \left(0, \frac{-1}{2}, \frac{1}{2} \right)$$

$$= \left(2, \frac{3}{2}, \frac{5}{2} \right).$$

The component of $(1, 2, 3)$ orthogonal to U is

$$\text{Proj}_{U^\perp}(1, 2, 3) = (1, 2, 3) - \text{Proj}_U(1, 2, 3)$$

$$= (1, 2, 3) - \left(2, \frac{3}{2}, \frac{5}{2} \right) = \left(-1, \frac{1}{2}, \frac{1}{2} \right).$$

It is easy to see that $\text{Proj}_{U^\perp}(1,2,3) = (-1, \frac{1}{2}, \frac{1}{2})$ is orthogonal to basis elements of U, and hence orthogonal to U.

Theorem 6.2.23. *Let V be an inner product space. Then for the subsets S, S_1, S_2 of V:*

(i) $S_1 \subseteq S_2 \Rightarrow S_2^\perp \subseteq S_1^\perp$.

(ii) $S^\perp = \langle S \rangle^\perp$.

(iii) $\langle S \rangle \subseteq S^{\perp\perp}$ *and equality holds if V is finite-dimensional.*

Proof. (i) Let $v \in S_2^\perp$. Then $\langle v, u \rangle = 0 \ \forall u \in S_2$. Since $S_1 \subseteq S_2$ $\langle v, u \rangle = 0 \ \forall u \in S_1$. This shows that $v \in S_1^\perp$, and hence $S_2^\perp \subseteq S_1^\perp$.

(ii) From (i), $S \subseteq \langle S \rangle \Rightarrow \langle S \rangle^\perp \subseteq S^\perp$. Now, let $v \in S^\perp$. Then $\langle v, u \rangle = 0 \ \forall u \in S$. Also, for $w = \sum a_i u_i \in \langle S \rangle$, where $a_i \in F$ and $u_i \in S$, $\langle v, w \rangle = \langle v, \sum a_i u_i \rangle = \sum \bar{a}_i \langle v, u_i \rangle = 0$. We have $v \in \langle S \rangle^\perp$ and hence $S^\perp \subseteq \langle S \rangle^\perp$. This proves that $S^\perp = \langle S \rangle^\perp$.

(iii) Let $v \in \langle S \rangle$. Then, for $w \in \langle S \rangle^\perp = S^\perp$, $\langle w, v \rangle = 0$. Since w is an arbitrary element, v is orthogonal to every element of $S^\perp$. Hence, $v \in S^{\perp\perp}$. Thus, $\langle S \rangle \subseteq S^{\perp\perp}$.

Let V be a finite-dimensional inner product space. Then $\langle S \rangle$, being a subspace of V satisfies $\langle S \rangle \oplus \langle S \rangle^\perp = V$. Hence, $\dim \langle S \rangle = \dim V - \dim \langle S \rangle^\perp$.

Also, $\langle S \rangle^\perp$ is a subspace of V, we have $\langle S \rangle^\perp \oplus \langle S \rangle^{\perp\perp} = V$, and hence $\dim \langle S \rangle^{\perp\perp} = \dim V - \dim \langle S \rangle^\perp$.

This shows that $\dim \langle S \rangle = \dim \langle S \rangle^{\perp\perp}$. Since $\langle S \rangle$ is a finite-dimensional vector space, $\langle S \rangle \cong \langle S \rangle^{\perp\perp}$.

But $\langle S \rangle \subseteq S^{\perp\perp} = \langle S \rangle^{\perp\perp}$ gives that $\langle S \rangle = \langle S \rangle^{\perp\perp}$.

From the above proof, it is clear that if W is a subspace of a finite-dimensional inner product space, then $W = W^{\perp\perp}$. □

ℹ️ Exercises

(6.2.1) Find the angle between $x = (1,0)$ and $y = (0,1)$ with respect to the following inner products:

    (i) $\langle x, y \rangle = x_1 y_1 + 2x_2 y_2$,

    (ii) $\langle x, y \rangle = 2x_1 y_1 + 5x_2 y_2$, and

    (iii) $\langle x, y \rangle = 4x_1 y_1 - x_1 y_2 - x_2 y_1 + 2x_2 y_2$,

    defined on $\mathbb{R}^2$.

(6.2.2) Find an orthonormal basis of $P_2(\mathbb{R})$ where the inner product is given by $\langle f, g \rangle = \int_{-1}^1 f(x)g(x)dx$.

(6.2.3) Find an orthonormal basis of $P_3(\mathbb{R})$ where the inner product is defined by $\langle f, g \rangle = \int_0^1 f(x)g(x)dx$.

(6.2.4) Show that the following sets of vectors are orthogonal and determine a corresponding orthonormal set:

    (i) $\{(2,-1,1),(1,1,-1),(0,1,1)\} \subseteq \mathbb{R}^3$;

    (ii) $\{(1,0,2,1),(-1,1,1,-1),(1,3,-1,1)\} \subseteq \mathbb{R}^4$.

(6.2.5) Show that the set of vectors:

    (i) $\{(1+i,1-i),(1-i,1+i)\}$ in $\mathbb{C}^2$ is orthogonal and find the corresponding orthonormal set.

    (ii) $\{(1-i,1+i,i),(-3+3i,2+2i,2i),(0,i,1-i)\}$ is orthogonal in $\mathbb{C}^3$ and find the corresponding orthonormal set.

(6.2.6) Find the orthonormal bases of $\mathbb{R}^3$ corresponding to the following bases:

     (i) $\{(0,1,2),(1,2,3),(3,1,1)\}$;

     (ii) $\{(1,3,3),(1,4,3),(1,3,4)\}$.

(6.2.7)   Find the orthonormal basis of $\mathbb{R}^4$ corresponding to the following bases:

     (i) $\{(1,1,-1,-2),(1,2,0,1),(0,2,1,3),(-1,1,2,6)\}$;

     (ii) $\{(1,2,3,1),(1,3,3,2),(2,4,3,3),(1,1,1,1)\}$.

(6.2.8)   Find the orthonormal basis of the subspace generated by the basis $\{(1,2,3,2),(0,1,1,3)\}$.

(6.2.9)   Find the orthonormal bases of $\mathbb{C}^3$ corresponding to the following bases:

     (i) $\{(1,0,0),(0,i,0),(1,1,i)\}$;

     (ii) $\{(3-i,2+2i,4),(2,2+4i,3),(1-i,-2i,-1)\}$.

(6.2.10) Find the orthonormal basis of $\mathbb{C}^2$ from its basis $\{(1,2i),(1,i)\}$.

(6.2.11) In the inner product space $C(R^{[0,1]})$ find the corresponding orthonormal basis of the subspace generate by the following set of vectors:

     (i) $\{1,x^2,x^3\}$;

     (ii) $\{1+x^2,x^3,x^5\}$.

(6.2.12) In the inner product space $C(\mathbb{R}^{[-1,1]})$ find the corresponding orthonormal bases of the subspace generated by the set of vectors:

     (i) $\{x,x^2,x^3\}$;

     (ii) $\{1-x^2,x^3,x^4\}$.

(6.2.13) In the vector space $\mathbb{R}^3$, find

     (i) $S_1^{\perp}$ if $S_1 = \langle(1,0,0)\rangle$;

     (ii) $S_2^{\perp}$ if $S_2 = \langle(-1,1,2)\rangle$;

     (iii) $S_3^{\perp}$ if $s_3 = \langle\{(0,1,2),(-1,2,0)\}\rangle$;

     (iv) $S_4^{\perp}$ if $S_4 = \langle\{(1,1,1),(0,-1,1)\}\rangle$.

(6.2.14) In the vector space $\mathbb{R}^4$, find

     (i) $S_1^{\perp}$ if $S_1 = \langle\{(1,2,3,0),(0,3,2,1)\}\rangle$;

     (ii) $S_2^{\perp}$ if $S_2 = \langle\{(1,1,1,1),(1,-1,0,0)\}\rangle$;

     (iii) $S_3^{\perp}$ if $S_3 = \langle\{(1,1,1,1),(1,1,-1,0),(1,-1,0,0)\}\rangle$.

(6.2.15) Let $V = \mathbb{R}^3$ be the inner product space. Then find $\mathrm{Proj}_U(v)$ and $\mathrm{Proj}_{U^{\perp}}(v)$ for each of the following:

     (i) $v = (3,-2,4)$, $U = \langle\{(\frac{2}{3},-\frac{2}{3},\frac{1}{3}),(\frac{1}{3},\frac{2}{3},\frac{2}{3})\}\rangle$;

     (ii) $v = (1,-3,6)$, $U = \langle\{(\frac{1}{\sqrt{6}},\frac{-2}{\sqrt{6}},\frac{1}{\sqrt{6}}),(\frac{2}{\sqrt{5}},\frac{1}{\sqrt{5}},0)\}\rangle$;

     (iii) $v = (-1,1,10)$, $U = \langle\{(1,2,-2),(\frac{2}{3},\frac{1}{3},\frac{2}{3})\}\rangle$;

     (iv) $v = (1,2,3)$, $U = \langle\{(1,1,-2),(\frac{2}{\sqrt{3}},\frac{2}{\sqrt{3}},\frac{2}{\sqrt{3}})\}\rangle$.

(6.2.16) In the inner product space $\mathbb{R}^4$, find $\mathrm{Proj}_U(v)$ and $\mathrm{Proj}_{U^{\perp}}(v)$ for each of the following:

     (i) $v = (1,2,-1,2)$, $U = \langle\{(\frac{1}{2},\frac{1}{2},\frac{1}{2},\frac{1}{2}),(\frac{1}{2},\frac{1}{2},\frac{-1}{2},\frac{-1}{2})\}\rangle$;

     (ii) $v = (1,0,2,3)$, $U = \langle\{(\frac{1}{\sqrt{6}},\frac{2}{\sqrt{6}},0,\frac{-1}{\sqrt{6}}),(\frac{1}{\sqrt{2}},0,0,\frac{1}{\sqrt{2}})\}\rangle$;

     (iii) $v = (1,2,3,4)$, $U = \langle\{(1,-1,2,-1),(-1,1,\frac{3}{2},1)\}\rangle$;

     (iv) $v = (3,2,1,0)$, $U = \langle\{(1,-1,2,-1),(1,2,0,-1)\}\rangle$.

6.3 The adjoint of a linear operator

By a linear operator, we mean a linear transformation that takes the vector space V to itself. Linear transformation $T : V \to F$, where V is a vector over the field F is called a linear functional.

Definition 6.3.1. Let $T : V \rightarrow V$ be a linear operator on an inner product space V. A linear operator $T^* : V \rightarrow V$ is called the adjoint of T if $\langle T(u), v \rangle = \langle u, T^*(v) \rangle$, for all $u, v \in V$.

Now, we discuss about the existence and uniqueness of the adjoint operator T^*.

Theorem 6.3.2. *Let $f : V \rightarrow F$ be a linear functional on a finite-dimensional inner product space V. Then there exists a unique vector $w \in V$ such that $f(v) = \langle v, w \rangle$ for all $v \in V$.*

Proof. Let $\dim(V) = n$. Then by the Gram–Schmidt orthogonalization process, we can find an orthonormal basis $\{v_1, v_2, \ldots, v_n\}$ of V. Since $\{v_1, v_2, \ldots, v_n\}$ is an orthonormal basis, every $v \in V$ can be written as $v = \langle v, v_1 \rangle v_1 + \langle v, v_2 \rangle v_2 + \cdots + \langle v, v_n \rangle v_n$. Then $f(v) = \langle v, v_1 \rangle f(v_1) + \langle v, v_2 \rangle f(v_2) + \cdots + \langle v, v_n \rangle f(v_n)$. Since $f(v_i)$, $1 \le i \le n$ are scalars from the property of inner product we have

$$f(v) = \langle v, \overline{f(v_1)} v_1 \rangle + \langle v, \overline{f(v_2)} v_2 \rangle + \cdots + \langle v, \overline{f(v_n)} v_n \rangle$$
$$= \langle v, \overline{f(v_1)} v_1 + \overline{f(v_2)} v_2 + \cdots + \overline{f(v_n)} v_n \rangle.$$

Take $w = \overline{f(v_1)} v_1 + \overline{f(v_2)} v_2 + \cdots + \overline{f(v_n)} v_n$. Then we have $f(v) = \langle v, w \rangle \ \forall v \in V$. For uniqueness, suppose for some $u, v \in V$,

$$f(v) = \langle v, w \rangle = \langle v, u \rangle \quad \forall v \in V.$$

Then

$$\langle v, w \rangle - \langle v, u \rangle = 0 \quad \forall v \in V$$
$$\Rightarrow \langle v, w - u \rangle = 0 \quad \forall v \in V.$$

Take $v = w - u$. Then

$$\langle w - u, w - u \rangle = 0 \Rightarrow w - u = 0 \Rightarrow w = u. \qquad \square$$

This is called the Riesz representation theorem for finite-dimensional spaces. The above result is not necessarily true for an infinite-dimensional inner product space.

From above theorem, it can be remarked that if $\langle v, w_1 \rangle = \langle v, w_2 \rangle$, $\forall v \in V$, then $w_1 = w_2$.

Example 6.3.3. Let $P(x)$ be the vector space of all polynomials over $\mathbb{R}$ with the inner product defined by

$$\langle p(x), q(x) \rangle = \int_0^1 p(x) q(x) dx, \quad \text{where } p(x), q(x) \in P(x),$$

Since $P(x)$ is an infinite-dimensional vector space, in this example we shall show that above result is not true.

Let $f : P(x) \to \mathbb{R}$ be a functional defined by $f(p(x)) = p(0)$. In particular, if $p(x) = x^2 - 2x + 1$, then $f(p(x)) = p(0) = 1$. Similarly, $f(x^n p(x)) = 0p(0) = 0$.

Suppose $f(p(x)) = \langle p(x), r(x) \rangle$, where $r(x)$ is any fixed polynomial in $P(x)$. Then $f(p(x)) = p(0) = \int_0^1 p(x)r(x)dx$, for all $p(x) \in P(x)$.

Hence,

$$f(x^{2n}r(x)) = 0 = \int_0^1 x^{2n}r(x)r(x)dx, \; n \geq 1,$$

$$\Rightarrow \int_0^1 x^{2n}(r(x))^2 dx = 0$$

$$\Rightarrow \langle x^n r(x), x^n r(x) \rangle = 0$$

$$\Rightarrow \|x^n r(x)\|^2 = 0$$

$$\Rightarrow x^n r(x) = 0.$$

Since $x^n \neq 0$, $r(x)$ is a zero polynomial. Then

$$f(p(x)) = \langle p(x), r(x) \rangle = \langle p(x), 0 \rangle = 0 \quad \forall p(x).$$

This shows that f is a zero functional, which is a contradiction that f is a nonzero functional. Thus, there is no $r(x) \in P(x)$ such that $f(p(x)) = \langle p(x), r(x) \rangle$.

Theorem 6.3.4. *Let $T : V \to V$ be a linear operator on a finite-dimensional inner product space V over F. Then T^* exists and is unique.*

Proof. To complete the proof, we need to find a unique linear operator $T^* : V \to V$ such that $\langle T(u), v \rangle = \langle u, T^*(v) \rangle \; \forall u, v \in V$. Consider the composite function $f_v \circ T : V \to F$, where $f_v : V \to F$ defined by $f_v(u) = \langle u, v \rangle$ is a linear functional. Then $f_v \circ T(u) = f_v(T(u)) = \langle T(u), v \rangle$. Hence, $f_v \circ T$ being the composition of linear transformations is a linear functional for all v. Then from above theorem, there is a unique element $\grave{v} \in V$, such that $f_v \circ T(u) = \langle u, \grave{v} \rangle$. That is, $\langle T(u), v \rangle = \langle u, \grave{v} \rangle$, for all $u \in V$.

Define $T^* : V \to V$ by $T^*(v) = \grave{v}$. Then $\langle T(u), v \rangle = \langle u, T^*(v) \rangle$, $\forall u, v \in V$. Now, we shall show that T^* is linear. Let $v_1, v_2 \in V$ and $a_1, a_2 \in F$. Then

$$\begin{aligned}\langle u, T^*(a_1 v_1 + a_2 v_2) \rangle &= \langle T(u), a_1 v_1 + a_2 v_2 \rangle \\ &= \overline{a_1}\langle T(u), v_1 \rangle + \overline{a_2}\langle T(u), v_2 \rangle \\ &= \overline{a_1}\langle u, T^*(v_1) \rangle + \overline{a_2}\langle u, T^*(v_2) \rangle \\ &= \langle u, a_1 T^*(v_1) \rangle + \langle u, a_2 T^*(v_2) \rangle \\ &= \langle u, a_1 T^*(v_1) + a_2 T^*(v_2) \rangle \quad \forall u \in V.\end{aligned}$$

Hence, $T^*(a_1 v_1 + a_2 v_2) = a_1 T^*(v_1) + a_2 T^*(v_2)$ and so T^* is linear.

For uniqueness, suppose

$$\langle u, T_1^*(v)\rangle = \langle u, T_2^*(v)\rangle, \quad \text{for all } u, v \in V.$$

Hence, $T_1^*(v) = T_2^*(v) \; \forall v$ and so $T_1^* = T_2^*$. □

Theorem 6.3.5. *Let T_1 and T_2 be linear operators on a finite-dimensional inner product space V and k is an scalar. Then:*
(i) $(T_1 + T_2)^* = T_1^* + T_2^*$,
(ii) $(kT_1)^* = \bar{k}T_1^*$,
(iii) $(T_1 T_2)^* = T_2^* T_1^*$,
(iv) $(T_1^*)^* = T_1$,
(v) $I^* = I$, *where I is the identity transformation,*
(vi) $0^* = 0$, *where 0 is the zero transformation.*

Proof. (i) Let $u, v \in V$. Then

$$\langle u, (T_1^* + T_2^*)(v)\rangle$$
$$= \langle u, T_1^*(v) + T_2^*(v)\rangle$$
$$= \langle u, T_1^*(v)\rangle + \langle u, T_2^*(v)\rangle$$
$$= \langle T_1(u), v\rangle + \langle T_2(u), v\rangle$$
$$= \langle T_1(u) + T_2(u), v\rangle$$
$$= \langle (T_1 + T_2)(u), v\rangle$$
$$= \langle u, (T_1 + T_2)^*(v)\rangle.$$

Since u, v are arbitrary, we have $(T_1 + T_2)^* = T_1^* + T_2^*$.
Similarly,
(ii)

$$\langle u, \bar{k}T_1^*(v)\rangle = k\langle u, T_1^*(v)\rangle = k\langle T_1(u), v\rangle$$
$$= \langle kT_1(u), v\rangle = \langle u, (kT_1)^*(v)\rangle$$
$$\Rightarrow (kT_1)^* = \bar{k}T_1^*.$$

(iii)

$$\langle u, T_2^* T_1^*(v)\rangle = \langle T_2(u), T_1^*(v)\rangle = \langle T_1 T_2(u), v\rangle$$
$$= \langle u, (T_1 T_2)^*(v)\rangle$$
$$\Rightarrow (T_1 T_2)^* = T_2^* T_1^*.$$

(iv)

$$\langle T_1(u), v\rangle = \langle u, T_1^*(v)\rangle = \overline{\langle T_1^*(v), u\rangle} = \overline{\langle v, (T_1^*)^* u\rangle}$$
$$= \langle (T_1^*)^* u, v\rangle$$
$$\Rightarrow (T_1^*)^* = T_1.$$

(v)
$$\langle u, v\rangle = \langle I(u), v\rangle = \langle u, I^*(v)\rangle$$
$$\Rightarrow \langle u, v\rangle = \langle u, I^*(v)\rangle$$
$$\Rightarrow \langle u, I(v)\rangle = \langle u, I^*(v)\rangle$$
$$\Rightarrow I^* = I.$$

(vi)
$$\langle u, 0^*(v)\rangle = \langle 0(u), v\rangle = \langle 0, v\rangle = 0 = \langle u, 0\rangle$$
$$= \langle u, 0(v)\rangle$$
$$\Rightarrow 0^* = 0. \qquad \square$$

Definition 6.3.6. A linear operator $T : V \rightarrow V$ on a finite-dimensional complex inner product space V is called

- self-adjoint or Hermitian (symmetric in case of real inner product) if $T^* = T$,
- skew-Hermitian (skew-symmetric in case of real inner product) if $T^* = -T$,
- unitary if $T^*T = TT^* = I$ (orthogonal in case of real inner product) and normal if $T^*T = TT^*$.

It is clear from the definition that Hermitian, skew-Hermitian and unitary operators are normal.

Theorem 6.3.7. *Let $T : V \rightarrow V$ be a linear operator on a finite-dimensional inner product space V and let $B = \{v_1, v_2, \ldots, v_n\}$ be an orthonormal basis of V. Then $[T^*]_{B,B} = [T]^*_{B,B}$. That is, the matrix of the adjoint of T is the adjoint (tranjugate) of the matrix of T.*

Proof. Since B is an orthonormal basis of V, any vector $v \in V$ can be written as $v = \sum_{i=1}^{n} \langle v, v_i\rangle v_i$.

Hence, $T(v_1) \in V$ can be written as

$$T(v_1) = \langle T(v_1), v_1\rangle v_1 + \langle T(v_1), v_2\rangle v_2 + \cdots + \langle T(v_1), v_n\rangle v_n,$$
$$\vdots$$
$$T(v_n) = \langle T(v_n), v_1\rangle v_1 + \langle T(v_n), v_2\rangle v_2 + \cdots + \langle T(v_n), v_n\rangle v_n.$$

Then

$$[T]_{B,B} = \begin{bmatrix} \langle T(v_1), v_1\rangle & \cdots & \langle T(v_n), v_1\rangle \\ \langle T(v_1), v_2\rangle & \cdots & \langle T(v_n), v_2\rangle \\ \vdots & & \\ \langle T(v_1), v_n\rangle & \cdots & \langle T(v_n), v_n\rangle \end{bmatrix}.$$

Thus, the ij-th entry of $[T]_{B,B} = \langle T(v_j), v_i\rangle$. Similarly, ij-th entry of $[T^*]_{B,B} = \langle T^*(v_j), v_i\rangle$.

Now, $\langle T^*(v_j), v_i \rangle = \overline{\langle v_i, T^*(v_j) \rangle} = \overline{\langle T(v_i), v_j \rangle}$, the complex conjugate of ji-th entry of $[T]_{B,B}$. Hence, $[T^*]_{B,B} = [T]^*_{B,B}$. □

Corollary 6.3.8. *Let T and S be linear operators on a finite- dimensional inner product space V and let $B = \{v_1, v_2, \ldots, v_n\}$ be an orthonormal basis of V. Then $[T]_{B,B} = [S]^*_{B,B}$ implies that $T = S^*$.*

Proof. From above theorem, $[T]_{B,B} = [S]^*_{B,B} = [S^*]_{B,B}$. Since matrix representation of linear operators T and S^* on V with respect to same bases are equal, $T = S^*$. □

Corollary 6.3.9. *Let T be a linear operator on a finite- dimensional inner product space V. Then for the orthonormal basis B, we have:*
(i)   *T is Hermitian if and only if $[T]_{B,B}$ is Hermitian,*
(ii) *T is skew-Hermitian if and only if $[T]_{B,B}$ is skew-Hermitian,*
(iii) *T is unitary if and only if $[T]_{B,B}$ is unitary,*
(iv) *T is normal if and only if $[T]_{B,B}$ is normal.*

Proof. (i)   Let T is Hermitian. Then $T = T^*$, and hence from Theorem 6.3.7, $[T]_{B,B} = [T^*]_{B,B} = [T]^*_{B,B}$. That is, $[T]_{B,B}$ is Hermitian.
Conversely, suppose $[T]_{B,B}$ is Hermitian. Then from Corollary 6.3.8,

$$[T]^*_{B,B} = [T]_{B,B} \Rightarrow T^* = T.$$

Similarly, remaining parts follow and left as an exercise. □

The following result establishes a beautiful connection between the notion of adjoint of a linear transformation with that of adjoint of a matrix.

Corollary 6.3.10. *Let A be an $n \times n$ matrix with entries in a field F. Then $L^*_A = L_{A^*}$ and $\langle AX, Y \rangle = \langle X, A^*Y \rangle \; \forall X, Y \in F^n$.*

Proof. We have that L_A and L_{A^*} are linear transformations on F^n. Also, if B is an standard basis of F^n, then $[L_A]_{B,B} = A$ and $[L_{A^*}]_{B,B} = A^*$, where A^* is the tranjugate of the matrix A.
Now from Theorem 6.3.7,

$$[L^*_A]_{B,B} = [L_A]^*_{B,B} = A^* = [L_{A^*}]_{B,B}, \quad \text{and hence} \quad L^*_A = L_{A^*}.$$

For column vectors X and Y in F^n,

$$\langle L_A(X), Y \rangle = \langle X, L^*_A(Y) \rangle = \langle X, L_{A^*}(Y) \rangle.$$

Since $L_A(X) = AX$ and $L_{A^*}(Y) = A^*Y$, we have $\langle AX, Y \rangle = \langle X, A^*Y \rangle \; \forall X, Y \in F^n$. □

By replacing operators T_1 and T_2 with n-square matrices A and B, respectively, we obtain Theorem 6.3.5 in matrix form. That is:
(i) $(A + B)^* = A^* + B^*$,
(ii) $(kA)^* = \bar{k}A^*$,

(iii) $(AB)^* = B^*A^*$,

(iv) $(A^*)^* = A$,

(v) $I_n^* = I_n$, where I_n is the identity matrix,

(vi) $0^* = 0$, where 0 is the zero matrix.

Theorem 6.3.11. *Let $T : V \to V$ be an operator on a finite- dimensional inner product space V over the field F. Then the following statements are equivalent:*

(i) *T is unitary (orthogonal),*

(ii) *$\langle T(u), T(v) \rangle = \langle u, v \rangle$, for all $u, v \in V$,*

(iii) *$\|T(u)\| = \|u\|$ $\forall u \in V$.*

Proof. (i) $\Rightarrow$ (ii)

Since T is unitary, $TT^* = T^*T = I$. Hence,

$$\langle T(u), T(v) \rangle = \langle u, T^*T(v) \rangle = \langle u, I(v) \rangle = \langle u, v \rangle, \quad \text{for all } u, v \in V.$$

(ii) $\Rightarrow$ (iii)

$$\|T(u)\|^2 = \langle T(u), T(u) \rangle = \langle u, u \rangle = \|u\|^2$$
$$\Rightarrow \|T(u)\| = \|u\| \quad \forall u \in V.$$

(iii) $\Rightarrow$ (i)

$$\|T(u)\|^2 = \|u\|^2 \Rightarrow \langle T(u), T(u) \rangle = \langle u, u \rangle$$
$$\Rightarrow \langle u, T^*T(u) \rangle = \langle u, I(u) \rangle \quad \forall u$$
$$\Rightarrow T^*T = I.$$

Since V is finite-dimensional $TT^* = I$. Hence, T is unitary. $\square$

If A is an n-square unitary (orthogonal) matrix, then from Corollary 6.3.10,

$$\langle AX, AY \rangle = \langle X, A^*AY \rangle = \langle X, Y \rangle \quad \forall X, Y \in F^n.$$

Hence, from above theorem we have the following.

Corollary 6.3.12. *Let A bc an n-square matrix. Then the following are equivalent:*

(i) *A is unitary (orthogonal).*

(ii) *$\langle AX, AY \rangle = \langle X, Y \rangle$ $\forall X, Y \in F^n$.*

(iii) *$\|AX\| = \|X\|$ $\forall X \in F^n$.*

Theorem 6.3.13. *Let V be a finite-dimensional inner product space. Then a linear operator $T : V \to V$ is unitary (orthogonal) if and only if it takes an orthonormal basis of V into an orthonormal basis.*

Proof. Let $\{v_1, v_2, \ldots, v_n\}$ be an orthonormal basis of V. If T is unitary, then

$$\langle T(v_i), T(v_j)\rangle = \langle v_i, v_j\rangle = \delta_{ij}.$$

Hence, the set $\{T(v_1), T(v_2), \ldots, T(v_n)\}$ forms an orthonormal basis of V.

Conversely, suppose that $\{v_1, v_2, \ldots, v_n\}$ and $\{T(v_1), T(v_2), \ldots, T(v_n)\}$ are orthonormal bases of V. Then for $u, v \in V$,

$$\langle T(u), T(v)\rangle = \left\langle T\left(\sum_{i=1}^{n} a_i v_i\right), T\left(\sum_{i=1}^{n} b_i v_i\right)\right\rangle$$
$$= \left\langle \sum_{i=1}^{n} a_i T(v_i), \sum_{i=1}^{n} b_i T(v_i)\right\rangle,$$

where $a_i = \langle u, v_i\rangle$ and $b_i = \langle v, v_i\rangle$.

Since $\{T(v_1), T(v_2), \ldots, T(v_n)\}$ is an orthonormal basis,

$$\langle T(v_i), T(v_j)\rangle = \delta_{ij}.$$

Hence, $\langle T(u), T(v)\rangle = \sum_{i=1}^{n} a_i \overline{b_i}$.

Also, $\langle u, v\rangle = \langle \sum_{i=1}^{n} a_i v_i, \sum_{i=1}^{n} b_i v_i\rangle = \sum_{i=1}^{n} a_i \overline{b_i}$. Thus, $\langle T(u), T(v)\rangle = \langle u, v\rangle$ implies that T is unitary. □

Corollary 6.3.14. *Any two n-dimensional complex inner product spaces are isomorphic as inner product spaces.*

Proof. Let U and V be two complex inner product spaces of dimension n. Then there exist orthonormal bases $\{u_1, u_2, \ldots, u_n\}$ and $\{v_1, v_2, \ldots, v_n\}$ of U and V, respectively. Define an isomorphism $T : U \to V$ such that $T(u_i) = v_i \; \forall i$. Then from the proof of the above theorem, we get that T preserves the inner product. Hence, T is an inner product space isomorphism. □

Since $\mathbb{C}^n$ is an n-dimensional complex inner product space, we have the following corollary.

Corollary 6.3.15. *An n-dimensional complex inner product space is isomorphic as an inner product space to $\mathbb{C}^n$.*

Theorem 6.3.16. *Let A be an n-square matrix. Then the following statements are equivalent:*
(i)   *A is an unitary (orthogonal) matrix.*
(ii) *The columns of A form an orthonormal set of vectors.*
(iii) *The rows of A form an orthonormal set of vectors.*

Proof. (i) ⇔ (ii)

Let $A = [X_1 \, X_2 \, \cdots \, X_n]$, where X_i are column vectors in F^n. Then

$$A^* = \begin{bmatrix} X_1^* \\ X_2^* \\ \vdots \\ X_n^* \end{bmatrix} \quad \text{and} \quad A^*A = \begin{bmatrix} X_1^*X_1 & X_1^*X_2 & \cdots & X_1^*X_n \\ X_2^*X_1 & X_2^*X_2 & \cdots & X_2^*X_n \\ \vdots & & & \\ X_n^*X_1 & X_n^*X_2 & \cdots & X_n^*X_n \end{bmatrix}.$$

Thus, $A^*A = I_n$ if and only if $X_i^*X_j = \delta_{ij}$. Hence, A is unitary if and only if columns of A form an orthonormal set of vectors.

(i) $\Leftrightarrow$ (iii)

Since $A^*A = I_n$ implies $AA^* = I_n$, we have (i) $\Leftrightarrow$ (iii). □

Example 6.3.17.

(a) Let $T : \mathbb{R}^2 \to \mathbb{R}^2$ be a linear transformation defined by $T(x,y) = (x, x+y)$.
Let us workout what $T^* : \mathbb{R}^2 \to \mathbb{R}^2$ is.
Let (s, t) be any vector in $\mathbb{R}^2$. Then $\langle T(x,y), (s,t) \rangle = \langle (x,y), T^*(s,t) \rangle$, for all $(x,y) \in \mathbb{R}^2$.

$$\langle (x, x+y), (s,t) \rangle = \langle (x,y), T^*(s,t) \rangle,$$
$$xs + xt + yt = \langle (x,y), T^*(s,t) \rangle,$$
$$(s+t)x + yt = \langle (x,y), T^*(s,t) \rangle,$$
$$\langle (x,y), (s+t,t) \rangle = \langle (x,y), T^*(s,t) \rangle, \quad \text{for all } (x,y) \in \mathbb{R}^2.$$

Hence, $T^*(s,t) = (s+t, t)$.
Let B be the standard orthonormal basis of $\mathbb{R}^2$. Then the matrix

$$[T]_{B,B} = \begin{bmatrix} 1 & 0 \\ 1 & 1 \end{bmatrix} \quad \text{and} \quad [T^*]_{B,B} = \begin{bmatrix} 1 & 1 \\ 0 & 1 \end{bmatrix}.$$

Thus, we have $[T^*]_{B,B} = [T]_{B,B}^* = [T]_{B,B}^t$.

(b) Let $T : \mathbb{C}^2 \to \mathbb{C}^2$ be a linear transformation defined by $T(x,y) = (3x + 5iy, ix - 2iy)$.
Let $(s,t) \in \mathbb{C}^2$ be any vector such that $\langle T(x,y), (s,t) \rangle = \langle (x,y), T^*(s,t) \rangle \; \forall (x,y) \in \mathbb{C}^2$.
Then

$$\langle (3x + 5iy, ix - 2iy), (s,t) \rangle$$
$$= (3x + 5iy)\bar{s} + (ix - 2iy)\bar{t}$$
$$= (3\bar{s} + i\bar{t})x + (5i\bar{s} - 2i\bar{t})y$$
$$= x(3\bar{s} + i\bar{t}) + y(5i\bar{s} - 2i\bar{t})$$
$$= x\overline{(3s - it)} + y\overline{(-5is + 2it)}$$
$$= \langle (x,y), (3s - it, -5is + 2it) \rangle$$
$$= \langle (x,y), T^*(s,t) \; \forall (x,y) \in \mathbb{C}^2 \rangle.$$

Hence,

$$T^*(s, t) = (3s - it, -5is + 2it) \quad \text{or} \quad T^*(x, y) = (3x - iy, -5ix + 2iy).$$

If B is the standard basis in $\mathbb{C}^2$, then

$$[T]_{B,B} = \begin{bmatrix} 3 & 5i \\ i & -2i \end{bmatrix} \quad \text{and} \quad [T]^*_{B,B} = \begin{bmatrix} 3 & -i \\ -5i & 2i \end{bmatrix}.$$

Thus, $[T^*]_{B,B} = [T]^*_{B,B}$.

Note that to compute T^* for a given linear operator T it is easy to compute the matrix $[T]_{B,B}$ with respect to standard orthonormal basis B and then compute T^* using the matrix $[T]^*_{B,B}$.

Example 6.3.18. Let $T_1 : \mathbb{C}^3 \to \mathbb{C}^3$ be a linear operator defined by $T_1(x, y, z) = (x + (1 - i)y + 2z, (1 + i)x + 3y + iz, 2x - iy)$. Then the matrix of T_1 with respect to the standard basis B of $\mathbb{C}^3$ is

$$[T_1]_{B,B} = \begin{bmatrix} 1 & 1-i & 2 \\ 1+i & 3 & i \\ 2 & -i & 0 \end{bmatrix}$$

is Hermitian. From the result that T is Hermitian if and only if the matrix $[T]_{B,B}$ is Hermitian, we have that T_1 is Hermitian.

Similarly, the operator T_2 on $\mathbb{C}^3$ defined by $T_2(x, y, z) = (ix, iz, iy)$ is skew-Hermitian and unitary both. The operator T_3 on $\mathbb{R}^3$ defined by $T_3(x, y, z) = (x + 2y + 3z, 2x + y + 4z, 3x + 4y)$ is symmetric. The operator T_4 on $\mathbb{R}^3$ defined by $T_4(x, y, z) = \frac{1}{3}(x + 2y + 2z, 2x + y - 2z, -2x + 2y - z)$ is orthogonal.

Theorem 6.3.19. *Let T be an operator on a complex inner product space V. Then T is Hermitian if and only if $\langle T(v), v \rangle \in \mathbb{R} \,\forall v \in V$.*

Proof. Suppose T is Hermitian. Then $\langle T(v), v \rangle = \langle v, T^*(v) \rangle = \langle v, T(v) \rangle = \overline{\langle T(v), v \rangle}$. This shows that $\langle T(v), v \rangle \in \mathbb{R} \,\forall v \in V$.

Conversely, suppose that $\langle T(v), v \rangle \in \mathbb{R} \,\forall v \in V$. Then $\langle T(v + w), v + w \rangle \in \mathbb{R} \,\forall v, w \in V$. Hence,

$$\langle T(v + w), v + w \rangle = \langle T(v) + T(w), v + w \rangle$$
$$= \langle T(v), v \rangle + \langle T(v), w \rangle + \langle T(w), v \rangle + \langle T(w), w \rangle \in \mathbb{R} \quad \forall v, w \in V.$$

Since $\langle T(v), v \rangle, \langle T(w), w \rangle \in \mathbb{R}$, we have

$$\langle T(v), w \rangle + \langle T(w), v \rangle \in \mathbb{R} \quad \forall v, w \in V. \tag{6.1}$$

Again, since $v + iw \in V$, we have $\langle T(v + iw), (v + iw) \rangle \in \mathbb{R}$. Then

$$\langle T(v + iw), (v + iw) \rangle$$
$$= \langle T(v), v \rangle + \langle T(v), iw \rangle + \langle iT(w), v \rangle + \langle iT(w), iw \rangle$$
$$= \langle T(v), v \rangle - i\langle T(v), w \rangle + i\langle T(w), v \rangle + i\bar{i}\langle T(w), w \rangle$$
$$= \langle T(v), v \rangle + \langle T(w), w \rangle + i(\langle T(w), v \rangle - \langle T(v), w \rangle) \in \mathbb{R}.$$

Since $\langle T(v), v \rangle + \langle T(w), w \rangle \in \mathbb{R}$, $i(\langle T(w), v \rangle - \langle T(v), w \rangle) \in \mathbb{R}$. This implies that

$$\langle T(w), v \rangle - \langle T(v), w \rangle \quad \forall v, w \in V \tag{6.2}$$

is purely imaginary.

Notice that if $z_1 = x_1 + iy_1$ and $z_2 = x_2 + iy_2$ are two complex numbers such that $z_1 + z_2 \in \mathbb{R}$ and $z_1 - z_2$ is purely imaginary, then $x_1 + x_2 + i(y_1 + y_2) \in \mathbb{R} \Rightarrow y_1 + y_2 = 0 \Rightarrow y_1 = -y_2$ and $x_1 - x_2 + i(y_1 - y_2)$ purely imaginary $\Rightarrow x_1 - x_2 = 0 \Rightarrow x_1 = x_2$.

Hence, $z_1 = x_1 + iy_1$ and $z_2 = x_1 - iy_1 \Rightarrow z_1 = \bar{z_2}$.

Thus, from equations (6.1) and (6.2) we have $\langle T(v), w \rangle = \overline{\langle T(w), v \rangle} = \langle v, T(w) \rangle$ $\forall v, w \in V$. Hence, T is Hermitian. $\qquad \square$

Corollary 6.3.20. *Let T be a Hermitian linear operator on a finite-dimensional complex inner product space V. Then:*
(i)   *$I + iT$ is invertible;*
(ii) *$I - iT$ is invertible;*
(iii) *$(I + iT)(I - iT)^{-1}$ is unitary.*

Proof. (i)   Since V is a finite-dimensional inner product space, it is sufficient to show that $I + iT$ is injective. Let $v \in \mathrm{Ker}(I + iT)$. Then $(I + iT)(v) = I(v) + iT(v) = v + iT(v) = 0 \Rightarrow v = -iT(v)$. Since $\langle v, v \rangle \in \mathbb{R}$,

$$\langle v, v \rangle = \langle -iT(v), v \rangle = -i\langle T(v), v \rangle \in \mathbb{R}.$$

Since T is Hermitian, from the above theorem we have that $\langle T(v), v \rangle \in \mathbb{R}$ $\forall v \in V$. Hence, $-i\langle T(v), v \rangle \in \mathbb{R} \Rightarrow \langle T(v), v \rangle = 0$. Then $\langle v, v \rangle = 0$, and so $v = 0$. Thus, $\mathrm{Ker}(I + iT) = \{0\}$. Hence, $I + iT$ is invertible.

(ii) As in part (i), we get $(I - iT)(v) = 0 \Rightarrow v = iT(v)$ and $\langle v, v \rangle = i\langle T(v), v \rangle$ $\forall v$. Hence, $\mathrm{Ker}(I - iT) = \{0\}$.

(iii) Since $I + iT$ and $I - iT$ are operators, we have

$$\left((I + iT)(I - iT)^{-1}\right)^* = \left((I - iT)^{-1}\right)^*(I + iT)^* = \left((I - iT)^*\right)^{-1}(I + iT)^*$$
$$= \left((I^* - \bar{i}T^*)\right)^{-1}(I^* + \bar{i}T^*) = (I + iT)^{-1}(I - iT).$$

Then

$$((I + iT)(I - iT)^{-1})^*(I + iT)(I - iT)^{-1} = (I + iT)^{-1}(I - iT)(I + iT)(I - iT)^{-1}$$
$$= (I + iT)^{-1}(I + iT)(I - iT)(I - iT)^{-1} = I.$$

It is easy to verify that $(I - iT)(I + iT) = (I + iT)(I - iT)$, which is used in the above step.

Similarly, $(I + iT)(I - iT)^{-1}((I + iT)(I - iT)^{-1})^* = I$. Hence, $(I + iT)(I - iT)^{-1}$ is unitary.

$\square$

Exercises

(6.3.1) Find the adjoint of $T : \mathbb{R}^3 \to \mathbb{R}^3$ defined by:
 (i) $T(x, y, z) = (x + 2y - z, -2x + z, x + 3y - 5z)$,
 (ii) $T(x, y, z) = (x + y, y + z, z + x)$,
 (iii) $T(x, y, z) = (x + y + z, x - y - z, 3y)$,
 (iv) $T(x, y, z) = (z, y, x)$.

(6.3.2) Find the adjoint of $T : \mathbb{C}^3 \to \mathbb{C}^3$ defined by:
 (i) $T(x, y, z) = (ix, -iy, z)$,
 (ii) $T(x, y, z) = (x + (1 + i)y - iz, (1 - i)y + (1 + i)z, 3ix - iz)$,
 (iii) $T(x, y, z) = ((1 + 2i)x + y + 2iz, -ix + (3 + 4i)z, -iy + iz)$,
 (iv) $T(x, y, z) = (x + iy, ix + 2iz, 5iy)$.

(6.3.3) If T is an invertible operator, then show that $(T^{-1})^* = (T^*)^{-1}$.

(6.3.4) Show that the operator $T : \mathbb{C}^3 \to \mathbb{C}^3$ defined by $T(x, y, z) = (2x + (3 + 2i)y - 4z, (3 - 2i)x + 5y + 6iz, -4x - 6iy + 3z)$ is Hermitian and iT defined by $iT(x, y, z) = T(ix, iy, iz)$ is skew-Hermitian.

(6.3.5) Show that the operators T_1 and T_2 on $\mathbb{C}^2$ defined by $T_1(x, y) = \frac{1}{2}(ix + \sqrt{3}y, \sqrt{3}x + iy)$ and $T_2(x, y) = \frac{1}{\sqrt{3}}(x + (1 + i)y, (1 - i)x - y)$ are unitary.

(6.3.6) Show that the operators T_1 and T_2 on $\mathbb{R}^3$ defined by $T_1(x, y, z) = (-2x + 5y + 4z, 5x + 7y + 5z, 4x + 5y + 3z)$ and $T_2(x, y, z) = (2y + 5z, -2x - 3z, -5x + 3y)$ are symmetric and skew-symmetric, respectively.

(6.3.7) Show that the operator T on $\mathbb{R}^3$ defined by $T(x, y, z) = (\frac{x}{3} + \frac{2}{3}y + \frac{2}{3}z, \frac{2}{3}x + \frac{y}{3} - \frac{2}{3}z, -\frac{2}{3}x + \frac{2}{3}y - \frac{1}{3}z)$ is orthogonal.

(6.3.8) Let V be a real inner product space. Then show that (i) the set of symmetric operators on V and (ii) the set of skew-symmetric operators on V form a subspace of the vector space of linear transformations on V.

(6.3.9) Let V be a complex inner product space. Then show that (i) the set of Hermitian operators on V and (ii) the set of skew-Hermitian operators on V are subgroups but not subspaces of vector space of linear transformations on V.

(6.3.10) Show that T is Hermitian if and only if iT is skew-Hermitian.

(6.3.11) Show that for a Hermitian linear operator T, kT is Hermitian if and only if k is real.

(6.3.12) Show that for a skew-Hermitian linear operator T, kT is skew-Hermitian if and only if k is purely imaginary.

(6.3.13) If T is a skew-Hermitian operator on a finite-dimensional complex inner product space, then show that $I + T$ and $I - T$ are invertible.

7 Determinant, eigenvalues and diagonalization of matrices

In this chapter, we delve into the fundamental concepts of determinants, eigenvalues and diagonalization of matrices, which have broad applications in mathematics, physics and engineering.

7.1 Determinant of a matrix

The determinant of an n-square matrix A is a number, which depends in a complicated way on the entries of A. The determinant of A is denoted by $\det(A)$ or $|A|$. Despite the complicated definition, the determinant of a matrix has some very remarkable properties, especially with regard to matrix multiplication, and row and column operations. Since the topic determinant is a part of the subject something called exterior algebra, which is beyond the scope of this book, we will not be able to give the proof of many of these remarkable properties here. So, we will write some results without their proofs.

The determinant of any square matrix is defined recursively, that is, the determinant of a large matrix is defined in terms of the determinant of smaller matrices. Before defining the determinant of a matrix, we give the definition of a submatrix.

Definition 7.1.1. Let $A = [a_{ij}]$ be an n-square matrix. Then the matrix A_{ij} of order $(n - 1) \times (n - 1)$ obtained from A by removing the i-th row and j-th column of A is called a submatrix of A.

For example, let

$$A = \begin{bmatrix} 1 & 2 & 3 \\ -1 & 0 & 2 \\ 3 & 4 & 2 \end{bmatrix},$$

be a 3-square matrix. Then

$$A_{11} = \begin{bmatrix} 0 & 2 \\ 4 & 2 \end{bmatrix}, \quad A_{12} = \begin{bmatrix} -1 & 2 \\ 3 & 2 \end{bmatrix}, \quad A_{13} = \begin{bmatrix} -1 & 0 \\ 3 & 4 \end{bmatrix},$$

$$A_{21} = \begin{bmatrix} 2 & 3 \\ 4 & 2 \end{bmatrix}, \quad A_{22} = \begin{bmatrix} 1 & 3 \\ 3 & 2 \end{bmatrix}, \quad A_{23} = \begin{bmatrix} 1 & 2 \\ 3 & 4 \end{bmatrix},$$

$$A_{31} = \begin{bmatrix} 2 & 3 \\ 0 & 2 \end{bmatrix}, \quad A_{32} = \begin{bmatrix} 1 & 3 \\ -1 & 2 \end{bmatrix} \quad \text{and} \quad A_{33} = \begin{bmatrix} 1 & 2 \\ -1 & 0 \end{bmatrix}$$

are submatrices of the matrix A.

Now, we define the determinant of a square matrix.

https://doi.org/10.1515/9783111516035-007

Definition 7.1.2. Let $A = [a_{ij}]$ be an n-square matrix. Then
(i) for $n = 1$, $A = [a_{11}]$ and $\det(A) = a_{11}$, and
(ii) for $n \geq 2$, $\det(A) = \sum_{j=1}^{n}(-1)^{1+j}a_{1j}\det(A_{1j})$.

For example, if $A = [6]$, then $\det(A) = 6$.
If

$$A = \begin{bmatrix} a_{11} & a_{12} \\ a_{21} & a_{22} \end{bmatrix},$$

then $\det(A) = (-1)^{1+1}a_{11}\det(A_{11}) + (-1)^{1+2}a_{12}\det(A_{12}) = a_{11}a_{22} - a_{12}a_{21}$. That is, if

$$A = \begin{bmatrix} a & b \\ c & d \end{bmatrix}$$

is any 2×2 matrix, then $\det(A) = ad - bc$.
If

$$A = \begin{bmatrix} 1 & 2 & 4 \\ -3 & 5 & 2 \\ 6 & 2 & 0 \end{bmatrix},$$

then

$$\det(A) = (-1)^{1+1}a_{11}\det(A_{11}) + (-1)^{1+2}a_{12}\det(A_{12}) + (-1)^{1+3}a_{13}\det(A_{13})$$

$$= (1)(1)\det\left(\begin{bmatrix} 5 & 2 \\ 2 & 0 \end{bmatrix}\right) + (-1)(2)\det\left(\begin{bmatrix} -3 & 2 \\ 6 & 0 \end{bmatrix}\right) + (1)(4)\det\left(\begin{bmatrix} -3 & 5 \\ 6 & 2 \end{bmatrix}\right)$$

$$= (1)(1)(-4) + (-1)(2)(-12) + (1)(4)(-36)$$

$$= -124.$$

The definition of the determinant is given by using the expansion along the first row of the matrix but the determinant of the matrix can be computed by expanding along any row or column of the matrix. That is clear from the following theorem whose proof is omitted.

Theorem 7.1.3. *Let $A = [a_{ij}]$ be an n-square matrix. Then*

$$\det(A) = \sum_{j=1}^{n}(-1)^{i+j}a_{ij}\det(A_{ij}) \quad \textit{(expansion along i-th row)}$$

$$= \sum_{i=1}^{n}(-1)^{i+j}a_{ij}\det(A_{ij}) \quad \textit{(expansion along j-th column)}.$$

The above theorem provides an amazing thing that the determinant of a matrix can be obtained by expanding along any row or column. In the following example, the determinant of the given matrix is computed along first row expansion and along third column expansion both.

Example 7.1.4. Let

$$A = \begin{bmatrix} 2 & 0 & -1 \\ 3 & 2 & 1 \\ -2 & 3 & 4 \end{bmatrix}.$$

Then the determinant of the matrix along:
(a) The first row expansion is

$$\det(A) = (-1)^{1+1}(2) \det\left(\begin{bmatrix} 2 & 1 \\ 3 & 4 \end{bmatrix}\right) + (-1)^{1+2}(0) \det\left(\begin{bmatrix} 3 & 1 \\ -2 & 4 \end{bmatrix}\right)$$

$$+ (-1)^{1+3}(-1) \det\left(\begin{bmatrix} 3 & 2 \\ -2 & 3 \end{bmatrix}\right)$$

$$= (1)(2)(5) + 0 + (1)(-1)(13)$$

$$= -3.$$

(b) The third column expansion is

$$\det(A) = (-1)^{1+3}(-1) \det\left(\begin{bmatrix} 3 & 2 \\ -2 & 3 \end{bmatrix}\right) + (-1)^{2+3}(1) \det\left(\begin{bmatrix} 2 & 0 \\ -2 & 3 \end{bmatrix}\right)$$

$$+ (-1)^{3+3}(4) \det\left(\begin{bmatrix} 2 & 0 \\ 3 & 2 \end{bmatrix}\right)$$

$$= (1)(-1)(13) + (-1)(1)(6) + (1)(4)(4)$$

$$= -3.$$

From Theorem 7.1.3, the following result immediately follows.

Corollary 7.1.5. *Let A be an n-square matrix. Then* $\det(A) = \det(A^t)$.

Example 7.1.6.
(a) Let $A = [a_{ij}]$ be an n-square matrix whose i-th row is zero. Then $\det(A) = 0$.
   By expansion along i-th row, $\det(A) = \sum_{j=1}^{n}(-1)^{i+j}a_{ij}\det(A_{ij})$. Since i-th row of A is zero ($a_{ij} = 0$) $\forall j$,

$$\det(A) = \sum_{j=1}^{n}(-1)^{i+j}(0)\det(A_{ij}) = 0.$$

Similarly, $\det(A) = 0$ if j-th column of A is zero.

Thus, the determinant of an n-square matrix A is zero if its at least one row or column is zero.

(b) Let

$$A = \begin{bmatrix} a_{11} & a_{12} & a_{13} & \cdots & a_{1n} \\ 0 & a_{22} & a_{23} & \cdots & a_{2n} \\ 0 & 0 & a_{33} & \cdots & a_{3n} \\ \vdots & & & & \\ 0 & 0 & 0 & \cdots & a_{nn} \end{bmatrix},$$

be an upper triangular matrix. Then to evaluate $\det(A)$ we use expansion along first column. Since there is only one nonzero entry in the first column $\det(A) = (-1)^{1+1} a_{11} \det(A_{11}) = a_{11} \det(A_{11})$, where

$$A_{11} = \begin{bmatrix} a_{22} & a_{23} & a_{24} & \cdots & a_{2n} \\ 0 & a_{33} & a_{34} & \cdots & a_{3n} \\ 0 & 0 & a_{44} & \cdots & a_{4n} \\ \vdots & & & & \\ 0 & 0 & 0 & \cdots & a_{nn} \end{bmatrix} = B \quad \text{(say)},$$

is an upper triangular matrix. Then $\det(A_{11}) = a_{22} \det(B_{11})$, where B_{11} is again an upper triangular sub matrix of A_{11}, and hence $\det(A) = a_{11}a_{22} \det(B_{11})$.

Proceeding in this fashion, we get $\det(A) = a_{11}a_{22} \cdots a_{nn}$.

In particular, let

$$A = \begin{bmatrix} 2 & 3 & 1 \\ 0 & 3 & -1 \\ 0 & 0 & 5 \end{bmatrix}.$$

Then $\det(A) = 2 \times 3 \times 5 = 30$.

If A is a lower triangular matrix, then A^t is an upper triangular. In this case, $\det(A) = \det(A^t) = $ Product of diagonal entries of A.

Thus, the determinant of an upper triangular matrix, lower triangular matrix or diagonal matrix is equal to the product of diagonal entries of the matrix.

(c) If I_n is the identity matrix, then $\det(I_n) = 1 \times 1 \times 1 \times \cdots \times 1 = 1$.

Theorem 7.1.7. *If B is the matrix obtained from an n-square matrix A by interchanging any two rows or columns, then $\det(B) = -\det(A)$.*

Proof. To prove the result, we use the principle of mathematical induction on n.

For $n = 2$, let

$$A = \begin{bmatrix} a_{11} & a_{12} \\ a_{21} & a_{22} \end{bmatrix}.$$

Then

$$B = \begin{bmatrix} a_{21} & a_{22} \\ a_{11} & a_{12} \end{bmatrix}.$$

Now, $\det(A) = a_{11}a_{22} - a_{12}a_{21}$ and $\det(B) = a_{12}a_{21} - a_{11}a_{22} = -(a_{11}a_{22} - a_{12}a_{21})$. Hence, $\det(B) = -\det(A)$. Thus, the result is true for $n = 2$.

Suppose the result is true for all l-square matrices. Let A be a $(l+1)$-square matrix and let B be the matrix obtained by interchanging any two rows or columns of A. To compute the determinant of B, we use the expansion along an unchanged k-th row of B. Then

$$\det(B) = \sum_{j=1}^{(l+1)} (-1)^{k+j} a_{kj} \det(B_{kj})$$

$$= (-1)^{k+1} a_{k1} \det(B_{k1}) + (-1)^{k+2} a_{k2} \det(B_{k2}) + \cdots + (-1)^{k+(l+1)} a_{k(l+1)} \det(B_{k(l+1)}).$$

Since B_{kj} are l-square submatrices of B, which are equal to l-square submatrices A_{kj} in the form of interchanged rows for all $j = 1, 2, \ldots, l+1$. Then by the induction hypothesis $\det(B_{kj}) = -\det(A_{kj})$.
Hence,

$$\det(B) = (-1)^{k+1} a_{k1}(-1) \det(A_{k1}) + (-1)^{k+2} a_{k2}(-1) \det(A_{k2}) + \cdots$$

$$+ (-1)^{k+(l+1)} a_{k(l+1)}(-1) \det(A_{k(l+1)})$$

$$= (-1) \sum_{j=1}^{(l+1)} (-1)^{k+j} a_{kj} \det(A_{kj})$$

$$= -\det(A).$$

Thus, the result is proved for $n = k + 1$. Hence, by the method of mathematical induction, the theorem is proved.

Similarly, the theorem can be proved for column interchanged matrices. □

For proper understanding the proof of the above theorem consider the matrix

$$A - \begin{bmatrix} 1 & 2 & 3 \\ 2 & 4 & -1 \\ 5 & 1 & 2 \end{bmatrix} \quad \text{and the matrix} \quad B = \begin{bmatrix} 5 & 1 & 2 \\ 2 & 4 & -1 \\ 1 & 2 & 3 \end{bmatrix},$$

which is obtained by interchanging the first and third rows of A. Since the second row of B remains unchanged, meaning the second rows of both matrices A and B are identical, we use the expansion along this row to calculate the determinant of B. Then

$$\det(B) = (-1)^{2+1}(2)\det(B_{21}) + (-1)^{2+2}(4)\det(B_{22}) + (-1)^{2+3}(-1)\det(B_{23})$$

$$= (-1)^{2+1}(2)\det\left(\begin{bmatrix} 1 & 2 \\ 2 & 3 \end{bmatrix}\right) + (-1)^{2+2}(4)\det\left(\begin{bmatrix} 5 & 2 \\ 1 & 3 \end{bmatrix}\right) + (-1)^{2+3}(-1)\det\left(\begin{bmatrix} 5 & 1 \\ 1 & 2 \end{bmatrix}\right)$$

$$= (-1)^{2+1}(2)(-1)\det\left(\begin{bmatrix} 2 & 3 \\ 1 & 2 \end{bmatrix}\right) + (-1)^{2+2}(4)(-1)\det\left(\begin{bmatrix} 1 & 3 \\ 5 & 2 \end{bmatrix}\right)$$

$$+ (-1)^{2+3}(-1)(-1)\det\left(\begin{bmatrix} 1 & 2 \\ 5 & 1 \end{bmatrix}\right)$$

$$= (-1)^{2+1}(2)(-1)\det(A_{21}) + (-1)^{2+2}(4)(-1)\det(A_{22}) + (-1)^{2+3}(-1)(-1)\det(A_{23})$$

$$= -[(-1)^{2+1}(2)\det(A_{21}) + (-1)^{2+2}(4)\det(A_{22}) + (-1)^{2+3}(-1)\det(A_{23})]$$

$$= -\det(A).$$

Thus, $\det(A) = -63$ and $\det(B) = 63$.

Corollary 7.1.8. *Let A be an n-square matrix with two identical rows (columns). Then* $\det(A) = 0$.

Proof. Suppose r-th and s-th rows (columns) of matrix A are identical. If we interchange r-th and s-th rows (columns) of A, A remains unchanged but from above theorem $\det(A) = -\det(A)$. Hence, $\det(A) = 0$. □

For example, if

$$A = \begin{bmatrix} 4 & -3 & 2 \\ 2 & 1 & 0 \\ 4 & -3 & 2 \end{bmatrix},$$

then $\det(A) = 0$.

Theorem 7.1.9. *Let A be an n-square matrix and let B be a matrix obtained from A by multiplying each entry of i-th row (column) of A by a scalar s. Then* $\det(B) = s\det(A)$.

Proof. The determinant of B along the i-th row expansion is

$$\det(B) = \sum_{j=1}^{n}(-1)^{i+j}sa_{ij}\det(B_{ij}) = s\sum_{j=1}^{n}(-1)^{i+j}a_{ij}\det(B_{ij}).$$

Since rows of A and B except i-th row are equal, $B_{ij} = A_{ij}$ $\forall i$.

Then $\det(B) = s\sum_{j=1}^{n}(-1)^{i+j}a_{ij}\det(A_{ij}) = s\det(A)$. The proof in the case when B is obtained from A by multiplying s in a column of A is similar. □

For example, let

$$A = \begin{bmatrix} 1 & 2 \\ 3 & 8 \end{bmatrix} \quad \text{and} \quad B = \begin{bmatrix} (6)(1) & (6)(2) \\ 3 & 8 \end{bmatrix} = \begin{bmatrix} 6 & 12 \\ 3 & 8 \end{bmatrix}.$$

Then

$$\det(B) = 12 = 6 \det\left(\begin{bmatrix} 1 & 2 \\ 3 & 8 \end{bmatrix}\right) = (6)(2) = 12.$$

Hence, $\det(B) = 6 \det(A)$.

Corollary 7.1.10. *Let A be an n-square matrix. Then $\det(sA) = s^n \det(A)$, where s is an scalar.*

Proof. The matrix sA is obtained by multiplying all the rows of A with s. Hence, from above theorem, $\det(sA) = (s)(s) \cdots (s) \det(A) = s^n \det(A)$. □

For example, let

$$A = \begin{bmatrix} 2 & 0 & -1 \\ 3 & 2 & 1 \\ -2 & 3 & 4 \end{bmatrix}.$$

Then $\det(A) = -3$ and $\det(5A) = 5^3 \det(A) = 125(-3) = -375$.

Example 7.1.11. Let

$$A = \begin{bmatrix} a_{11} & a_{12} & a_{13} \\ a_{21} & a_{22} & a_{23} \\ a_{31} & a_{32} & a_{33} \end{bmatrix} \quad \text{and} \quad B = \begin{bmatrix} a_{11} & a_{12} & a_{13} \\ (a_{21} + la_{11}) & (a_{22} + la_{12}) & (a_{23} + la_{13}) \\ a_{31} & a_{32} & a_{33} \end{bmatrix}$$

be obtained by multiplying l in first row and adding it to second row of A.
Then the determinant of B along the second row expansion is

$$\det(B) = (-1)^{2+1}(a_{21} + la_{11}) \det(B_{21}) + (-1)^{2+2}(a_{22} + la_{12}) \det(B_{22})$$
$$+ (-1)^{2+3}(a_{23} + la_{13}) \det(B_{23})$$
$$= [(-1)a_{21} \det(B_{21}) + a_{22} \det(B_{22}) + (-1)(a_{23} \det(B_{23})] + l[(-1)a_{11} \det(B_{21})$$
$$+ a_{12} \det(B_{22}) + (-1)(a_{13} \det(B_{23}))]$$
$$= \left[(-1)a_{21} \det\left(\begin{bmatrix} a_{12} & a_{13} \\ a_{32} & a_{33} \end{bmatrix}\right) + a_{22} \det\left(\begin{bmatrix} a_{11} & a_{13} \\ a_{31} & a_{33} \end{bmatrix}\right)\right.$$
$$\left. + (-1)(a_{23} \det\left(\begin{bmatrix} a_{11} & a_{12} \\ a_{31} & a_{32} \end{bmatrix}\right)\right]$$
$$+ l\left[(-1)a_{11} \det\left(\begin{bmatrix} a_{12} & a_{13} \\ a_{32} & a_{33} \end{bmatrix}\right) + a_{12} \det\left(\begin{bmatrix} a_{11} & a_{13} \\ a_{31} & a_{33} \end{bmatrix}\right)\right.$$
$$\left. + (-1)(a_{13} \det\left(\begin{bmatrix} a_{11} & a_{12} \\ a_{31} & a_{32} \end{bmatrix}\right)\right]$$

$$= \det(A) + l \det \left(\begin{bmatrix} a_{11} & a_{12} & a_{13} \\ a_{11} & a_{12} & a_{13} \\ a_{31} & a_{32} & a_{33} \end{bmatrix} \right) \quad \text{along second row expansion}$$

$$= \det(A) + l(0)$$

$$= \det(A).$$

The following theorem is the generalization of the above example.

Theorem 7.1.12. *Let A be an n-square matrix and let B be the matrix obtained from A by multiplying a row (or column) by a scalar l and then adding it to another row (or column). Then* $\det(B) = \det(A)$.

Proof. Suppose B is obtained by multiplying l in r-th row and adding it to s-th row of $A = [a_{ij}]$. Then the s-th row of B will be

$$(a_{s1} + la_{r1}) + (a_{s2} + la_{r2}) + \cdots + (a_{sn} + la_{rn}),$$

and remaining rows will be same as of A.

Computing $\det(B)$ along the s-th row expansion, we get

$$\det(B) = \sum_{j=1}^{n} (-1)^{s+j} b_{sj} \det(B_{sj})$$

$$= \sum_{j=1}^{n} (-1)^{s+j} (a_{sj} + la_{rj}) \det(B_{sj}).$$

Since remaining rows of B are the same as rows of A, $B_{sj} = A_{sj}$ for all j. Then

$$\det(B) = \sum_{j=1}^{n} (-1)^{s+j} (a_{sj} + la_{rj}) \det(A_{sj})$$

$$= \sum_{j=1}^{n} (-1)^{s+j} a_{sj} \det(A_{sj}) + l \sum_{j=1}^{n} (-1)^{s+j} a_{rj} \det(A_{sj})$$

$$= \det(A) + l \det A_1,$$

where A_1 is obtained from A by replacing s-th row of A by r-th row. Then $\det(A_1) = 0$. Hence, $\det(B) = \det(A)$.

Similarly, the theorem can be proved for column cases, also. □

In the following examples, we use above results to compute the determinant of matrices.

Example 7.1.13. Let

$$A = \begin{bmatrix} -2 & 3 & 0 & 1 \\ 9 & -2 & 0 & 1 \\ 1 & 3 & -2 & -1 \\ 4 & 1 & 2 & 6 \end{bmatrix}.$$

We compute the determinant of A by converting into the upper triangular matrix, using elementary row operations.

Then

$$\det(A) = \det\left(\begin{bmatrix} -2 & 3 & 0 & 1 \\ 9 & -2 & 0 & 1 \\ 1 & 3 & -2 & -1 \\ 4 & 1 & 2 & 6 \end{bmatrix}\right)$$

$$= -\det\left(\begin{bmatrix} 1 & 3 & -2 & -1 \\ 9 & -2 & 0 & 1 \\ -2 & 3 & 0 & 1 \\ 4 & 1 & 2 & 6 \end{bmatrix}\right) \quad \text{(from } (R_1 \leftrightarrow R_3) \text{ and Theorem 7.1.7)}$$

$$= -\det\left(\begin{bmatrix} 1 & 3 & -2 & -1 \\ 0 & -29 & 18 & 10 \\ 0 & 9 & -4 & -1 \\ 0 & -11 & 10 & 10 \end{bmatrix}\right)$$

(from $R_2 \to R_2 - 9R_1, R_3 \to R_3 + 2R_1, R_4 \to R_4 - 4R_1$ and Theorem 7.1.12)

$$= (-1)(-29)\det\left(\begin{bmatrix} 1 & 3 & -2 & -1 \\ 0 & 1 & \frac{-18}{29} & \frac{-10}{29} \\ 0 & 9 & -4 & -1 \\ 0 & -11 & 10 & 10 \end{bmatrix}\right)$$

(from $R_2 \longleftrightarrow \frac{-1}{29}R_2$ and Theorem 7.1.9)

$$= 29\det\left(\begin{bmatrix} 1 & 3 & -2 & -1 \\ 0 & 1 & \frac{-18}{29} & \frac{-10}{29} \\ 0 & 0 & \frac{46}{29} & \frac{61}{29} \\ 0 & 0 & \frac{92}{29} & \frac{180}{29} \end{bmatrix}\right) \quad \text{(from } R_3 \to R_3 - 9R_2, R_4 \to R_4 + 11R_2)$$

$$= 29\det\left(\begin{bmatrix} 1 & 3 & -2 & -1 \\ 0 & 1 & \frac{-18}{29} & \frac{-10}{29} \\ 0 & 0 & \frac{46}{29} & \frac{61}{29} \\ 0 & 0 & 0 & \frac{58}{29} \end{bmatrix}\right) \quad \text{(from } R_4 \to R_4 - 2R_3)$$

$$= 29 \times 1 \times 1 \times \frac{46}{29} \times \frac{58}{29} = 92.$$

Example 7.1.14.

$$\det\left(\begin{bmatrix} 1 & a & 1 & b \\ a & 1 & b & 1 \\ 1 & b & 1 & a \\ b & 1 & a & 1 \end{bmatrix}\right)$$

$$= \det\left(\begin{bmatrix} 2+a+b & 2+a+b & 2+a+b & 2+a+b \\ a & 1 & b & 1 \\ 1 & b & 1 & a \\ b & 1 & a & 1 \end{bmatrix}\right) \quad \text{(from } R_1 \to R_1 + R_2 + R_3\text{)}$$

$$= (2+a+b)\det\left(\begin{bmatrix} 1 & 1 & 1 & 1 \\ a & 1 & b & 1 \\ 1 & b & 1 & a \\ b & 1 & a & 1 \end{bmatrix}\right)$$

$$= (2+a+b)\det\left(\begin{bmatrix} 1 & 1 & 1 & 1 \\ 0 & 1-a & b-a & 1-a \\ 0 & b-1 & 0 & a-1 \\ 0 & 1-b & a-b & 1-b \end{bmatrix}\right)$$

$$\text{(from } R_2 \to R_2 - aR_1, R_3 \to R_3 - R_1, R_4 \to R_4 - bR_1\text{)}$$

$$= (2+a+b)\det\left(\begin{bmatrix} 1-a & b-a & 1-a \\ b-1 & 0 & a-1 \\ 1-b & a-b & 1-b \end{bmatrix}\right) \quad \text{(by expansion along the first column)}$$

$$= (2+a+b)(a-b)\det\left(\begin{bmatrix} 1-a & -1 & 1-a \\ b-1 & 0 & a-1 \\ 1-b & 1 & 1-b \end{bmatrix}\right) \quad \text{(from } C_2 \to \frac{1}{a-b}C_2\text{)}$$

$$= (2+a+b)(a-b)\det\left(\begin{bmatrix} 2-a-b & 0 & 2-a-b \\ b-1 & 0 & a-1 \\ 1-b & 1 & 1-b \end{bmatrix}\right) \quad \text{(from } R_1 \to R_1 + R_3\text{)}$$

$$= (2+a+b)(a-b)(2-a-b)\det\left(\begin{bmatrix} 1 & 0 & 1 \\ b-1 & 0 & a-1 \\ 1-b & 1 & 1-b \end{bmatrix}\right) \quad \text{(from } R_1 \to \frac{1}{2-a-b}R_1\text{)}$$

$$= (2+a+b)(a-b)(2-a-b)(-1)^{3+2}(1)\det\left(\begin{bmatrix} 1 & 1 \\ b-1 & a-1 \end{bmatrix}\right)$$

(by expansion along C_2)

$$= (2+a+b)(a-b)(2-a-b)(-1)\{(a-1)-(b-1)\}$$

$$= (a-b)^2(a+b+2)(a+b-2).$$

Example 7.1.15. In this example, we compute the determinants of the three possible elementary row (column) matrices.

(a) Let E_1 be the elementary row (column) matrix obtained by interchanging any two rows (columns) of I_n. Then $\det(E_1) = -\det(I_n) = -1$.
(b) Let E_2 be the elementary matrix obtained by multiplying scalar l in the i-th row (column) of I_n. Then $\det(E_2) = l\det(I_n) = l(1) = l$.
(c) If E_3 is the elementary matrix obtained by multiplying r-th row (column) of I_n by l and adding it to s-th row (column) of I_n, then by Theorem 7.1.12, $\det(E_3) = \det(I_n) = 1$.

Thus, the determinants of elementary matrices are nonzero.

Theorem 7.1.16. *If A is an n-square matrix and E is an n-square elementary matrix, then* $\det(EA) = \det(E)\det(A)$.

Proof. We prove the result for all of the three types of elementary row matrices. Let E_1 be the elementary matrix obtained by interchanging two rows of I_n. Then E_1A is the matrix resulting from interchanging the corresponding two rows of A. Hence, $\det(E_1A) = -\det(A) = (-1)\det(A) = \det(E_1)\det(A)$.

Let E_2 be the elementary matrix obtained by multiplying scalar l to the r-th row of I_n. Then E_2A is the matrix resulting from multiplying l to the r-th row of A. Hence, $\det(E_2A) = l\det(A) = \det(E_2)\det(A)$.

Finally, let E_3 be the elementary matrix obtained by multiplying r-th row by l and adding it to s-th row of I_n. Then E_3A is the matrix resulting from multiplying r-th row by l and adding to s-th row of A. Hence, $\det(E_3A) = (1)\det(A) = \det(E_3)\det(A)$.

Similarly, we can prove the result for all three elementary column matrices, and hence the theorem is proved for all elementary matrices. □

Theorem 7.1.17. *An n-square matrix A is singular if and only if* $\det(A) = 0$.

Proof. Let A be an n-square singular matrix and let B be the matrix, which is the reduced row echelon form of A. Then

$$B = E_rE_{r-1}E_{r-2}\cdots E_2E_1A, \quad \text{where } E_1, E_2, \ldots, E_r \text{ are elementary row matrices.}$$

Hence,

$$\det(B) = \det(E_rE_{r-1}E_{r-2}\cdots E_2E_1A)$$
$$= \det(E_r)\det(E_{r-1})\det(E_{r-2})\cdots\det(E_2)\det(E_1)\det(A).$$

Since A is singular and B is the reduced row echelon form of A, B must contain some zero rows. Hence, $\det(B) = 0$.

Also, a determinant of elementary matrices are nonzero, and we have that

$$0 = \det(E_r)\det(E_{r-1})\det(E_{r-2})\cdots\det(E_2)\det(E_1)\det(A).$$

Hence, $\det(A) = 0$.

Conversely, suppose that $\det(A) = 0$. We shall prove the contrapositive. Suppose A is nonsingular. Then by elementary row operations, A can be reduced to the identity matrix I_n. Suppose there exist elementary matrices $E_1, E_2, \ldots, E_{s-1}, E_s$, such that

$$E_s E_{s-1} \cdots E_2 E_1 A = I_n.$$

Then

$$\det(E_s E_{s-1} \cdots E_2 E_1 A) = \det(I_n),$$
$$\det(E_s) \det(E_{s-1}) \cdots \det(E_2) \det(E_1) \det(A) = 1.$$

Since the determinant of elementary matrices is nonzero,

$$\det(A) = \frac{1}{\det(E_s) \det(E_{s-1}) \cdots \det(E_2) \det(E_1)} \neq 0.$$

This proves that A is singular. $\square$

Equivalently, we have that an n-square matrix A is invertible if and only if $\det(A) \neq 0$.

Theorem 7.1.18. *Let A and B be n-square matrices. Then $\det(AB) = \det(A) \det(B)$.*

Proof. We prove the theorem by satisfying the result for both of the cases, singular and nonsingular of A.

Suppose that A is a singular matrix. Then $\det(A) = 0$. Since A is singular, AB is singular. Then $\det(AB) = 0 = (0) \det(B) = \det(A) \det(B)$.

Now, suppose that A is a nonsingular matrix. Then by the Gauss–Jordan method, we have

$$I_n = E_s E_{s-1} \cdots E_2 E_1 A, \quad \text{where } E_1, E_2, \ldots, E_s \text{ are elementary matrices.}$$

Then

$$A = E_1^{-1} E_2^{-1} \cdots E_s^{-1}.$$

Since the inverse of an elementary matrix is also an elementary matrix,

$$\det(AB) = \det(E_1^{-1} E_2^{-1} \cdots E_s^{-1} B)$$
$$= \det(E_1^{-1}) \det(E_2^{-1}) \cdots \det(E_s^{-1}) \det(B)$$
$$= \det(E_1^{-1} E_2^{-1} \cdots E_s^{-1}) \det(B)$$
$$= \det(A) \det(B). \qquad \square$$

Definition 7.1.19. Let $A = [a_{ij}]$ be an n-square matrix. Then $C_{ij} = (-1)^{i+j} \det(A_{ij})$ is called the cofactor of a_{ij}.

Example 7.1.20. Let

$$A = \begin{bmatrix} 2 & 0 & -1 \\ 3 & 2 & 1 \\ -2 & 3 & 4 \end{bmatrix}.$$

Then

$$C_{11} = (-1)^{1+1} \det(A_{11}) = (-1)^2 \det\left(\begin{bmatrix} 2 & 1 \\ 3 & 4 \end{bmatrix}\right) = 5,$$

$$C_{12} = (-1)^{1+2} \det(A_{12}) = (-1)^3 \det\left(\begin{bmatrix} 3 & 1 \\ -2 & 4 \end{bmatrix}\right) = -14,$$

$$C_{13} = (-1)^{1+3} \det(A_{13}) = (-1)^4 \det\left(\begin{bmatrix} 3 & 2 \\ -2 & 3 \end{bmatrix}\right) = 13,$$

$$C_{21} = (-1)^{2+1} \det(A_{21}) = (-1)^3 \det\left(\begin{bmatrix} 0 & -1 \\ 3 & 4 \end{bmatrix}\right) = -3,$$

$$C_{22} = (-1)^{2+2} \det(A_{22}) = (-1)^4 \det\left(\begin{bmatrix} 2 & -1 \\ 3=-2 & 4 \end{bmatrix}\right) = 6,$$

$$C_{23} = (-1)^{2+3} \det(A_{23}) = (-1)^5 \det\left(\begin{bmatrix} 2 & 0 \\ -2 & 3 \end{bmatrix}\right) = -6,$$

$$C_{31} = (-1)^{3+1} \det(A_{31}) = (-1)^4 \det\left(\begin{bmatrix} 0 & -1 \\ 2 & 1 \end{bmatrix}\right) = 2,$$

$$C_{32} = (-1)^{3+2} \det(A_{32}) = (-1)^5 \det\left(\begin{bmatrix} 2 & -1 \\ 3 & 1 \end{bmatrix}\right) = -5,$$

$$C_{33} = (-1)^{3+3} \det(A_{33}) = (-1)^6 \det\left(\begin{bmatrix} 2 & 0 \\ 3 & 2 \end{bmatrix}\right) = 4,$$

are cofactors of A.

Let

$$A = \begin{bmatrix} a_{11} & a_{12} & a_{13} \\ a_{21} & a_{22} & a_{23} \\ a_{31} & a_{32} & a_{33} \end{bmatrix}$$

be a 3-square matrix.
Then the expression

$$(-1)^{2+1} a_{21} A_{21} + (-1)^{2+2} a_{22} A_{22} + (-1)^{2+3} a_{23} A_{23}$$
$$= a_{21} C_{21} + a_{22} C_{22} + a_{23} C_{23},$$

represents the $\det(A)$, obtained by expansion along the second row.

In the above expression, if we replace a_{21}, a_{22}, a_{23} by a_{11}, a_{12}, a_{13}, respectively, we get

$$(-1)^{2+1} a_{11} A_{21} + (-1)^{2+2} a_{12} A_{22} + (-1)^{2+3} a_{13} A_{23},$$

which is the determinant of the matrix

$$A_1 = \begin{bmatrix} a_{11} & a_{12} & a_{13} \\ a_{11} & a_{12} & a_{13} \\ a_{31} & a_{32} & a_{33} \end{bmatrix},$$

computed along the second row. Since first and second rows of matrix A_1 are identical, $\det(A_1) = 0$. Hence,

$$0 = (-1)^{2+1} a_{11} A_{21} + (-1)^{2+2} a_{12} A_{22} + (-1)^{2+3} a_{13} A_{23}$$
$$= a_{11} C_{21} + a_{12} C_{22} + a_{13} C_{23}.$$

Similarly, we get

$$\sum_{k=1}^{3} a_{ik} C_{jk} = \begin{cases} \det(A), & \text{if } i = j, \\ 0, & \text{if } i \neq j. \end{cases}$$

In view of above discussion, we have the following theorem.

Theorem 7.1.21. *Let $A = [a_{ij}]$ be an n-square matrix. Then*
(a)

$$\sum_{k=1}^{n} a_{ik} C_{jk} = \begin{cases} \det(A), & \text{if } i = j, \\ 0, & \text{if } i \neq j, \end{cases}$$

in case of expansion along a row,
(b)

$$\sum_{k=1}^{n} a_{ki} C_{kj} = \begin{cases} \det(A), & \text{if } i = j, \\ 0, & \text{if } i \neq j, \end{cases}$$

in case of expansion along a column.

Definition 7.1.22. Let $A = [a_{ij}]$ be an n-square matrix. Then transpose of the matrix

$$\begin{bmatrix} C_{11} & C_{12} & \cdots & C_{1n} \\ C_{21} & C_{22} & \cdots & C_{2n} \\ \vdots & & & \\ C_{n1} & C_{n2} & \cdots & C_{nn} \end{bmatrix},$$

where C_{ij} are cofactors of A is called the adjoint of A, denoted by adj(A). Thus, adj(A) = $[C_{ij}]^t$.

Example 7.1.23. Consider the matrix

$$A = \begin{bmatrix} 2 & 0 & -1 \\ 3 & 2 & 1 \\ -2 & 3 & 4 \end{bmatrix}$$

given in Example 7.1.20. Then

$$\text{adj}(A) = \begin{bmatrix} C_{11} & C_{12} & C_{13} \\ C_{21} & C_{22} & C_{23} \\ C_{31} & C_{32} & C_{33} \end{bmatrix}^t$$

$$= \begin{bmatrix} 5 & -14 & 13 \\ -3 & 6 & -6 \\ 2 & -5 & 4 \end{bmatrix}^t$$

$$= \begin{bmatrix} 5 & -3 & 2 \\ -14 & 6 & -5 \\ 13 & -6 & 4 \end{bmatrix}.$$

Theorem 7.1.24. *Let $A = [a_{ij}]$ be an n-square matrix. Then A adj(A) = adj(A)A = det(A)I_n.*

Proof. Given that $A = [a_{ij}]$, an n-square matrix. Then

$$\text{adj}(A) = \begin{bmatrix} C_{11} & C_{12} & \cdots & C_{1n} \\ C_{21} & C_{22} & \cdots & C_{2n} \\ \vdots & & & \\ C_{n1} & C_{n2} & \cdots & C_{nn} \end{bmatrix}^t = \begin{bmatrix} C_{11} & C_{21} & \cdots & C_{n1} \\ C_{12} & C_{22} & \cdots & C_{n2} \\ \vdots & & & \\ C_{1n} & C_{2n} & \cdots & C_{nn} \end{bmatrix}.$$

Hence,

$$A \text{ adj}(A) = \begin{bmatrix} a_{11} & a_{12} & \cdots & a_{1n} \\ a_{21} & a_{22} & \cdots & a_{2n} \\ \vdots & & & \\ a_{n1} & a_{n2} & \cdots & a_{nn} \end{bmatrix} \begin{bmatrix} C_{11} & C_{21} & \cdots & C_{n1} \\ C_{12} & C_{22} & \cdots & C_{n2} \\ \vdots & & & \\ C_{1n} & C_{2n} & \cdots & C_{nn} \end{bmatrix}$$

$$= \begin{bmatrix} \sum_{j=1}^{n} a_{1j}C_{1j} & \sum_{j=1}^{n} a_{1j}C_{2j} & \cdots & \sum_{j=1}^{n} a_{1j}C_{nj} \\ \sum_{j=1}^{n} a_{2j}C_{1j} & \sum_{j=1}^{n} a_{2j}C_{2j} & \cdots & \sum_{j=1}^{n} a_{2j}C_{nj} \\ \vdots & & & \\ \sum_{j=1}^{n} a_{nj}C_{1j} & \sum_{j=1}^{n} a_{nj}C_{2j} & \cdots & \sum_{j=1}^{n} a_{nj}C_{nj} \end{bmatrix}$$

$$= \begin{bmatrix} \det(A) & 0 & \cdots & 0 \\ 0 & \det(A) & \cdots & 0 \\ \vdots & & & \\ 0 & 0 & \cdots & \det(A) \end{bmatrix}$$

$$= \det(A) \begin{bmatrix} 1 & 0 & \cdots & 0 \\ 0 & 1 & \cdots & 0 \\ \vdots & & & \\ 0 & 0 & \cdots & 1 \end{bmatrix}$$

$$= \det(A) I_n.$$

Similarly, we can prove that $\mathrm{adj}(A)A = \det(A)I_n$. ☐

Corollary 7.1.25. *If* $\det(A) \neq 0$, *then* A *is invertible and* $A^{-1} = \frac{1}{\det(A)} \mathrm{adj}(A)$.

Proof. From the above theorem, we have $A \, \mathrm{adj}(A) = \det(A)I_n$. Since $\det(A) \neq 0$, $A \frac{\mathrm{adj}(A)}{\det(A)} = I_n$. Hence, $A^{-1} = \frac{1}{\det(A)} \mathrm{adj}(A)$. ☐

Example 7.1.26. Consider Example 7.1.23. Then

$$A = \begin{bmatrix} 2 & 0 & -1 \\ 3 & 2 & 1 \\ -2 & 3 & 4 \end{bmatrix} \quad \text{and} \quad \mathrm{adj}(A) = \begin{bmatrix} 5 & -3 & 2 \\ -14 & 6 & -5 \\ 13 & -6 & 4 \end{bmatrix}.$$

Hence, from the result $A^{-1} = \frac{1}{\det(A)} \mathrm{adj}(A)$, we have

$$A^{-1} = \frac{1}{-3} \begin{bmatrix} 5 & -3 & 2 \\ -14 & 6 & -5 \\ 13 & -6 & 4 \end{bmatrix} = \begin{bmatrix} \frac{-5}{3} & 1 & \frac{-2}{3} \\ \frac{14}{3} & -2 & \frac{5}{3} \\ \frac{-13}{3} & 2 & \frac{-4}{3} \end{bmatrix},$$

where $\det(A) = -3$.

Theorem 7.1.27. *If* A *and* B *are* n-*square matrices such that* $AB = I_n$, *then* $BA = I_n$.

Proof. Since $\det(AB) = \det(A)\det(B) = 1$, $\det(A) \neq 0$, $\det(B) \neq 0$. Hence, A and B are invertible matrices. Since B is an invertible matrix, there exists a matrix C such that $BC = I_n$.

Now, $BA = (BA)I_n = (BA)(BC) = B(AB)C = BI_nC = BC = I_n$. ☐

Thus, from the above theorem it is clear that to prove that a matrix B is the inverse of a matrix A, it is sufficient to show that $AB = I_n$ except $AB = BA = I_n$.

Example 7.1.28. If A and B are n-square similar matrices, then $\det(A) = \det(B)$.

Since A and B are similar matrices, there exists a nonsingular matrix P such that $A = P^{-1}BP$. Then $\det(A) = \det(P^{-1}BP) = \det(P^{-1})\det(B)\det(P) = \det(B)\det(P^{-1})\det(P) = \det(B)\det(P^{-1}P) = \det(B)\det(I_n) = \det(B)$.

Example 7.1.29. If A is an n-square invertible matrix, then:
(i) $\det(\text{adj}(A)) = (\det(A))^{n-1}$, and
(ii) $\text{adj}(\text{adj}(A)) = (\det(A))^{n-2}A$.

(i) From Theorem 7.1.24, we have

$$A\,\text{adj}(A) = \text{adj}(A)A = \det(A)I_n = \begin{bmatrix} \det(A) & 0 & \cdots & 0 \\ 0 & \det(A) & \cdots & 0 \\ \vdots & & & \\ 0 & 0 & \cdots & \det(A) \end{bmatrix}.$$

Hence,

$$\det(A(\text{adj}(A))) = \det(A)\det(\text{adj}(A))$$

$$= \det\left(\begin{bmatrix} \det(A) & 0 & \cdots & 0 \\ 0 & \det(A) & \cdots & 0 \\ \vdots & & & \\ 0 & 0 & \cdots & \det(A) \end{bmatrix} \right)$$

$$= (\det(A))^n.$$

Thus, $\det(\text{adj}(A)) = (\det(A))^{n-1}$.

(ii) Denote $\text{adj}(A) = B$. From Theorem 7.1.24, we have

$$\text{adj}(B)B = \det(B)I_n,$$

which implies

$$\text{adj}(B)BA = \det(B)I_nA$$
$$\Rightarrow \text{adj}(B)\,\text{adj}(A)A = \det(B)A$$
$$\Rightarrow \text{adj}(B)(\det(A)I_n) = \det(B)A$$
$$\Rightarrow \det(A)\,\text{adj}(B) = \det(\text{adj}(A))A = (\det(A))^{n-1}A.$$

Therefore, $\text{adj}(B) = (\det(A))^{n-2}A$, and hence

$$\text{adj}(\text{adj}(A)) = (\det(A))^{n-2}A.$$

Example 7.1.30. Let A be an n-square skew-symmetric matrix. In this example, we discuss the determinant of A.
Since A is skew-symmetric, $A^t = -A$.
Thus,

$$\det(A^t) = \det(-A) = (-1)^n \det(A).$$

Since $\det(A^t) = \det(A)$, from the above equation we have

$$\det(A)(1 - (-1)^n) = 0.$$

Thus, either $\det(A) = 0$, or $(-1)^n = 1$. In other words, the determinant of a skew-symmetric matrix is 0 whenever its dimension is odd.

If the dimension of a skew-symmetric matrix is even, then we cannot say anything about its determinant.

---

i Exercises

(7.1.1)   Find the determinants of the following matrices:

$$A = \begin{bmatrix} 1 & 4 & 7 \\ 2 & 5 & 8 \\ 3 & 6 & 9 \end{bmatrix}, \qquad B = \begin{bmatrix} 1 & 3 & 0 \\ 0 & 5 & 1 \\ 4 & -1 & 2 \end{bmatrix},$$

$$C = \begin{bmatrix} 2 & 0 & -1 & -2 \\ 3 & 2 & 1 & -1 \\ -2 & 3 & 4 & -4 \\ 3 & 2 & 4 & 2 \end{bmatrix}, \qquad D = \begin{bmatrix} 1 & 0 & -1 & -2 \\ 0 & 2 & 1 & 0 \\ 5 & 3 & -2 & -4 \\ 0 & 0 & -4 & 2 \end{bmatrix}.$$

(7.1.2)   If

$$A = \begin{bmatrix} 1 & 1 & 1 \\ a & b & c \\ a^2 & b^2 & c^2 \end{bmatrix},$$

then show that $\det(A) = (a - b)(b - c)(c - a)$.

(7.1.3)   If

$$A = \begin{bmatrix} (y + z)^2 & x^2 & x^2 \\ y^2 & (z + x)^2 & y^2 \\ z^2 & z^2 & (x + y)^2 \end{bmatrix},$$

then show that $\det(A) = 2xyz(x + y + z)^3$.

(7.1.4)   Show that the determinant $\det(A) = (x - d)(x - g)(x - k)(x - m)$, where

$$A = \begin{bmatrix} x & d & d & d & d \\ a & x & g & g & g \\ b & e & x & k & k \\ c & f & h & x & m \\ 1 & 1 & 1 & 1 & 1 \end{bmatrix}.$$

(7.1.5)   Evaluate the determinant of the following matrices:

$$A = \begin{bmatrix} 1 & 1 & 1 \\ a + b & b + c & c + a \\ a^2 + b^2 & b^2 + c^2 & c^2 + a^2 \end{bmatrix}, \qquad B = \begin{bmatrix} 1 & 1 & 1 & 1 \\ x & y & z & u \\ x^2 & y^2 & z^2 & u^2 \\ x^3 + yzu & y^3 + zux & z^3 + uxy & u^3 + xyz \end{bmatrix}.$$

(7.1.6)   Let $A = [a_{ij}]$, where $a_{ij} = i + j$ and

$$B = [b_{ij}], \quad \text{where } b_{ij} = \begin{cases} 1, & \text{if } i + j \text{ is odd,} \\ 0, & \text{otherwise} \end{cases}$$

be matrices of order 4×4. Then find the determinant of the following matrices:

(i) A (ii) B (iii) $3A - 5B$ (iv) AB (v) BA (vi) A^2B (vii) AB^2.

(7.1.7)   Let $A = [a_{ij}]$, where $a_{ij} = (-1)^{i+j}$ and

$$B = [b_{ij}], \quad \text{where } b_{ij} = \begin{cases} 1, & \text{if } i + j \text{ is even,} \\ 0, & \text{otherwise} \end{cases}$$

be matrices of order 4×4. Then find the determinant of the following matrices:

(i) A (ii) B (iii) $3A - 5B$ (iv) AB (v) BA (vi) A^2B (vii) AB^2.

(7.1.8)   Find all the values of t such that the matrix

$$A = \begin{bmatrix} 1 & 0 & 4 \\ 0 & 4+t & -2 \\ 0 & 1 & t \end{bmatrix}$$

is nonsingular.

(7.1.9)   Find all the values of a so that the matrix

$$A = \begin{bmatrix} a & 2 & 0 \\ a^2 & 3 & -1 \\ 1 & 1 & 1 \end{bmatrix}$$

is singular.

(7.1.10) Prove that the determinant of an n-square real skew-symmetric matrix is zero if n is odd.

(7.1.11) Let A and B be n-square matrices. Test whether $\det(A + B) = \det(A) + \det(B)$ is true. If so, then give a proof. If not, then give a counterexample.

(7.1.12) Find the cofactor, adjoint and also the inverse of the following matrices:

$$A = \begin{bmatrix} 3 & -1 & -2 \\ 0 & 0 & -1 \\ 3 & -5 & 0 \end{bmatrix}, \quad B = \begin{bmatrix} 1 & 3 & 0 \\ 0 & 5 & 1 \\ 4 & -1 & 2 \end{bmatrix}.$$

(7.1.13) If the matrix A is invertible, then show that adj(A) is also invertible. Is the converse true? Support your finding.

(7.1.14) Suppose that A is a 3-square invertible matrix with determinant 5. Find the det(adj(A)).

(7.1.15) Suppose that A is a 4-square invertible matrix such that det(adj(A)) = 27. Find the determinant of A.

(7.1.16) Show that the determinant of an orthogonal matrix is ±1.

7.2 Eigenvalues and eigenvectors (of linear transformations and matrices)

Eigenvalues and eigenvectors provide a powerful framework for understanding the properties and behavior of matrices and linear transformations. They offer valuable insights into stability, dynamics and structural properties, making them indispensable tools in the study of linear algebra and its applications.

Understanding eigenvalues and eigenvectors is foundational in linear algebra and has broad implications across diverse fields.

Definition 7.2.1. Let V be a vector space over a field F and let $T : V \to V$ be a linear transformation. A scalar $\lambda \in F$ is called an eigenvalue of T if their exists a nonzero vector $x \in V$ such that $T(x) = \lambda x$.

If λ is an eigenvalue of T, then any nonzero vector $x \in V$ satisfying $T(x) = \lambda x$ is called an eigenvector of T associated to λ.

Eigenvalues are also called characteristic values or characteristic roots or latent roots or proper values.

Example 7.2.2.
(a) Let $I : V \to V$ be the identity linear transformation. Then $I(x) = x, \forall x \in V$. This shows that 1 is the only eigenvalue of I and all nonzero vectors of V are eigenvectors associated to eigenvalue 1.
(b) The zero linear transformation $0 : V \to V$ is defined by $0(x) = 0 = 0.x, \forall x \in V$. Hence, 0 is the only eigenvalue of the zero linear transformation and every nonzero vector of V is eigenvector associated to 0.
(c) Let $T : \mathbb{R}^2 \to \mathbb{R}^2$ be a linear transformation, which rotates every vector counterclockwise by $\frac{\pi}{3}$. This rotation in $\mathbb{R}^2$ moves every vector in such a way that no vector is taken to a scaling of itself. In this case, T has no eigenvalue, and hence no eigenvector. In general, rotation linear transformation is defined as $T(x, y) = (x \cos \theta - y \sin \theta, x \sin \theta + y \cos \theta)$.

Theorem 7.2.3. *Let V be a vector space over a field F and let $T : V \to V$ be a linear transformation. Then:*
(i)   *for all $k \neq 0$, kx is an eigenvector of T associated to eigenvalue λ if x is an eigenvector of T associated to λ;*
(ii) *eigenvalue λ, associated with an eigenvector x is unique;*
(iii) *$\lambda \in F$ is an eigenvalue of T if and only if $T - \lambda I$, where I is the identity linear transformation is singular;*
(iv) *T is singular if and only if one of its eigenvalue is zero.*

Proof. (i) $T(kx) = kT(x) = k\lambda x = \lambda(kx)$. This implies that kx is an eigenvector of T associated to eigenvalue λ.

(ii) Suppose $T(x) = \lambda_1 x = \lambda_2 x$. Then $0 = T(0) = T(x - x) = T(x) - T(x) = \lambda_1 x - \lambda_2 x = (\lambda_1 - \lambda_2)x \Rightarrow \lambda_1 - \lambda_2 = 0$ or $x = 0$. Since x is an eigenvector, $x \neq 0$. Hence, $\lambda_1 - \lambda_2 = 0 \Rightarrow \lambda_1 = \lambda_2$.

(iii) If λ is an eigenvalue of T, then there exists a nonzero vector $x \in V$ such that $T(x) = \lambda x = \lambda I(x) \Rightarrow T(x) - \lambda I(x) = 0 \Rightarrow (T - \lambda I)(x) = 0 \Rightarrow T - \lambda I$ is singular.

Conversely, suppose that $T - \lambda I$ is singular. Then there exists a nonzero vector $x \in V$ such that $(T - \lambda I)(x) = 0 \Rightarrow T(x) - \lambda I(x) = 0 \Rightarrow T(x) = \lambda x$. Hence, λ is an eigenvalue of T.

(iv) It follows from (iii) by taking $\lambda = 0$. □

It is shown in Theorem 5.3.2 that if $A = [a_{ij}]$ is an n-square matrix with entries in a field F, then there exists a linear transformation $L_A : F^n \to F^n$ defined by $L_A(X) = AX$ such that $[L_A]_{B,B} = A$, where B is the standard ordered basis of F^n. Suppose $\lambda \in F$ is an eigenvalue of L_A, then there exists a nonzero column vector $X \in F^n$ such that $L_A(X) = AX = \lambda X$. Accordingly, we have the following definition.

Definition 7.2.4. Let A be an n-square matrix with entries in a field F. Then scalar $\lambda \in F$ is called an eigenvalue of A if there exists a nonzero column vector $X \in F^n$ such that $AX = \lambda X$.

For an eigenvalue λ, any nonzero column vector X satisfying $AX = \lambda X$ is called an eigenvector associated to λ.

If X is an eigenvector of A associated to eigenvalue λ, then $A(kX) = kAX = k\lambda X = \lambda(kX)$, and shows that kX is also an eigenvector associated to λ, $k \neq 0 \in F$. Also, every eigenvalue λ of A associated with an eigenvector is unique.

Theorem 7.2.5. *Let A be an n-square matrix with entries in a field F. Then the scalar $\lambda \in F$ is an eigenvalue of A if and only if $|A - \lambda I_n| = 0$, where $|A - \lambda I_n|$ denotes the determinant of the matrix $(A - \lambda I_n)$.*

Proof. Since $\lambda \in F$ is an eigenvalue of A, there exists a nonzero column vector $X \in F^n$ such that $AX = \lambda X$. Then $AX = \lambda X = \lambda I_n(X) \Rightarrow (A - \lambda I_n)X = 0$, a homogeneous system of n equations in n variables. This system has a nontrivial solution X if and only if $|A - \lambda I_n| = 0$. Hence, λ is an eigenvalue of A if and only if $|A - \lambda I_n| = 0$. □

Definition 7.2.6. Let $A = [a_{ij}]$ be an n-square matrix. Then

$$|A - \lambda I_n| = \begin{vmatrix} a_{11} - \lambda & a_{12} & \cdots & a_{1n} \\ a_{21} & a_{22} - \lambda & \cdots & a_{2n} \\ \vdots & & & \\ a_{n1} & a_{n2} & \cdots & a_{nn} - \lambda \end{vmatrix}$$

is an n-degree polynomial in λ. This polynomial is called the characteristic polynomial of A and is denoted by $P_A(\lambda)$. By the characteristic equation of A, we mean the equation $P_A(\lambda) = 0$.

From above theorem, it is clear that the eigenvalues of A are the roots of the characteristic equation. Thus, an n-square matrix may have at most n distinct eigenvalues.

From the properties of determinant, we have the following result.

Theorem 7.2.7. *Let A be an n-square matrix. Then the characteristic equation of A is given by $P_A(\lambda) = \lambda^n - a_1\lambda^{n-1} + a_2\lambda^{n-2} + \cdots + (-1)^r a_r\lambda^{n-r} + \cdots + (-1)^n a_n = 0$, where a_r is the sum of principal r-minors of A.*

Theorem 7.2.8. *Similar matrices have the same characteristic polynomial, and hence the same eigenvalues also.*

Proof. Let A and B be similar n-square matrices. Then there exists a nonsingular matrix P such that $B = P^{-1}AP$.

Hence,

$$P_B(\lambda) = |B - \lambda I_n| = |P^{-1}AP - \lambda I_n| = |P^{-1}AP - \lambda P^{-1}P| = |P^{-1}AP - P^{-1}\lambda P|$$
$$= |P^{-1}(A - \lambda I_n)P| = |P^{-1}||A - \lambda I_n||P| = |A - \lambda I_n| = P_A(\lambda).$$

Since $P_A(\lambda) = 0$ and $P_B(\lambda) = 0$ are characteristic equations and eigenvalues are roots of characteristic equations, A and B have the same eigenvalues. □

Since matrices representing a linear transformation T on a finite-dimensional vector space V with respect to different choice a bases are similar, the above theorem enables us to define the characteristic polynomial of T.

Definition 7.2.9. Let T be a linear transformation on a finite-dimensional vector space V over a field F. Then the characteristic polynomial of T is defined as the characteristic polynomial of any matrix representing T.

Example 7.2.10. Let the matrix

$$A = \begin{bmatrix} 2 & 1 & 1 \\ 1 & 2 & 1 \\ 0 & 0 & 1 \end{bmatrix}.$$

Then the characteristic equation of A is $|A - \lambda I_3| = 0$.

$$\begin{vmatrix} 2-\lambda & 1 & 1 \\ 1 & 2-\lambda & 1 \\ 0 & 0 & 1-\lambda \end{vmatrix} = 0 \Rightarrow (1-\lambda)((2-\lambda)^2 - 1) = 0$$

$$\Rightarrow (1-\lambda)(4 + \lambda^2 - 4\lambda - 1) = 0 \Rightarrow (1-\lambda)(\lambda - 3)(\lambda - 1) = 0$$

$$\Rightarrow (\lambda - 3)(\lambda - 1)^2 = 0.$$

Since eigenvalues of a matrix are the roots of its characteristic equation, 1, 1, 3 are the eigenvalues of the matrix A.

Alternatively, from Theorem 7.2.7,

$$P_A(\lambda) = \lambda^3 - a_1\lambda^2 + a_2\lambda - a_3 = 0,$$

where

$a_1 = a_{11} + a_{22} + a_{33} = \text{tr}(A) = 5,$

$a_2 = $ the sum of 2-minors of $A = \begin{vmatrix} 2 & 1 \\ 0 & 1 \end{vmatrix} + \begin{vmatrix} 2 & 1 \\ 0 & 1 \end{vmatrix} + \begin{vmatrix} 2 & 1 \\ 1 & 2 \end{vmatrix} = 2 + 2 + 3 = 7,$ and

$a_3 = $ the sum of 3-minors of $A = |A| = 3.$

Then $P_A(\lambda) = \lambda^3 - 5\lambda^2 + 7\lambda - 3 = 0 \Rightarrow (\lambda - 3)(\lambda - 1)^2 = 0.$

Thus, the characteristic equation of a matrix can be obtained by anyone method from the above two.

Example 7.2.11. Let

$$A = \begin{bmatrix} 2 & k \\ -k & 0 \end{bmatrix}, \quad k > 0$$

is a real number. Then

$$|A - \lambda I_2| = \begin{vmatrix} -\lambda & k \\ -k & -\lambda \end{vmatrix} = 0 \Rightarrow \lambda^2 + k^2 = 0.$$

If A is representing the linear transformation $T : \mathbb{R}^2 \to \mathbb{R}^2$ with respect to the standard ordered basis of $\mathbb{R}^2$, then T has no eigenvalues. But if A is representing the linear transformation $T : \mathbb{C}^2 \to \mathbb{C}^2$ with respect to the standard ordered basis of $\mathbb{C}^2$, then T has $\pm ki$ eigenvalues. This shows that A has no eigenvalues in $\mathbb{R}$ but has the two eigenvalues ki and $-ki$ in $\mathbb{C}$. Therefore, we must be careful about the field involved during the calculation of eigenvalues of a matrix.

Theorem 7.2.12. *Let A be an n-square matrix with entries in a field F. Then the following properties hold:*
(i) *A is singular if and only if zero is an eigenvalue of A.*
(ii) *A is nonsingular if and only if all the eigenvalues of A are nonzero. If λ is an eigenvalue of A, then λ^{-1} is an eigenvalue of A^{-1}.*
(iii) *If λ is an eigenvalue of A, then the polynomial $f(A) = a_0 I_n + a_1 A + a_2 A^2 + \cdots + a_r A^r$, which is an n-square matrix has $f(\lambda) = a_0 + a_1\lambda + a_2\lambda^2 + \cdots + a_r\lambda^r$, as an eigenvalue, where $a_0, a_1, a_2, \ldots, a_r \in F$.*
(iv) *The sum of the eigenvalues of A is equal to $\text{tr}(A)$ and the product of eigenvalues of A is equal to $|A|$.*
(v) *A and A^t have the same eigenvalues.*

Proof. (i) If A is singular, then the homogeneous system of linear equations $AX = 0$ has a nontribal solution. Then there exists a nonzero column vector X_1 such that $AX_1 = 0 = 0.X_1$. Hence, 0 is an eigenvalue of A.

Conversely, if 0 is an eigenvalue of A, then there exists $X \neq 0$ such that $AX = 0X = 0$. That is the homogeneous system $AX = 0$ has a nontrivial solution. Hence, $|A| = 0$ and so A is singular.

(ii) Suppose A is a nonsingular matrix. Then from (i) it is clear that if λ is an eigenvalue of A, then $\lambda \neq 0$ if and only if A is nonsingular. If λ is an eigenvalue of A, then there exists a nonzero vector X such that $AX = \lambda X$. Since A is nonsingular, A^{-1} exists. Then $AX = \lambda X \Rightarrow A^{-1}AX = A^{-1}\lambda X \Rightarrow X = \lambda A^{-1}X \Rightarrow A^{-1}X = \lambda^{-1}X$. That is, λ^{-1} is an eigenvalue of A^{-1}.

(iii) Let λ be an eigenvalue of A. Then there exists a nonzero vector X such that $AX = \lambda X$. Then $A^2X = A(AX) = A\lambda X = \lambda(AX) = \lambda\lambda X = \lambda^2 X$. This shows that λ^2 is an eigenvalue of A^2. Similarly, $A^nX = \lambda^nX$, $kAX = k\lambda X = (k\lambda)X$ and eigenvalues of identity matrix is 1. Hence,

$$
\begin{aligned}
f(A)X &= a_0 I_n X + a_1 AX + a_2 A^2 X + \cdots + a_r A^r X \\
&= a_0 X + a_1 \lambda X + a_2 \lambda^2 X + \cdots + a_r \lambda^r X \\
&= (a_0 + a_1 \lambda + a_2 \lambda^2 + \cdots + a_r \lambda^r)X \\
&= f(\lambda)X.
\end{aligned}
$$

This proves that $f(\lambda)$ is the eigenvalue of $f(A)$.

(iv) From Theorem 7.2.7, the characteristic equation of A is $P_A(\lambda) = \lambda^n - a_1\lambda^{n-1} + a_2\lambda^{n-2} + \cdots + (-1)^r a_r \lambda^{n-r} + \cdots + (-1)^n a_n = 0$. Since eigenvalues of A are the roots of its characteristic equation, the sum of the roots of $P_A(\lambda) = 0 =$ the sum of the eigenvalues of $A = -(-a_1) = a_1 = a_{11} + a_{22} + \cdots + a_{nn} = \mathrm{tr}(A)$.

Similarly, the product of roots of equation $P_A(\lambda) = 0 =$ product of eigenvalues of $A = (-1)^n a_n = a_n$, if n is even and $-(-1)^n a_n = a_n$ if n is odd, where $a_n = |A|$.

(v) $|A - \lambda I_n| = |(A - \lambda I_n)^t| = |A^t - \lambda I_n|$. This shows that the characteristic polynomials of A and A^t are the same, and hence A and A^t have the same eigenvalues. □

Example 7.2.13. Let

$$
A = \begin{bmatrix} 1 & 0 & -1 \\ 1 & 2 & 1 \\ 2 & 2 & 3 \end{bmatrix}.
$$

Then the characteristic equation $P_A(\lambda) = 0$ gives

$$
\begin{vmatrix} 1-\lambda & 0 & -1 \\ 1 & 2-\lambda & 1 \\ 2 & 2 & 3-\lambda \end{vmatrix} = 0 \Rightarrow \lambda^3 - 6\lambda^2 + 11\lambda - 6 = 0 \Rightarrow (\lambda - 1)(\lambda - 2)(\lambda - 3) = 0.
$$

Thus, the eigenvalues of A are 1, 2, 3.

Let

$$X = \begin{bmatrix} x_1 \\ x_2 \\ x_3 \end{bmatrix}$$

be the eigenvector associated to eigenvalue λ of A. Then

$$(A - \lambda I_3)X = 0, \quad \text{or} \quad \begin{bmatrix} 1-\lambda & 0 & -1 \\ 1 & 2-\lambda & 1 \\ 2 & 2 & 3-\lambda \end{bmatrix} \begin{bmatrix} x_1 \\ x_2 \\ x_3 \end{bmatrix} = \begin{bmatrix} 0 \\ 0 \\ 0 \end{bmatrix}.$$

Now, we put $\lambda = 1, 2$ and 3 in the above equation and find associated eigenvectors of A.

For $\lambda = 1$,

$$\begin{bmatrix} 0 & 0 & -1 \\ 1 & 1 & 1 \\ 2 & 2 & 2 \end{bmatrix} \begin{bmatrix} x_1 \\ x_2 \\ x_3 \end{bmatrix} = \begin{bmatrix} 0 \\ 0 \\ 0 \end{bmatrix}.$$

Then

$$-x_3 = 0, \quad x_1 + x_2 + x_3 = 0, \quad 2x_1 + 2x_2 + x_3 = 0, \quad \text{or}$$
$$x_3 = 0, \quad x_1 + x_2 = 0.$$

Take $x_1 = k$. Then $x_2 = -k$.
Hence,

$$\begin{bmatrix} x_1 \\ x_2 \\ x_3 \end{bmatrix} = \begin{bmatrix} k \\ -k \\ 0 \end{bmatrix} = k \begin{bmatrix} 1 \\ -1 \\ 0 \end{bmatrix}.$$

Thus,

$$\begin{bmatrix} 1 \\ -1 \\ 0 \end{bmatrix}$$

is an eigenvector corresponding to eigenvalue $\lambda = 1$.

For $\lambda = 2$,

$$\begin{bmatrix} -1 & 0 & -1 \\ 1 & 0 & 1 \\ 2 & 2 & 1 \end{bmatrix} \begin{bmatrix} x_1 \\ x_2 \\ x_3 \end{bmatrix} = \begin{bmatrix} 0 \\ 0 \\ 0 \end{bmatrix}.$$

Then

$$-x_1 - x_3 = 0, \quad x_1 + x_3 = 0, \quad 2x_1 + 2x_2 + x_3 = 0, \quad \text{or}$$
$$x_1 + x_3 = 0, \quad 2x_1 + 2x_2 + x_3 = 0.$$

Take $x_3 = k$. Then $x_1 = -k$ and $x_2 = \frac{k}{2}$.
Hence,

$$\begin{bmatrix} x_1 \\ x_2 \\ x_3 \end{bmatrix} = \begin{bmatrix} -k \\ \frac{k}{2} \\ k \end{bmatrix} = k \begin{bmatrix} -1 \\ \frac{1}{2} \\ 1 \end{bmatrix}.$$

Thus,

$$\begin{bmatrix} -1 \\ \frac{1}{2} \\ 1 \end{bmatrix}$$

is an eigenvector corresponding to eigenvalue $\lambda = 2$.
For $\lambda = 3$,

$$\begin{bmatrix} -2 & 0 & -1 \\ 1 & -1 & 1 \\ 2 & 2 & 0 \end{bmatrix} \begin{bmatrix} x_1 \\ x_2 \\ x_3 \end{bmatrix} = \begin{bmatrix} 0 \\ 0 \\ 0 \end{bmatrix}.$$

Then

$$-2x_1 - x_3 = 0, \quad x_1 - x_2 + x_3 = 0, \quad 2x_1 + 2x_2 = 0, \quad \text{or}$$
$$x_2 = -x_1, \quad x_3 = -2x_1.$$

Take $x_1 = k$.
Hence,

$$\begin{bmatrix} x_1 \\ x_2 \\ x_3 \end{bmatrix} = \begin{bmatrix} k \\ -k \\ -2k \end{bmatrix} = k \begin{bmatrix} 1 \\ -1 \\ -2 \end{bmatrix}.$$

Thus,

$$\begin{bmatrix} 1 \\ -1 \\ -2 \end{bmatrix}$$

is an eigenvector corresponding to eigenvalue $\lambda = 3$.

Example 7.2.14. Let

$$A = \begin{bmatrix} 2 & 1 & 1 \\ 1 & 2 & 1 \\ 0 & 0 & 1 \end{bmatrix}.$$

Then the characteristic equation $P_A(\lambda) = |A - \lambda I_3| = 0$ is

$$\begin{vmatrix} 2 - \lambda & 1 & 1 \\ 1 & 2 - \lambda & 1 \\ 0 & 0 & 1 - \lambda \end{vmatrix} = 0 \Rightarrow (1 - \lambda)^2 (3 - \lambda) = 0.$$

Hence the eigenvalues of A are 1, 1, 3. To find the eigenvector associated to eigenvalues of A, we need to solve the following equation for $\lambda = 1, 3$:

$$\begin{bmatrix} 2 - \lambda & 1 & 1 \\ 1 & 2 - \lambda & 1 \\ 0 & 0 & 1 - \lambda \end{bmatrix} \begin{bmatrix} x_1 \\ x_2 \\ x_3 \end{bmatrix} = \begin{bmatrix} 0 \\ 0 \\ 0 \end{bmatrix}.$$

For, $\lambda = 1$, we have $x_1 + x_2 + x_3 = 0 \Rightarrow x_3 = -x_1 - x_2$. Take $x_1 = k_1$ and $x_2 = k_2$. Then

$$\begin{bmatrix} x_1 \\ x_2 \\ x_3 \end{bmatrix} = \begin{bmatrix} k_1 \\ k_2 \\ -k_1 - k_2 \end{bmatrix} = k_1 \begin{bmatrix} 1 \\ 0 \\ -1 \end{bmatrix} + k_2 \begin{bmatrix} 0 \\ 1 \\ -1 \end{bmatrix}.$$

Thus,

$$\begin{bmatrix} 1 \\ 0 \\ -1 \end{bmatrix} \quad \text{and} \quad \begin{bmatrix} 0 \\ 1 \\ -1 \end{bmatrix}$$

are eigenvectors associated to eigenvalue 1.

Similarly, for $\lambda = 3$ we have

$$-x_1 + x_2 + x_3 = 0, \quad x_3 = 0, \quad \text{or}$$
$$x_1 = x_2, \quad x_3 = 0.$$

Take $x_1 = k$.
Then

$$\begin{bmatrix} x_1 \\ x_2 \\ x_3 \end{bmatrix} = \begin{bmatrix} k \\ k \\ 0 \end{bmatrix} = k \begin{bmatrix} 1 \\ 1 \\ 0 \end{bmatrix}.$$

Hence,

$$\begin{bmatrix} 1 \\ 1 \\ 0 \end{bmatrix}$$

is an eigenvector associated to eigenvalue $\lambda = 3$.

Example 7.2.15. Let

$$A = \begin{bmatrix} 1 & 1 & 1 \\ 0 & 1 & 1 \\ 0 & 0 & 1 \end{bmatrix}.$$

Then the characteristic equation $|A - \lambda I_3| = 0 \Rightarrow (1 - \lambda)^3 = 0 \Rightarrow \lambda = 1, 1, 1$.
Now, we find the eigenvector of A associated to eigenvalue 1. Then

$$\begin{bmatrix} 0 & 1 & 1 \\ 0 & 0 & 1 \\ 0 & 0 & 0 \end{bmatrix} \begin{bmatrix} x_1 \\ x_2 \\ x_3 \end{bmatrix} = \begin{bmatrix} 0 \\ 0 \\ 0 \end{bmatrix}$$

$$\Rightarrow x_2 + x_3 = 0, x_3 = 0 \Rightarrow x_1 = k, x_2 = 0, x_3 = 0.$$

Hence, eigenvectors of A associated to eigenvalue 1 are of the form

$$\begin{bmatrix} k \\ 0 \\ 0 \end{bmatrix} = k \begin{bmatrix} 1 \\ 0 \\ 0 \end{bmatrix}.$$

In this case, eigenvalue 1 is repeated three times but there is only one linearly independent associated eigenvector.

Example 7.2.16.
(a) Let

$$A = \begin{bmatrix} a_{11} & a_{12} & a_{13} \\ 0 & a_{22} & a_{23} \\ 0 & 0 & a_{33} \end{bmatrix}$$

be an upper triangular matrix. Then the characteristic equation

$$P_A(\lambda) = \begin{vmatrix} a_{11} - \lambda & a_{12} & a_{13} \\ 0 & a_{22} - \lambda & a_{23} \\ 0 & 0 & a_{33} - \lambda \end{vmatrix} = (a_{11} - \lambda)(a_{22} - \lambda)(a_{33} - \lambda) = 0 \Rightarrow \lambda = a_{11}, a_{22}, a_{33}.$$

Thus, in general if A is an n-square triangular matrix, then all the diagonal elements of A are its eigenvalues.

(b) Let

$$D = \begin{bmatrix} d_1 & 0 & 0 \\ 0 & d_2 & 0 \\ 0 & 0 & d_3 \end{bmatrix}$$

be a diagonal matrix. Then $P_D(\lambda) = (d_1 - \lambda)(d_2 - \lambda)(d_3 - \lambda) = 0 \Rightarrow \lambda = d_1, d_2, d_3$ are eigenvalues of D. Vectors $e_1 = (1, 0, 0)$, $e_2 = (0, 1, 0)$ and $e_3 = (0, 0, 1)$ are eigenvectors associated to eigenvalues d_1, d_2 and d_3, respectively.

Similarly, for an n-square diagonal matrix $D = \text{diag}(d_1, d_2, \ldots, d_n)$, eigenvalues are $d_1, d_2, \ldots, d_n$ and $e_1, e_2, \ldots, e_n$ are associated eigenvectors, respectively, where $e_i = (0, 0, \ldots, 0, 1, 0, \ldots, 0) \ \forall 1 \le i \le n$.

(c) Let A be an idempotent matrix. Then $A^2 = A$. If λ is an eigenvalue of A, then there exists a nonzero vector X such that $AX = \lambda X$. Also, $A^2 X = A(AX) = A(\lambda X) = \lambda(AX) = \lambda^2 X$.

Hence, $A^2 X = AX \Rightarrow \lambda^2 X = \lambda X \Rightarrow (\lambda^2 - \lambda)X = 0$. Since $X \ne 0$, $\lambda^2 - \lambda = 0 \Rightarrow \lambda = 0, 1$. Thus, eigenvalues of an idempotent matrix are $0, 1$.

Similarly, it is easy to prove that the eigenvalues of a nilpotent matrix are 0.

(d) Let $A = [a_{ij}]$ be an n-square matrix with entries in a field F such that $\sum_{j=1}^{n} a_{ij} = k$, for all $1 \le i \le n$. Then k will be an eigenvalue of A and

$$\begin{bmatrix} 1 \\ 1 \\ \vdots \\ 1 \end{bmatrix}$$

will be its associated eigenvector.

Example 7.2.17. Let A be a matrix of order 3×3 such that $|A - 2I_3| = 0$, $\text{tr}(A) = 10$ and $|A| = 30$. Then we find the eigenvalues of A.

From Theorem 7.2.5, we have that $\lambda \in F$ is an eigenvalue of A if and only if $|A - \lambda\lambda I_n| = 0$. Hence, $|A - 2I_3| = 0 \Rightarrow 2$ is an eigenvalue of A. Let $\lambda_1 = 2, \lambda_2, \lambda_3$ be the eigenvalues of A. Then

$$\lambda_1 + \lambda_2 + \lambda_3 = \text{tr}(A)$$
$$\Rightarrow 2 + \lambda_2 + \lambda_3 = 10$$
$$\Rightarrow \lambda_2 + \lambda_3 = 8$$
$$\Rightarrow \lambda_3 = 8 - \lambda_2, \quad \text{and}$$
$$\lambda_1 \lambda_2 \lambda_3 = |A| \Rightarrow 2\lambda_2\lambda_3 = 30 \Rightarrow \lambda_2\lambda_3 = 15.$$

Substituting $\lambda_3 = 8 - \lambda_2$, in $\lambda_2\lambda_3 = 15$ we get

$$\lambda_2(8 - \lambda_2) = 15$$
$$\Rightarrow \lambda_2^2 - 8\lambda_2 + 15 = 0$$
$$\Rightarrow (\lambda_2 - 5)(\lambda_2 - 3) = 0$$
$$\Rightarrow \lambda_2 = 5, \text{ or } 3.$$

If $\lambda_2 = 5$, then $\lambda_3 = 8 - 5 = 3$. If $\lambda_2 = 3$, then $\lambda_3 = 8 - 3 = 5$. Thus, eigenvalues of A are 2, 3, 5.

Alternatively, we have the following. Let $\lambda_1, \lambda_2, \lambda_3$ be the eigenvalues of A. Then from Theorem 7.2.12(iii), $\lambda_1 - 2, \lambda_2 - 2, \lambda_3 - 2$ will be eigenvalues of matrix $A - 2I_3$. Since $|A - 2I_3| = 0$, at least one eigenvalue of $A - 2I_3$ will be zero. Suppose $\lambda_1 - 2 = 0$, then $\lambda_1 = 2$. Hence, $\lambda_1\lambda_2\lambda_3 = 30$ and $\lambda_1 + \lambda_2 + \lambda_3 = 10$ give the result.

Example 7.2.18. Let

$$A = \begin{bmatrix} 1 & -2 & 3 \\ 0 & -2 & 4 \\ 0 & 0 & -1 \end{bmatrix}.$$

Then find the eigenvalues of $2A^3 + 5A^2 - 3A + 2I_3$.

Since A is an upper triangular matrix, eigenvalues of A are 1, −2, −3, the principal diagonal of A.

Then from Theorem 7.2.12 (iii), eigenvalues of $2A^3 + 5A^2 - 3A + 2I_3$ are

$$2(1)^3 + 5(1)^2 - 3(1) + 2 \times 1 = 6,$$
$$2(-2)^3 + 5(-2)^2 - 3(-2) + 2 \times 1 = 12, \quad \text{and}$$
$$2(-1)^3 + 5(-1)^2 - 3(-1) + 2 \times 1 = 8.$$

Example 7.2.19. Let A and B be n-square matrices with entries in a field F. Then AB and BA have the same eigenvalues.

If $\lambda = 0$ is an eigenvalue of AB, then AB is singular. Hence, $0 = |AB| = |BA| \Rightarrow BA$ is a singular matrix. Then $\lambda = 0$ is an eigenvalue of BA, also.

Now, let $\lambda \neq 0$ be an eigenvalue of AB. Then there exists a nonzero column vector X such that $ABX = \lambda X$.

Again, $(BA)(BX) = B(ABX) = B\lambda X = \lambda(BX)$ implies that λ is an eigenvalue of BA with the associated eigenvector $BX \neq 0$.

If $BX = 0$, then $ABX = 0 \Rightarrow \lambda X = 0$. Since $\lambda \neq 0$, $\lambda X = 0 \Rightarrow X = 0$, which is a contradiction that X is an eigenvector.

Example 7.2.20. Let

$$A = \begin{bmatrix} 1 & 2 & 1 \\ 1 & 5 & -2 \\ 2 & -3 & 5 \end{bmatrix}.$$

Then find the eigenvalues of A.

Since the sum of each row is 4, 4 is an eigenvalue of A.

Now, suppose $\lambda_1 = 4, \lambda_2, \lambda_3$ are eigenvalues of A. Then $\mathrm{tr}(A) = \lambda_1 + \lambda_2 + \lambda_3 \Rightarrow 11 = 4 + \lambda_2 + \lambda_3 \Rightarrow \lambda_2 + \lambda_3 = 7$.

Also, $\lambda_1\lambda_2\lambda_3 = |A| \Rightarrow 4\lambda_2\lambda_3 = 3$.

Solving the above two equations, we get $\lambda_2 = \frac{7+\sqrt{37}}{2}$ and $\lambda_3 = \frac{7-\sqrt{37}}{2}$.

Example 7.2.21. Let

$$A = \begin{bmatrix} 1 & r & r^2 & r^3 & r^4 \\ 0 & r & r^2 & r^3 & r^4 \\ 0 & 0 & r^2 & r^3 & r^4 \\ 0 & 0 & 0 & r^3 & r^4 \\ 0 & 0 & 0 & 0 & r^4 \end{bmatrix},$$

where $r = e^{\frac{2\pi i}{5}}$ is a fifth root of unity. Then we calculate the trace of the matrix $I_5 + A + A^2$ with the help of the eigenvalues of A.

Since A is an upper triangular matrix, eigenvalues of A are $1, r, r^2, r^3, r^4$. Hence, eigenvalues of $I_5 + A + A^2$ are $1 + 1 + 1 = 3, 1 + r + r^2, 1 + r^2 + r^4, 1 + r^3 + r^6$ and $1 + r^4 + r^8$.

Now, trace of $(I_5 + A + A^2)$ = sum of the eigenvalues. Hence,

$$\mathrm{tr}(I_5 + A + A^2) = 3 + (1 + r + r^2) + (1 + r^2 + r^4) + (1 + r^3 + r^6) + (1 + r^4 + r^8)$$
$$= 3 + (1 + r + r^2 + r^3 + r^4) + (1 + r^2 + r^4 + r^6 + r^8) + 2$$
$$= 5 + (1 + r + r^2 + r^3 + r^4) + (1 + r^2 + r^4 + r + r^3).$$

Since r is the fifth root of unity, $1 + r + r^2 + r^3 + r^4 = 0$. Hence, from the above equation, $\mathrm{tr}(I_5 + A + A^2) = 5$.

Exercises

(7.2.1) Determine eigenvalues and their associated eigenvectors of the following matrices:

$$A = \begin{bmatrix} 1 & 0 & -1 \\ 1 & 2 & 1 \\ 2 & 2 & 3 \end{bmatrix}, \quad B = \begin{bmatrix} 3 & 10 & 5 \\ -2 & -3 & -4 \\ 3 & 5 & 7 \end{bmatrix}, \quad C = \begin{bmatrix} 6 & -2 & 2 \\ -2 & 3 & -1 \\ 2 & -1 & 3 \end{bmatrix}.$$

(7.2.2)   Let

$$A = \begin{bmatrix} 6 & -2 & 2 \\ -2 & 3 & -1 \\ 2 & -1 & 3 \end{bmatrix}.$$

Then find the eigenvalues of

(i) A^5 (ii) $5A^3$ (iii) $(A - 3I_3)^2$ (iv) $3A^3 - 2A^2 + A - I_3$

(v) $5A^4 + 4A^3 - 6A^2 + 2A - 7I_3$ (vi) A^t.

(7.2.3)   For the matrices A, B, C given in exercise (7.2.1), find $\text{tr}(A^4)$, $\text{tr}(B^5)$ and $\text{tr}(C^{10})$.

(7.2.4)   Let A be a 3-square matrix such that $|2A - 6I_3| = 0$, $\text{tr}(A) = 15$ and $|A| = 105$. Then find the sum of squares of eigenvalues of A.

(7.2.5)   Let A be a 3-square matrix such that $|A - I_3| = 0$, $\text{tr}(A) = 13$ and $|A| = 32$. Then find the eigenvalues of A.

(7.2.6)   Consider λ as an eigenvalue of square matrices A and B, both of size n, corresponding to the common eigenvector x. Show that: (a) 2λ is an eigenvalue of $A+B$ corresponding to x, and (b) λ^2 is an eigenvalue of AB corresponding to x.

(7.2.7)   Suppose that 1, −1, 2, −2 are eigenvalues of a 4-square matrix A. If $B = A^4 - 3A^2 + 3I_4$, then find $\det(A + B)$, $\det(A - B)$, $\det(B^2)$, $\text{tr}(A + B)$, $\text{tr}(A - B)$ and $\text{tr}(B^2)$.

7.3 Diagonalization

We know that any computation involving diagonal matrices is quite simple in comparison to other matrices. For a given linear transformation T on a finite-dimensional vectors space V, the existence of an ordered basis B for V, such that $[T]_{B,B}$ is a diagonal matrix is called a diagonalization problem. In this section, we deal with this problem.

Let $T : V \to V$ be a linear transformation and let $B = \{v_1, v_2, \ldots, v_n\}$ be a basis of V, such that $T(v_i) = \lambda_i v_i$, $1 \le i \le n$, where λ_i are scalars. Then

$$T(v_1) = \lambda_1 v_1 = \lambda_1 v_1 + 0v_2 + 0v_3 + \cdots + 0v_n,$$
$$T(v_2) = \lambda_2 v_2 = 0v_1 + \lambda_2 v_2 + 0v_3 + \cdots + 0v_n,$$

$$\vdots$$

$$T(v_n) = \lambda_n v_n = 0v_1 + 0v_2 + 0v_3 + \cdots + \lambda_n v_n,$$

and hence

$$[T]_{B,B} = \begin{bmatrix} \lambda_1 & 0 & 0 & \cdots & 0 \\ 0 & \lambda_2 & 0 & \cdots & 0 \\ \vdots & \vdots & & & \\ 0 & 0 & \cdots & 0 & \lambda_n \end{bmatrix}$$

is a diagonal matrix.

Since all v_i, $1 \le i \le n$, being basis elements are nonzero vectors, all λ_i, $1 \le i \le n$ are eigenvalues of T and v_i are there associated eigenvectors.

Thus, from this illustration it is clear that the concepts of eigenvalues and eigenvectors discussed in the previous section are naturally attached with diagonalization problems. Thus, we have the following definition.

Definition 7.3.1. Let V be a finite-dimensional vector space over a field F. Then a linear transformation $T : V \to V$ is said to be diagonalizable if there is a basis of V with respect to which the matrix of T is a diagonal matrix.

Or equivalently, T is diagonalizable if and only if eigenvectors of T form a basis of V.

In Corollary 5.2.4, if we replace V by F^n, T by L_A, where A is an n-square matrix and basis B_1 by standard ordered basis of F^n, then we get the result that A is diagonalizable if and only if L_A is diagonalizable. In view of this result, we have the following definition for a matrix.

Definition 7.3.2. An n-square matrix A with entries in a field F is said to be diagonalizable if it is similar to a diagonal matrix. Or equivalently, A is diagonalizable if and only if eigenvectors of A form of basis of F^n.

Theorem 7.3.3. *The set $\{x_1, x_2, \ldots, x_k\}$ of eigenvectors associated to distinct eigenvalues $\lambda_1, \lambda_2, \ldots, \lambda_k$ of linear transformation T on a finite-dimensional vector space V is linearly independent. Or equivalently, eigenvectors associated to distinct eigenvalues of a matrix A are linearly independent.*

Proof. To prove the result, we use the method of mathematical induction on the number of distinct eigenvalues of T.

Suppose $T : V \to V$ has only one eigenvalue λ_1. Then there exist a nonzero eigenvector x associated to λ. Since $x \ne 0$, $\{x\}$ is linearly independent. Thus, the result is true for $k = 1$.

Suppose that the statement holds for $k = r$ distinct eigenvalues. Let $\{x_1, x_2, \ldots, x_r, x_{r+1}\}$ represent the set of eigenvectors corresponding to the distinct eigenvalues $\lambda_1, \lambda_2, \ldots, \lambda_r, \lambda_{r+1}$.

Now, for the scalars $\{a_1, a_2, \ldots, a_r, a_{r+1}\}$, assume that

$$a_1 x_1 + a_2 x_2 + \cdots + a_r x_r + a_{r+1} x_{r+1} = 0. \tag{7.1}$$

Then

$$T(a_1 x_1 + a_2 x_2 + \cdots + a_r x_r + a_{r+1} x_{r+1}) = T(0),$$
$$a_1 T(x_1) + a_2 T(x_2) + \cdots + a_r T(x_r) + a_{r+1} T(x_{r+1}) = 0,$$
$$a_1 \lambda_1 x_1 + a_2 \lambda_2 x_2 + \cdots + a_r \lambda_r x_r + a_{r+1} \lambda_{r+1} x_{r+1} = 0. \tag{7.2}$$

Applying $(7.2) - \lambda_{r+1}(7.1)$, we have

$$a_1 (\lambda_1 - \lambda_{r+1}) x_1 + a_2 (\lambda_2 - \lambda_{r+1}) x_2 + \cdots + a_r (\lambda_r - \lambda_{r+1}) x_r = 0.$$

It is assumed that $\{x_1, x_2, \ldots, x_r\}$ are linearly independent,

$$a_1(\lambda_1 - \lambda_{r+1}) = a_2(\lambda_2 - \lambda_{r+1}) = \cdots = a_r(\lambda_r - \lambda_{r+1}) = 0.$$

Since all λ_r are distinct, $a_1 = a_2 = \cdots = a_r = 0$. Then from (7.1), $a_{r+1}x_{r+1} = 0$. Since $x_{r+1} \neq 0$, $a_{r+1} = 0$. Hence, the set $\{x_1, x_2, \ldots, x_r, x_{r+1}\}$ is linearly independent. Thus, by the induction hypothesis the result is true for all distinct eigenvalues. □

With the help of this result, we obtain the following important corollary related to diagonalization of a linear transformation or equivalently, diagonalization of a matrix.

Corollary 7.3.4. *Let V be an n-dimensional vectors space over a field F. Then the linear transformation $T : V \to V$ having n distinct eigenvalues is diagonalizable.*

Proof. Suppose $x_1, x_2, \ldots, x_n$ are n eigenvectors associated to distinct eigenvalues $\lambda_1, \lambda_2, \ldots, \lambda_n$ of T. Then from the above theorem the set $\{x_1, x_2, \ldots, x_n\} \subset V$ is linearly independent. Since $\dim(V) = n$, the set of eigenvectors of T form a basis of V. Hence, T is diagonalizable. Also, from $T(x_i) = \lambda_i x_i$, $1 \le i \le n$, the matrix of T related to the basis $\{x_1, x_2, \ldots, x_n\}$ is the diagonal matrix $\mathrm{diag}(\lambda_1, \lambda_2, \ldots, \lambda_n)$. □

For example, the matrix

$$\begin{bmatrix} 1 & 2 & 5 \\ 0 & 2 & 6 \\ 0 & 0 & 4 \end{bmatrix}$$

is diagonalizable.

Remark. Converse of the above result is not true. Since I_3 is a diagonal matrix but 1 is the only eigenvalue of I_3 repeated 3 times.

The fundamental theorem of algebra states that every polynomial has a root in the field of complex numbers. Thus, the characteristic equation $|A - \lambda I_n| = 0$ has n roots, including repetition, over the field of complex numbers. That is, if $\lambda_1, \lambda_2, \ldots, \lambda_r$ are distinct roots of the characteristic equation, then

$$0 = |A - \lambda I_n| = (\lambda - \lambda_1)^{a_1}(\lambda - \lambda_2)^{a_2} \cdots (\lambda - \lambda_r)^{a_r}, \quad \text{where } a_1 + a_2 + \cdots + a_r = n.$$

We call $a_1, a_2, \ldots, a_r$ the algebraic multiplicities of $\lambda_1, \lambda_2, \ldots, \lambda_r$, respectively.

Definition 7.3.5. The algebraic multiplicity of an eigenvalue λ is the number of times λ is repeated as a root of the characteristic polynomial. We use the notation $\mathrm{Mul}_a(\lambda)$ for algebraic multiplicity of λ.

Definition 7.3.6. Let λ be an eigenvalue of linear transformation $T : V \to V$. Then $V(\lambda) = \{x \in V : T(x) = \lambda x\}$, the set of all eigenvectors associated to λ including zero vector is a subspace of V, called the λ-eigenspace.

The dimension of the λ-eigenspace $V(\lambda)$ is called the geometric multiplicity of the eigenvalue λ and is denoted by $\mathrm{Mul}_g(\lambda)$.

Equivalently, if λ is an eigenvalue of an n-square matrix A with entries in a field F, then $F^n(\lambda)$ = the set of all eigenvectors associated to λ with the zero column vector is a subspace of F^n, called the λ-eigenspace.

Hence, the geometric multiplicity,

$$\mathrm{Mul}_g(\lambda) = \dim(F^n(\lambda)) = \dim(N(A - \lambda I_n)) = n - \mathrm{rank}(A - \lambda I_n).$$

Example 7.3.7. Consider Example 7.2.14. We have $\mathrm{Mul}_a(1) = 2$ and $\mathrm{Mul}_a(3) = 1$.
Now,

$$\mathrm{rank}(A - I_3) = \mathrm{rank}\left(\begin{bmatrix} 1 & 1 & 1 \\ 1 & 1 & 1 \\ 0 & 0 & 0 \end{bmatrix}\right) = 1.$$

Hence, $\mathrm{Mul}_g(1) = 3 - \mathrm{rank}(A - I_3) = 2$.
Similarly, $\mathrm{Mul}_g(3) = 3 - \mathrm{rank}(A - 3I_3) = 3 - 2 = 1$.
Now, let

$$B = \begin{bmatrix} 1 & 2 & 3 \\ 0 & 2 & 3 \\ 0 & 0 & 2 \end{bmatrix}.$$

Then $|B - \lambda I_3| = (1 - \lambda)(2 - \lambda)^2 = 0$.
Here, $\mathrm{Mul}_a(1) = 1$ and $\mathrm{Mul}_a(2) = 2$.
Now,

$$\mathrm{Mul}_g(1) = 3 - \mathrm{rank}(B - I_3) = 3 - 2 = 1, \quad \text{and}$$
$$\mathrm{Mul}_g(2) = 3 - \mathrm{rank}(B - 2I_3) = 3 - 2 = 1.$$

Thus, the algebraic multiplicity of an eigenvalue λ may not be equal to its geometric multiplicity. Note that if $\mathrm{Mul}_a(\lambda) = 1$, then $\mathrm{Mul}_g(\lambda) = 1$ (see Corollary 7.3.10).

Theorem 7.3.8. *Let T be a linear transformation on an n-dimensional vectors space V and let $\lambda_1, \lambda_2, \ldots, \lambda_k$ be its distinct eigenvalues. Then T is diagonalizable if and only if $\mathrm{Mul}_g(\lambda_1) + \mathrm{Mul}_g(\lambda_2) + \cdots + \mathrm{Mul}_g(\lambda_k) = n$.*

Proof. Suppose $\mathrm{Mul}_g(\lambda_1) + \mathrm{Mul}_g(\lambda_2) + \cdots + \mathrm{Mul}_g(\lambda_k) = n$. Then we have

$$\dim(V(\lambda_1)) + \dim(V(\lambda_2)) + \cdots + \dim(V(\lambda_k)) = n$$
$$\Rightarrow V(\lambda_1) \oplus V(\lambda_2) \oplus \cdots \oplus V(\lambda_k) = V.$$

That is, V is the direct sum of eigenspaces of T.

If B_i denotes the basis of λ_i-eigenspace for all i, then $B = \bigcup_{i=1}^{k} B_i$ is a basis of V consisting of eigenvectors of T. Hence, T is diagonalizable.

Conversely, suppose T is diagonalizable. Then eigenvectors of T form a basis of V. Hence, in view of the result that the associated eigenvectors to distinct eigenvalues are linearly independent, $V(\lambda_1) \oplus V(\lambda_2) \oplus \cdots \oplus V(\lambda_k) = V$.

Therefore, $\text{Mul}_g(\lambda_1) + \text{Mul}_g(\lambda_2) + \cdots + \text{Mul}_g(\lambda_k) = n$. □

Equivalently, the above result can also be proved for a matrix.

Theorem 7.3.9. *If λ is an eigenvalue of a linear transformation on a finite-dimensional vector space, then $\text{Mul}_g(\lambda) \leq \text{Mul}_a(\lambda)$.*

Proof. Let T be a linear transformation on an n-dimensional vectors space V and let λ be an eigenvalue of geometric multiplicity r. Then $\dim V(\lambda) = r$.

Suppose $\{v_1, v_2, \ldots, v_r\}$ is a basis of $V(\lambda)$. Extending it to a basis of V, let $B = \{v_1, v_2, \ldots, v_r, v_{r+1}, \ldots, v_n\}$ be an ordered basis of V. Then matrix of T related to the basis B is a block matrix

$$[T]_{B,B} = \begin{bmatrix} \lambda I_r & A \\ 0 & B \end{bmatrix}.$$

Then the characteristic polynomial $P_{[T]_{B,B}}(t) = P_{\lambda I_r}(t) P_B(t) = (\lambda - t)^r P_B(t)$, where $P_B(t)$ is a polynomial of degree $n - r$. It follows that the algebraic multiplicity of λ is greater than or equal to r. Hence, $\text{Mul}_g(\lambda) \leq \text{Mul}_a(\lambda)$. □

Corollary 7.3.10. *If algebraic multiplicity of an eigenvalue λ is one, then $\text{Mul}_g(\lambda) = \text{Mul}_a(\lambda) = 1$.*

Proof. Since eigenvector x associated to an eigenvalue λ is nonzero, $\dim(V(\lambda)) \geq 1$. Hence, $\text{Mul}_g(\lambda) \geq 1$. But $1 \leq \text{Mul}_g(\lambda) \leq \text{Mul}_a(\lambda) = 1 \Rightarrow 1 \leq \text{Mul}_g(\lambda) \leq 1 \Rightarrow \text{Mul}_g(\lambda) = 1$. □

Corollary 7.3.11. *Let V be an n-dimensional vector space. Then the linear transformation $T : V \to V$ is diagonalizable if and only if for every eigenvalue λ of T, $\text{Mul}_g(\lambda) = \text{Mul}_a(\lambda)$.*

Proof. Let $\lambda_1, \lambda_2, \ldots, \lambda_k$ be distinct eigenvalues of T such that $\text{Mul}_a(\lambda_i) = \text{Mul}_g(\lambda_i)$ $\forall 1 \leq i \leq k$. Then $\sum_{i=1}^{k} \text{Mul}_a(\lambda_i) = \sum_{i=1}^{k} \text{Mul}_g(\lambda_i) = n$. Hence, the result follows from Theorems 7.3.8 and 7.3.9. □

Example 7.3.12. Recall the matrix

$$A = \begin{bmatrix} 2 & 1 & 1 \\ 1 & 2 & 1 \\ 0 & 0 & 1 \end{bmatrix}$$

given in Example 7.2.14. Then $P_A(\lambda) = (1 - \lambda)^2(3 - \lambda) = 0 \Rightarrow \lambda = 1, 3$ are eigenvalues of A. Also, A has three linearly independent eigenvectors, hence A is diagonalizable.

Alternatively, from Example 7.3.7 we have $\text{Mul}_g(1) + \text{Mul}_g(3) = 2 + 1 = 3$. Hence, A is diagonalizable.

Next, recall the matrix B in Example 7.3.7,

$$B = \begin{bmatrix} 1 & 2 & 3 \\ 0 & 2 & 3 \\ 0 & 0 & 2 \end{bmatrix}.$$

Then $P_B(\lambda) = (1 - \lambda)(2 - \lambda)^2 = 0$, and $\text{Mul}_g(1) + \text{Mul}_g(2) = 1 + 1 = 2 \neq 3$. Hence, B is not diagonalizable.

Note that an n-square matrix A is diagonalizable if the sum of the geometric multi-plicities of its eigenvalues equals n. If this sum is not equal to n, then A is not diagonaliz-able. Thus, determining whether A is diagonalizable does not require computing all of its eigenvectors.

We have that if A is a diagonal matrix, then A is similar to a diagonal matrix. That is, there exists a nonsingular matrix P such that $P^{-1}AP = D$. The following theorem provides the solution of the problem of determining such a matrix P.

Theorem 7.3.13. *Let A be an n-square diagonalizable matrix and let P be a matrix whose columns are eigenvectors of A. Then $P^{-1}AP = D$ is the diagonal matrix whose diagonal elements are eigenvalues of A.*

Proof. Since A is diagonalizable, A has n linearly independent eigenvectors.

Let $X_1, X_2, \ldots, X_n$ be the eigenvectors associated to eigenvalues $\lambda_1, \lambda_2, \ldots, \lambda_n$ of A, re-spectively.

Let $P = [X_1 \ X_2 \ \cdots \ X_n]$. Since columns of P are linearly independent, P is invertible. Now,

$$AP = \begin{bmatrix} AX_1 & AX_2 & \cdots & AX_n \end{bmatrix} = \begin{bmatrix} \lambda_1 X_1 & \lambda_2 X_2 & \cdots & \lambda_n X_n \end{bmatrix}$$

$$= \begin{bmatrix} X_1 & X_2 & \cdots & X_n \end{bmatrix} \begin{bmatrix} \lambda_1 & 0 & 0 & \cdots & 0 \\ 0 & \lambda_2 & 0 & \cdots & 0 \\ & & \vdots & & \\ 0 & 0 & \cdots & 0 & \lambda_n \end{bmatrix}$$

$$= PD.$$

Then $AP = PD \Rightarrow P^{-1}AP = D$. $\qquad \square$

Remark.

1. If x is an eigenvector, then kx is also an eigenvector where $k \neq 0$. Hence, the matrix P is not unique.
2. Eigenvalues $\lambda_1, \lambda_2, \ldots, \lambda_n$ need not be distinct.

3. To visualize above computation of matrix product, consider the matrix

$$A = \begin{bmatrix} a_{11} & a_{12} & a_{13} \\ a_{21} & a_{22} & a_{23} \\ a_{31} & a_{32} & a_{33} \end{bmatrix}$$

and let

$$X_1 = \begin{bmatrix} x_1 \\ x_2 \\ x_3 \end{bmatrix}, \quad X_2 = \begin{bmatrix} y_1 \\ y_2 \\ y_3 \end{bmatrix}, \quad X_3 = \begin{bmatrix} z_1 \\ z_2 \\ z_3 \end{bmatrix}$$

be the eigenvectors associated to eigenvalues $\lambda_1, \lambda_2, \lambda_3$ of A, respectively.
Then

$$P = \begin{bmatrix} X_1 & X_2 & X_3 \end{bmatrix} = \begin{bmatrix} x_1 & y_1 & z_1 \\ x_2 & y_2 & z_2 \\ x_3 & y_3 & z_3 \end{bmatrix}.$$

Now,

$$AP = \begin{bmatrix} a_{11} & a_{12} & a_{13} \\ a_{21} & a_{22} & a_{23} \\ a_{31} & a_{32} & a_{33} \end{bmatrix} \begin{bmatrix} x_1 & y_1 & z_1 \\ x_2 & y_2 & z_2 \\ x_3 & y_3 & z_3 \end{bmatrix}$$

$$= \begin{bmatrix} a_{11}x_1 + a_{12}x_2 + a_{13}x_3 & a_{11}y_1 + a_{12}y_2 + a_{13}y_3 & a_{11}z_1 + a_{12}z_2 + a_{13}z_3 \\ a_{21}x_1 + a_{22}x_2 + a_{23}x_3 & a_{21}y_1 + a_{22}y_2 + a_{23}y_3 & a_{21}z_1 + a_{22}z_2 + a_{23}z_3 \\ a_{31}x_1 + a_{32}x_2 + a_{33}x_3 & a_{31}y_1 + a_{32}y_2 + a_{33}y_3 & a_{31}z_1 + a_{32}z_2 + a_{33}z_3 \end{bmatrix}.$$

Note that

$$AX_1 = \begin{bmatrix} a_{11} & a_{12} & a_{13} \\ a_{21} & a_{22} & a_{23} \\ a_{31} & a_{32} & a_{33} \end{bmatrix} \begin{bmatrix} x_1 \\ x_2 \\ x_3 \end{bmatrix} = \begin{bmatrix} a_{11}x_1 + a_{12}x_2 + a_{13}x_3 \\ a_{21}x_1 + a_{22}x_2 + a_{23}x_3 \\ a_{31}x_1 + a_{32}x_2 + a_{33}x_3 \end{bmatrix}$$

is equal to the first column of the matrix AP. Similarly, AX_2 and AX_3 are the second and third columns of AP.
Hence, $AP = [AX_1 \ AX_2 \ AX_3]$.
Again,

$$[\lambda_1 X_1 \ \lambda_2 X_2 \ \lambda_3 X_3] = \begin{bmatrix} \lambda_1 x_1 & \lambda_2 y_1 & \lambda_3 z_1 \\ \lambda_1 x_2 & \lambda_2 y_2 & \lambda_3 z_2 \\ \lambda_1 x_3 & \lambda_2 y_3 & \lambda_3 z_3 \end{bmatrix} = \begin{bmatrix} x_1 & y_1 & z_1 \\ x_2 & y_2 & z_2 \\ x_3 & y_3 & z_3 \end{bmatrix} \begin{bmatrix} \lambda_1 & 0 & 0 \\ 0 & \lambda_2 & 0 \\ 0 & 0 & \lambda_3 \end{bmatrix} = PD.$$

Example 7.3.14. Let

$$A = \begin{bmatrix} 2 & 2 & 1 \\ 1 & 3 & 1 \\ 1 & 2 & 2 \end{bmatrix}.$$

Then the characteristic equation of A is $\lambda^3 - 7\lambda^2 + 11\lambda - 5 = 0$, and hence $\lambda = 5, 1, 1$ are eigenvalues of A.

Eigenvectors associated to eigenvalues 5, 1, 1 are

$$X_1 = \begin{bmatrix} 1 \\ 1 \\ 1 \end{bmatrix}, \quad X_2 = \begin{bmatrix} 1 \\ 0 \\ -1 \end{bmatrix}, \quad X_3 = \begin{bmatrix} 2 \\ -1 \\ 0 \end{bmatrix},$$

respectively.

Since the number of linearly independent eigenvectors of 3-square matrix A is 3, matrix A is diagonalizable.

Consider

$$P = [X_1 \quad X_2 \quad X_3] = \begin{bmatrix} 1 & 1 & 2 \\ 1 & 0 & -1 \\ 1 & -1 & 0 \end{bmatrix},$$

then

$$P^{-1} = \begin{bmatrix} \frac{1}{4} & \frac{1}{2} & \frac{1}{4} \\ \frac{1}{4} & \frac{1}{2} & \frac{-3}{4} \\ \frac{1}{4} & \frac{-1}{4} & \frac{1}{4} \end{bmatrix}.$$

Hence, $P^{-1}AP = D$, where

$$D = \begin{bmatrix} 5 & 0 & 0 \\ 0 & 1 & 0 \\ 0 & 0 & 1 \end{bmatrix}$$

is the diagonal matrix.

$P^{-1}AP = D \Rightarrow A = PDP^{-1}$.

Then

$$A^2 = (PDP^{-1})(PDP^{-1}) = PDP^{-1}PDP^{-1} = PDI_nDP^{-1} = PD^2P^{-1}.$$

Similarly, $A^n = PD^nP^{-1}$.

Hence,

$$A^3 = PD^3P^{-1} = \begin{bmatrix} 1 & 1 & 2 \\ 1 & 0 & -1 \\ 1 & -1 & 0 \end{bmatrix} \begin{bmatrix} 125 & 0 & 0 \\ 0 & 1 & 0 \\ 0 & 0 & 1 \end{bmatrix} \begin{bmatrix} \frac{1}{4} & \frac{1}{2} & \frac{1}{4} \\ \frac{1}{4} & \frac{1}{2} & \frac{-3}{4} \\ \frac{1}{4} & \frac{-1}{4} & \frac{1}{4} \end{bmatrix}$$

$$= \begin{bmatrix} 32 & 62 & 31 \\ 31 & 63 & 31 \\ 31 & 62 & 32 \end{bmatrix}.$$

Example 7.3.15.

(a) Let

$$A = \begin{bmatrix} 5 & 3 & -2 \\ 0 & 5 & 1 \\ 0 & 0 & 5 \end{bmatrix}.$$

The characteristic equation of A is $(\lambda - 5)^3 = 0$. Since $\text{Mul}_a(5) = 3 \neq \text{Mul}_g(5) = 1$, A is not diagonalizable.

Alternatively, if A is diagonalizable, then there exists an invertible matrix P such that

$$P^{-1}AP = \begin{bmatrix} 5 & 0 & 0 \\ 0 & 5 & 0 \\ 0 & 0 & 5 \end{bmatrix} = 5I_3.$$

Then $A = P5I_3P^{-1} = 5I_3PP^{-1} = 5I_3$, which is a contradiction. Hence, A is not diagonalizable.

(b) Let A be an n-square matrix with entries in a field F. Then A is diagonalizable if and only if $A + kI_n$ is diagonalizable for all $k \in F$.

If A is diagonalizable, then $A = PDP^{-1}$, where D is a diagonal matrix and P is invertible.

Now, $A + kI_n = PDP^{-1} + kI_n = PDP^{-1} + kI_nPP^{-1} = P(D + kI_n)P^{-1} = P\acute{D}P^{-1}$, where $\acute{D} = D + kI_n$ is a diagonal matrix.

Conversely, if $A + kI_n$ is diagonalizable, then $A + kI_n = PDP^{-1} \Rightarrow A = PDP^{-1} - kI_n = PDP^{-1} - kI_nPP^{-1} = P(D - kI_n)P^{-1} = P\acute{D}P^{-1}$, where $\acute{D} = D - kI_n$ is a diagonal matrix.

Example 7.3.16. Let A be an n-square matrix. If λ_1 and λ_2 are eigenvalues of A and X_1 and X_2 are associated eigenvectors of λ_1 and λ_2, respectively, then we shall show that $A^r(c_1X_1 + c_2X_2) = c_1\lambda_1^rX_1 + c_2\lambda_2^rX_2$, where c_1 and c_2 are scalars.

We have $AX_1 = \lambda_1X_1$. Then $A^2X_1 = A(AX_1) = A(\lambda_1X_1) = \lambda_1(AX_1) = \lambda_1(\lambda_1X_1) = \lambda_1^2X_1$.

Similarly, $A^rX_1 = \lambda_1^rX_1$ and $A^rX_2 = \lambda_2^rX_2$.

Hence, $A^r(c_1X_1 + c_2X_2) = c_1A^rX_1 + c_2A^rX_2 = c_1\lambda_1^rX_1 + c_2\lambda_2^rX_2$.

In particular, if A is a 3-square matrix and -2 and 3 are eigenvalues with associated eigenvectors

$$\begin{bmatrix} 1 \\ 2 \\ 0 \end{bmatrix} \quad \text{and} \quad \begin{bmatrix} 0 \\ 1 \\ -1 \end{bmatrix},$$

respectively, then we can compute

$$A^4 \begin{bmatrix} -1 \\ 0 \\ -2 \end{bmatrix}.$$

Suppose

$$X_1 = \begin{bmatrix} 1 \\ 2 \\ 0 \end{bmatrix} \quad \text{and} \quad X_2 = \begin{bmatrix} 0 \\ 1 \\ -1 \end{bmatrix}.$$

Then

$$\begin{bmatrix} -1 \\ 0 \\ -2 \end{bmatrix} = (-1) \begin{bmatrix} 1 \\ 2 \\ 0 \end{bmatrix} + 2 \begin{bmatrix} 0 \\ 1 \\ -1 \end{bmatrix} = (-1)X_1 + 2X_2.$$

Hence,

$$A^4 \begin{bmatrix} -1 \\ 0 \\ -2 \end{bmatrix} = A^4((-1)X_1 + 2X_2)$$

$$= (-1)A^4 X_1 + 2A^4 X_2$$

$$= (-1)(-2)^4 X_1 + 2(3^4)X_2$$

$$= -16 \begin{bmatrix} 1 \\ 2 \\ 0 \end{bmatrix} + 162 \begin{bmatrix} 0 \\ 1 \\ -1 \end{bmatrix} = \begin{bmatrix} -16 \\ 130 \\ -162 \end{bmatrix}.$$

Example 7.3.17. Let J be a 21-square matrix with all the entries equal to 1. In this example, we shall show that J and $J - I_{21}$ are diagonalizable matrices.

We have rank(J) = 1 ≠ 21, hence 0 is an eigenvalue of J.

Also, since row sum of J is 21, 21 is an eigenvalue of J.

$$\text{Mul}_g(0) = 21 - \text{rank}(J - 0I_{21}) = 21 - \text{rank}(J) = 20 \quad \text{and}$$
$$\text{Mul}_g(21) = \text{Mul}_a(21) = 1.$$

Thus, $\text{Mul}_g(0) + \text{Mul}_g(21) = 21$, and hence J is diagonalizable.

Also, we have $\text{Mul}_a(0) = 20$ and $\text{Mul}_a(21) = 1$. This implies eigenvalues of J are $0, 0, \ldots, 0$ (20 times), and 21.

Hence, eigenvalues of $J - I_{21}$ are $-1, -1, \ldots, -1$ (20 times), and 20.

$$\mathrm{Mul}_g(-1) = 21 - \mathrm{rank}[(J - I_{21}) - (-1)I_{21}] = 21 - \mathrm{rank}(J) = 20 \quad \text{and}$$
$$\mathrm{Mul}_g(20) = \mathrm{Mul}_a(20) = 1.$$

Thus, $\mathrm{Mul}_g(-1) + \mathrm{Mul}_g(20) = 21$, and hence $J - I_{21}$ is diagonalizable.

In general, the n-square matrix J with all the entries equal to 1 is diagonalizable, and hence $J - I_n$ is diagonalizable.

Exercises

(7.3.1) Show that a strictly upper triangular matrix is nondiagonalizable.

(7.3.2) Let P be an n-square matrix and Q be a m-square matrix. Then show that the block diagonal matrix

$$R = \begin{bmatrix} P & 0 \\ 0 & Q \end{bmatrix}$$

is diagonalizable if and only if P and Q are diagonalizable.

(7.3.3) Find eigenvalues and their eigenvectors of the differential linear transformation $D : P_3(x) \to P_3(x)$, where $P_3(x)$ is a polynomial space over R.

(7.3.4) Show that any triangular matrix with distinct principal diagonals is diagonalizable.

(7.3.5) Check whether the following matrices are diagonalizable or not:

$$A = \begin{bmatrix} 2 & 1 & 1 \\ 2 & 3 & 2 \\ 3 & 3 & 4 \end{bmatrix}, \quad B = \begin{bmatrix} 2 & -2 & 2 \\ 1 & 1 & 1 \\ 1 & 3 & -1 \end{bmatrix}, \quad C = \begin{bmatrix} 1 & 1 & 0 \\ 0 & 1 & 0 \\ 0 & 0 & 1 \end{bmatrix},$$

$$D = \begin{bmatrix} -3 & -7 & -5 \\ 2 & 4 & 3 \\ 1 & 2 & 2 \end{bmatrix}, \quad E = \begin{bmatrix} 0 & 1 & 1 \\ 1 & 0 & 1 \\ 1 & 1 & 0 \end{bmatrix}, \quad F = \begin{bmatrix} 1 & 1 & 1 \\ 1 & 1 & 1 \\ 1 & 1 & 1 \end{bmatrix}.$$

(7.3.6) In the above example, if any matrix A is diagonalizable, then find the matrix P such that $P^{-1}AP$ becomes a diagonal matrix.

(7.3.7) Find the matrix P such that PAP^{-1} is in diagonal form, where

$$A = \begin{bmatrix} 1 & 1 & 1 \\ 0 & 2 & 1 \\ -4 & 4 & 3 \end{bmatrix}.$$

Hence, calculate A^6.

(7.3.8) Let A be a 3-square singular matrix such that 2 and 3 are eigenvalues of A. Then show that the matrix $A^3 + 2A + I_3$ is diagonalizable.

(7.3.9) Let A be a 3-square singular matrix. If 2 and 5 are eigenvalues of A, then show that $A^3 - 2A + I_3$ is diagonalizable.

(7.3.10) Let -3 and 2 be two eigenvalues and

$$X_1 = \begin{bmatrix} 1 \\ -1 \\ 2 \end{bmatrix} \quad \text{and} \quad X_2 = \begin{bmatrix} 2 \\ 1 \\ 0 \end{bmatrix}$$

are associated eigenvectors, respectively, of a matrix A of order 3×3. Then compute A^3Y, where

$$Y = \begin{bmatrix} -1 \\ -2 \\ 2 \end{bmatrix}$$

is the linear combination of X_1 and X_2.

(7.3.11)   Let A be a 4-square matrix. If -1 and 3 are eigenvalues and

$$X_1 = \begin{bmatrix} 1 \\ 1 \\ 0 \\ 2 \end{bmatrix} \quad \text{and} \quad X_2 = \begin{bmatrix} 0 \\ 0 \\ 2 \\ -1 \end{bmatrix}$$

are associated eigenvectors of A, respectively. Then compute A^3Y, where

$$Y = \begin{bmatrix} -2 \\ -2 \\ 2 \\ -5 \end{bmatrix}$$

is the linear combination of X_1 and X_2.

(7.3.12)   If $\lambda_1, \lambda_2, \ldots, \lambda_n$ are distinct eigenvalues and $X_1, X_2, \ldots, X_n$ are associated eigenvectors of an n-square real matrix A, respectively. Then show that A^rX can be computed for any $X \in \mathbb{R}^n$, even A is unknown.

(7.3.13)   Find geometric multiplicities of each eigenvalues and a basis for each eigenspace of the matrix

$$A = \begin{bmatrix} 2 & 1 & 1 \\ 1 & 2 & 1 \\ 0 & 0 & 1 \end{bmatrix}.$$

(7.3.14)   Suppose that A is a diagonalizable matrix. If $P_A(\lambda) = \lambda(\lambda - 2)^2(\lambda + 3)^3(\lambda - 5)^4$ is the characteristic polynomial of A, then find the size of A and the dimension of eigenspaces $V(2)$ and $V(5)$.

(7.3.15)   Suppose that A is a diagonalizable matrix. If $P_A(\lambda) = \lambda(\lambda - 1)^2(\lambda - 2)^3(\lambda - 3)^5$ is the characteristic polynomial of A, then find the size of A and the dimension of eigenspaces $V(2)$ and $V(3)$.

(7.3.16)   Let A be a 5-square real matrix with trace 21. If 3 and 4 are eigenvalues of A, each with algebraic multiplicity 2, then show that A is invertible.

(7.3.17)   Let J be a 31-square matrix with all the entries equal to 1. Then show that the matrix $J - I_{31}$ is invertible and diagonalizable.

7.4 Eigenvalues and eigenvectors of some special matrices

Recall that if

$$X = \begin{bmatrix} x_1 \\ x_2 \\ \vdots \\ x_n \end{bmatrix} \quad \text{and} \quad Y = \begin{bmatrix} y_1 \\ y_2 \\ \vdots \\ y_n \end{bmatrix}$$

are column vectors in $\mathbb{C}^n$, then the inner product

$$\langle X, Y \rangle = x_1\overline{y_1} + x_2\overline{y_2} + \cdots + x_n\overline{y_n} = Y^*X, \quad \text{where } Y^* = \overline{Y}^t.$$

Also, an n-square matrix $A \in M_n(\mathbb{C})$ is called Hermitian if $A^* = A$.

Now, we prove some remarkable results on eigenvalues and eigenvectors of Hermitian, skew-Hermitian and unitary matrices.

Theorem 7.4.1. *All the eigenvalues of a Hermitian matrix are real.*

Proof. Let $A \in M_n(\mathbb{C})$ be a Hermitian matrix. Then $A^* = A$.

If λ is an eigenvalue of A, then there exists a nonzero column vector X, such that $AX = \lambda X$.

Now, $\lambda X^*X = X^*\lambda X = X^*(AX) = (X^*A)X = (X^*A^*)X = (AX)^*X = (\lambda X)^*X = \overline{\lambda}X^*X$.

Since $X \neq 0$, $X^*X = \|X\|^2 \neq 0$.

Hence, $\lambda\|X\|^2 = \overline{\lambda}\|X\|^2 \Rightarrow \lambda = \overline{\lambda} \Rightarrow \lambda$ is a real number. ☐

It is known that not every real matrix necessarily has an eigenvalue, but the following result shows that every symmetric matrix does have at least one eigenvalue.

Corollary 7.4.2. *All the eigenvalues of a symmetric matrix are real.*

Proof. The result follows from the fact that every symmetric matrix can be considered as a complex Hermitian matrix. ☐

Corollary 7.4.3. *All the eigenvalues of a skew-Hermitian matrix are either zero or purely imaginary.*

Proof. Let A be a skew-Hermitian matrix and λ be its eigenvalue. We know that A is skew-Hermitian if and only if iA is Hermitian. Hence, λ is an eigenvalue of A if and only if $i\lambda$ is an eigenvalue of iA. This shows that $i\lambda$ is real. It follows that λ is zero or purely imaginary. ☐

Corollary 7.4.4. *There is no nonzero eigenvalue of a real skew-symmetric matrix.*

Proof. Let A be an n-square real skew-symmetric matrix. Then A is complex skew-Hermitian, and hence all the nonzero eigenvalues of a are purely imaginary. λ is an eigenvalue of A is equivalent to say that λ is an eigenvalue of the linear transformation $L_A : \mathbb{R}^n \longrightarrow \mathbb{R}^n$. Then $L_A(X) = \lambda X$ is possible only when $\lambda \in \mathbb{R}$. Thus, there is no nonzero eigenvalue of A. ☐

Theorem 7.4.5. *If λ is an eigenvalue of an unitary matrix U, then $|\lambda| = 1$.*

Proof. From Corollary 6.3.12, we have

$$\langle UX, UX \rangle = \langle X, X \rangle$$

$$\Rightarrow \langle \lambda X, \lambda X \rangle = \|X\|^2$$

$$\Rightarrow \lambda\bar{\lambda}\langle X,X\rangle = \|X\|^2$$
$$\Rightarrow |\lambda|^2\|X\|^2 = \|X\|^2.$$

Since $X \neq 0$, $|\lambda|^2 = 1$. Hence, $|\lambda| = 1$. □

Corollary 7.4.6. *If λ is an eigenvalue of an orthogonal matrix O, then $\lambda = \pm 1$.*

Proof. Since every orthogonal matrix is an unitary matrix, $|\lambda| = 1$. But $\lambda \in \mathbb{R}$, and hence $\lambda = \pm 1$. □

Theorem 7.4.7. *Let $A \in M_n(\mathbb{C})$ and let λ_1, λ_2 be distinct eigenvalues of A. Then their associated eigenvectors X_1, X_2 are orthogonal if:*
(i)   *A is Hermitian;*
(ii) *A is skew-Hermitian;*
(iii) *A is unitary.*

Proof. (i)   If A is Hermitian, then $A^* = A$ and eigenvalues λ_1, λ_2 are real. Hence,

$$\lambda_1 X_1^* X_2 = (\overline{\lambda_1}X_1)^* X_2 = (\lambda_1 X_1)^* X_2 = (AX_1)^* X_2$$
$$= X_1^* A^* X_2 = X_1^* A X_2 = X_1^* \lambda_2 X_2 = \lambda_2 X_1^* X_2$$
$$\Rightarrow (\lambda_1 - \lambda_2)X_1^* X_2 = 0 \Rightarrow \lambda_1 - \lambda_2 = 0 \text{ or } X_1^* X_2 = 0.$$

Since $\lambda_1 \neq \lambda_2$, $X_1^* X_2 = 0$. That is, $\langle X_2, X_1 \rangle = 0$.
(ii)   If A is skew-Hermitian, then $A^* = -A$ and all the eigenvalues of A are either zero or purely imaginary. Since $\lambda_1 \neq \lambda_2$, at least one is nonzero. Suppose $\lambda_1 \neq 0$. Then λ_1 is purely imaginary, and hence $\overline{\lambda_1} = -\lambda_1$.
Now,

$$\lambda_1 X_1^* X_2 = (\overline{\lambda_1}X_1)^* X_2 = -(\lambda_1 X_1)^* X_2 = -(AX_1)^* X_2 = -X_1^* A^* X_2$$
$$= -X_1^* (-A)X_2 = X_1^* A X_2 = X_1^* \lambda_2 X_2 = \lambda_2 X_1^* X_2$$
$$\Rightarrow (\lambda_1 - \lambda_2)X_1^* X_2 = 0 \Rightarrow X_1^* X_2 = 0.$$

(iii)   If A is an unitary matrix, then $\langle AX_1, AX_2 \rangle = \langle X_1, X_2 \rangle$.
Hence,

$$\langle X_1, X_2 \rangle = \langle \lambda_1 X_1, \lambda_2 X_2 \rangle = \lambda_1\overline{\lambda_2}\langle X_1, X_2 \rangle \Rightarrow (1 - \lambda_1\overline{\lambda_2})\langle X_1, X_2 \rangle = 0.$$

Since $|\lambda| = 1 \Rightarrow \lambda\bar{\lambda} = 1 \Rightarrow \bar{\lambda} = \lambda^{-1}$, we have $\lambda_1 \neq \lambda_2 \Rightarrow \frac{\lambda_1}{\lambda_2} \neq 1 \Rightarrow \lambda_1\lambda_2^{-1} \neq 1 \Rightarrow \lambda_1\overline{\lambda_2} \neq 1$. This shows that $\langle X_1, X_2 \rangle = 0$. □

We have that Hermitian, skew-Hermitian and unitary operators are normal operators, and we discuss here some nice results and properties related to eigenvalues and eigenvectors of normal operators.

Theorem 7.4.8. *Let T be a normal operator such that $T(v) = \lambda v$. Then $T^*(v) = \bar{\lambda} v$, where v is a vector and λ is a scalar.*

Proof. Let $T : V \to V$ be a normal operator. To prove the result, we have to show that $\|T^*(v) - \bar{\lambda} v\| = 0$.

Consider

$$\langle T^*(v) - \bar{\lambda} v, T^*(v) - \bar{\lambda} v \rangle$$

$$= \langle T^*(v), T^*(v) \rangle + \langle -\bar{\lambda} v, T^*(v) \rangle + \langle T^*(v), -\bar{\lambda} v \rangle + \langle -\bar{\lambda} v, -\bar{\lambda} v \rangle$$

$$= \langle v, TT^*(v) \rangle - \bar{\lambda} \langle v, T^*(v) \rangle - \lambda \langle T^*(v), v \rangle + \bar{\lambda} \lambda \langle v, v \rangle$$

$$= \langle v, TT^*(v) \rangle - \bar{\lambda} \langle T(v), v \rangle - \lambda \langle v, T(v) \rangle + \bar{\lambda} \lambda \langle v, v \rangle$$

$$= \langle v, T^*T(v) \rangle - \bar{\lambda} \langle \lambda v, v \rangle - \lambda \langle v, \lambda v \rangle + \lambda \bar{\lambda} \langle v, v \rangle$$

$$= \langle T(v), \lambda v \rangle - \bar{\lambda} \lambda \langle v, v \rangle - \lambda \bar{\lambda} \langle v, v \rangle + \lambda \bar{\lambda} \langle v, v \rangle$$

$$= \lambda \bar{\lambda} \langle v, v \rangle - \bar{\lambda} \lambda \langle v, v \rangle$$

$$= 0.$$

Then

$$\left\| T^*(v) - \bar{\lambda} v \right\|^2 = 0 \Rightarrow \left\| T^*(v) - \bar{\lambda} v \right\| = 0 \Rightarrow T^*(v) - \bar{\lambda} v = 0 \Rightarrow T^*(v) = \bar{\lambda} v.$$

In particular, we have the following.

If T is Hermitian, then $T = T^*$, and hence $T(v) = T^*(v) \Rightarrow \lambda v = \bar{\lambda} v \Rightarrow (\lambda - \bar{\lambda})v = 0$. Since $v \neq 0$, $\lambda = \bar{\lambda} \Rightarrow \lambda$ is real.

If T is skew-Hermitian, then $T^* = -T$. Hence, $T^*(v) = -T(v) \Rightarrow \bar{\lambda} v = -\lambda v \Rightarrow (\bar{\lambda}+\lambda)v = 0 \Rightarrow \bar{\lambda} = -\lambda$. This shows that $\lambda = 0$ or purely imaginary.

Similarly, if T is Unitary, then $TT^* = T^*T = I$. Hence,

$$T^*(v) = \bar{\lambda} v \Rightarrow TT^*(v) = T(\bar{\lambda} v) \Rightarrow I(V) = \bar{\lambda} T(v) \Rightarrow v = \bar{\lambda} \lambda v \Rightarrow (1 - \bar{\lambda} \lambda)v = 0$$

$$\Rightarrow \bar{\lambda} \lambda = 1 \Rightarrow |\lambda|^2 = 1 \Rightarrow |\lambda| = 1.$$

□

Also, we have that the linear operator L_A is normal if and only if matrix A is normal. Replacing T by L_A in the above theorem, we obtain a similar result for normal matrices.

Corollary 7.4.9. *Eigenvectors associated to two distinct eigenvalues of a normal operator are orthogonal.*

Proof. Let v_1 and v_2 be associated eigenvectors of distinct eigenvalues λ_1 and λ_2 of a normal operator T, respectively.

Then $T(v_1) = \lambda_1 v_1$ and $T(v_2) = \lambda_2 v_2$. Also, from the above theorem, $T^*(v_1) = \bar{\lambda}_1 v_1$ and $T^*(v_2) = \bar{\lambda}_2 v_2$. Then

$$\lambda_1 \langle v_1, v_2 \rangle = \langle \lambda_1 v_1, v_2 \rangle = \langle T(v_1), v_2 \rangle = \langle v_1, T^*(v_2) \rangle = \langle v_1, \overline{\lambda_2} v_2 \rangle = \lambda_2 \langle v_1, v_2 \rangle.$$

This implies $(\lambda_1 - \lambda_2) \langle v_1, v_2 \rangle = 0$.

Since $\lambda_1 - \lambda_2 \neq 0$, $\langle v_1, v_2 \rangle = 0$. Hence, v_1 and v_2 are orthogonal. $\square$

Accordingly, this result is also true for all Hermitian, skew-Hermitian and unitary operators (matrices).

Note that if B is an orthonormal basis consisting of eigenvectors of an operator T, then the matrix $[T]_{B,B}$ is diagonal. Since a diagonal matrix is normal, $[T]_{B,B}$ is normal, and hence T is a normal operator.

We now prove an important theorem known as spectral theorem for normal operators that the converse of the above discussion is also true.

Theorem 7.4.10. *Let T be a normal operator on a finite-dimensional complex inner product space V. Then T is diagonalizable.*

Proof. We prove the result, using the method of mathematical induction on the dimension of V. Let $\dim(V) = n$. If $\dim(V) = 1$, take any nonzero vector $v \in V$ and then $B = \{v_1\}$, where $v_1 = \frac{v}{\|v\|}$ is an orthonormal basis of V. Since $\dim(V) = 1$, $V = \{\alpha v : \alpha \in \mathbb{C}\}$. Hence, v is an eigenvector of T. Thus, the set of eigenvectors of T form an orthonormal basis of V and so T is diagonalizable. Hence, the result is true for $\dim(V) = 1$.

Suppose the result is true for all normal operators on vector spaces of dimension less than n. Since T has at least one eigenvalue λ_1 in the case of a complex inner product space, let us take a unit vector $v_1 \in V$ such that $T(v_1) = \lambda_1 v_1$.

Let U be the subspace of V generated by the eigenvector v_1. Then $V = U \oplus U^{\perp}$. Since $\dim(U) = 1$, $\dim(U^{\perp}) = n - 1$.

Now, we shall show that the restriction of T and T^* on $U^{\perp}$ are also operators on $U^{\perp}$ and $T^* /_{U^{\perp}}$ is an adjoint of $T /_{U^{\perp}}$.

Let $v_2 \in U^{\perp}$. Then $\langle T(v_2), v_1 \rangle = \langle v_2, T^*(v_1) \rangle = \langle v_2, \overline{\lambda_1} v_1 \rangle = \lambda_1 \langle v_2, v_1 \rangle$. Since $\langle v_2, v_1 \rangle = 0$, $\langle T(v_2), v_1 \rangle = 0$. This shows that $T(v_2) \in U^{\perp}$.

Similarly,

$$\langle T^*(v_2), v_1 \rangle = \langle v_2, T(v_1) \rangle = \langle v_2, \lambda_1 v_1 \rangle = \overline{\lambda_1} \langle v_2, v_1 \rangle = 0 \quad \text{implies} \quad T^*(v_2) \in U^{\perp}.$$

Also, since $U^{\perp} \subset V$, $T /_{U^{\perp}}$ and $T^* /_{U^{\perp}}$ are adjoint to each other. Since $\dim(U^{\perp}) = n - 1$, by the induction hypothesis $U^{\perp}$ has an orthonormal basis $\{v_2, v_3, \ldots, v_n\}$ consisting of eigenvectors of $T /_{U^{\perp}}$. Clearly, these eigenvectors are also eigenvectors of T. Then adding v_1 to this basis, we get $\{v_1, v_2, v_3, \ldots, v_n\}$, an orthonormal basis of V. Since each $v_i, 1 \leq i \leq n$ is an eigenvector of T, and T is diagonalizable. $\square$

Corollary 7.4.11. *Hermitian, skew-Hermitian and unitary operators on a finite-dimensional complex inner product space V are diagonalizable.*

Proof. All the operators are normal, and hence diagonalizable. $\square$

Corollary 7.4.12. *Hermitian, skew-Hermitian and unitary complex matrices are diagonalizable.*

Proof. Let A be an n-square Hermitian (skew-Hermitian, unitary) matrix. Then A is Hermitian (skew-Hermitian, unitary) if and only if the linear operator $L_A : \mathbb{C}^n \to \mathbb{C}^n$ is Hermitian (skew-Hermitian, unitary). Hence, the result follows from the above corollary. □

We have proved that the normal operator on a finite-dimensional complex inner product space is diagonalizable. Now, we provide an example of a normal operator on a real inner product space, which is normal but not diagonalizable.

Example 7.4.13. Let $T : \mathbb{R}^3 \to \mathbb{R}^3$ be an operator defined by $T(x,y,z) = (-y - 2z, x - 3z, 2x + 3y)$. Then $T^*(x,y,z) = (y + 2z, -x + 3z, -2x - 3y)$, and hence $TT^* = T^*T$ (verify). Now, we shall so that T is not diagonalizable. Let B be standard basis of $\mathbb{R}^3$. Then

$$[T]_{B,B} = \begin{bmatrix} 0 & -1 & -2 \\ 1 & 0 & -3 \\ 2 & 3 & 0 \end{bmatrix}.$$

The characteristic equation of T = Characteristic equation of the matrix $[T]_{B,B}$ = $\lambda^3 + 14\lambda = 0$. Then $0, \pm i\sqrt{14}$ are eigenvalues of T. We have $\mathrm{Mul}_g(0) = 1$ but there is no vector $(x,y,z) \in \mathbb{R}^3$ such that $T(x,y,z) = \pm i\sqrt{14}(x,y,z) \in \mathbb{R}^3$. Thus, the set of eigenvectors of T is not a basis of $\mathbb{R}^3$. Hence, T is not diagonalizable. However, if T is defined on the complex inner product space $\mathbb{C}^3$ by $T(x,y,z) = (-y-2z, x-3z, 2x+3y)$, where $(x,y,z) \in \mathbb{C}^3$. Then T has distinct eigenvalues $0, \pm i\sqrt{14}$ and so T is diagonalizable.

By noting that all the eigenvalues of a Hermitian operator are real, and hence its characteristic equation splits over the reals, we prove the spectral theorem for Hermitian operators defined on a real or complex inner product space.

Theorem 7.4.14. *A Hermitian operator and a finite-dimensional inner product space (real or complex) is diagonalizable.*

Proof. Let $\dim(V) = n$. We use the method of mathematical induction on n. If $n = 1$, then we get an orthonormal basis $\{v\}$ of V, where $v = \frac{u}{\|u\|}$ for any nonzero vector $u \in V$. That is eigenvector v forms an orthonormal basis of V. Hence, the Hermitian operator T on V is diagonalizable.

Now suppose that the result is true for $\dim(V) < n$. Since all the eigenvectors of a Hermitian operator are real, we can find a real eigenvalue λ_1 and a nonzero eigenvector v_1. Also, $T^* = T$. After that, repeating the rest of the proof of the spectral theorem for normal operators we obtain the proof of this theorem. □

Corollary 7.4.15. *Every Hermitian and symmetric matrix is diagonalizable.*

Proof. Let A be an n-square matrix. Then we have that A is Hermitian if and only if $L_A : \mathbb{C}^n \to \mathbb{C}^n$ is Hermitian and A is symmetric if and only if $L_A : \mathbb{R}^n \to \mathbb{R}^n$ is symmetric. Hence, the result follows from above theorem. $\square$

Corollary 7.4.16. *Let A be a Hermitian matrix. Then there exists an unitary matrix U such that $U^* A U$ is a diagonal matrix. Similarly, if A is a symmetric matrix, then there exists an orthogonal matrix O such that $O^t A O$ is a diagonal matrix.*

Proof. Let A be an n-square Hermitian matrix. Then the linear transformation $L_A : \mathbb{C}^n \to \mathbb{C}^n$ is Hermitian. From Theorem 7.4.14, there exists an orthonormal basis $\{X_1, X_2, \ldots, X_n\}$ of $\mathbb{C}^n$, where each column vector $X_i \in \mathbb{C}^n, 1 \le i \le n$ is an eigenvector of L_A. That is, $L_A(X_i) = AX_i = \lambda_i X_i$ for some $\lambda_i \in \mathbb{C}^n$. Define the matrix $U = [X_1 \ X_2 \ \cdots \ X_n]$, where eigenvectors $X_1, X_2, \ldots, X_n$ are as columns of U. Since all the column vectors are orthonormal, U is an unitary matrix.

Then

$$U^* A U = \begin{bmatrix} X_1^* \\ X_2^* \\ \vdots \\ X_n^* \end{bmatrix} A \begin{bmatrix} X_1 & X_2 & \cdots & X_n \end{bmatrix}$$

$$= \begin{bmatrix} X_1^* \\ X_2^* \\ \vdots \\ X_n^* \end{bmatrix} \begin{bmatrix} AX_1 & AX_2 & \cdots & AX_n \end{bmatrix}$$

$$= \begin{bmatrix} X_1^* \\ X_2^* \\ \vdots \\ X_n^* \end{bmatrix} \begin{bmatrix} \lambda_1 X_1 & \lambda_2 X_2 & \cdots & \lambda_n X_n \end{bmatrix}$$

$$= \begin{bmatrix} \lambda_1 X_1^* X_1 & \lambda_2 X_1^* X_2 & \cdots & \lambda_n X_1^* X_n \\ \lambda_1 X_2^* X_1 & \lambda_2 X_2^* X_2 & \cdots & \lambda_n X_2^* X_n \\ \vdots & & & \\ \lambda_1 X_n^* X_1 & \lambda_2 X_n^* X_2 & \cdots & \lambda_n X_n^* X_n \end{bmatrix}$$

$$= \begin{bmatrix} \lambda_1 & 0 & 0 & \cdots & 0 \\ 0 & \lambda_2 & 0 & \cdots & 0 \\ \vdots & & & & \\ 0 & 0 & 0 & \cdots & \lambda_n \end{bmatrix},$$

where $X_i^* X_j$ represents the standard inner product in $\mathbb{C}^n$. Since $X_1, X_2, \ldots, X_n$ are orthonormal, $X_i^* X_j = \delta_{ij}$.

Similarly, if A is symmetric then $L_A : \mathbb{R}^n \to \mathbb{R}^n$ is symmetric. We get an orthonormal basis $\{X_1, X_2, \ldots, X_n\}$ of $\mathbb{R}^n$, and hence the orthogonal matrix $O = [X_1\ X_2\ \cdots\ X_n]$. Since $X_i \in \mathbb{R}^n\ \forall i$, the inner product is defined as $\langle X_i, X_j \rangle = X_j^t X_i$. Then we get $O^t A O$ a diagonal matrix. □

Since O is an orthogonal matrix $O^t = O^{-1}$, and hence $O^t A O = O^{-1} A O = D$, a diagonal matrix containing corresponding eigenvalues in the diagonal. This special type of diagonalization for symmetric matrices is called orthogonally diagonalization.

Remark. Once we have identified the unique eigenvalues, we proceed to determine their associated eigenspaces. Using the Gram–Schmidt process, we construct an orthonormal basis for each eigenspace. Consequently, this yields an orthonormal basis containing eigenvectors. Notably, this method extends to the diagonalization of both Hermitian and real symmetric matrices.

Exercises

(7.4.1) Find orthogonal matrices O, P, Q and R such that the matrices $O^t A O, P^t B P, Q^t C Q$ and $R^t D R$ are diagonal, where

$$A = \begin{bmatrix} -6 & 2 & 1 \\ 2 & -6 & 1 \\ 1 & 1 & -5 \end{bmatrix}, \quad B = \begin{bmatrix} -3 & 2 & -4 \\ 2 & -6 & -2 \\ -4 & -2 & -3 \end{bmatrix},$$

$$C = \begin{bmatrix} -1 & 6 & -4 \\ 6 & -2 & 2 \\ -4 & 2 & 3 \end{bmatrix}, \quad D = \begin{bmatrix} -3 & 2 & -4 \\ 2 & -6 & -2 \\ -4 & -2 & -3 \end{bmatrix}$$

are real symmetric matrices.

(7.4.2) Using the properties of eigenvalues show that the determinant of a Hermitian matrix is always real.

(7.4.3) Let A be a 101-square skew-Hermitian matrix. Using the properties of eigenvalues of A, show that the determinant of A is either 0, or it is purely imaginary.

(7.4.4) Let A be a 100-square skew-Hermitian matrix. Using the properties of eigenvalues of A show that the determinant of A is purely real.

(7.4.5) Let A be an n-square Hermitian matrix. Then using the properties of eigenvalues of A show that $iI_n + A$ is invertible.

(7.4.6) Let A be an n-square skew-symmetric (skew-Hermitian). Then using the properties of eigenvalues of A show that $I_n + A$ is invertible.

(7.4.7) Show that all eigenvalues of A^*A are real, and A^*A is diagonalizable.

(7.4.8) Let A be a real matrix. Show that $A^t A$ is diagonalizable.

(7.4.9) Show that for every real matrix A, there exist an orthogonal matrix O such that $AA^t = ODO^t$, where D is a diagonal matrix.

7.5 Cayley–Hamilton theorem and the minimal polynomial

The celebrated Cayley–Hamilton theorem is one of the most fundamental results in basic linear algebra. Before stating and proving this important theorem, we explain some elementary ideas used in the proof.

Let

$$
A = \begin{bmatrix} 1 & 2 & 3 \\ -1 & 0 & 2 \\ 0 & 3 & 4 \end{bmatrix}
$$

be a 3-square matrix. Then

$$
A - \lambda I_3 = \begin{bmatrix} 1-\lambda & 2 & 3 \\ -1 & -\lambda & 2 \\ 0 & 3 & 4-\lambda \end{bmatrix}
$$

and

$$
\operatorname{adj}(A - \lambda I_3) = \begin{bmatrix} \lambda^2 - 4\lambda - 6 & 2\lambda + 1 & 3\lambda + 4 \\ -\lambda + 4 & \lambda^2 - 5\lambda + 4 & 2\lambda - 5 \\ -3 & 3\lambda - 3 & \lambda^2 - \lambda + 2 \end{bmatrix}
$$

$$
= \lambda^2 \begin{bmatrix} 1 & 0 & 0 \\ 0 & 1 & 0 \\ 0 & 0 & 1 \end{bmatrix} + \lambda \begin{bmatrix} -4 & 2 & 3 \\ -1 & -5 & 2 \\ 0 & 3 & -1 \end{bmatrix} + \begin{bmatrix} -6 & 1 & 4 \\ 4 & 4 & -5 \\ -3 & -3 & 2 \end{bmatrix}
$$

$$
= \lambda^2 B_2 + \lambda B_1 + B_0,
$$

where B_2, B_1, B_0 are above 3×3 matrices.

Here, we observe that the elements of $\operatorname{adj}(A - \lambda I_3)$ are cofactors of the matrix $A - \lambda I_3$, and hence are polynomials in λ of degree at most 2 and $\operatorname{adj}(A - \lambda I_3)$ is written as $\lambda^2 B_2 + \lambda B_1 + B_0$, where B_2, B_1, B_0 are 3×3 matrices, which are independent of λ. This shows that if A is any n-square matrix with entries in a field F, then $\operatorname{adj}(A - \lambda I_n) = \lambda^{n-1} B_{n-1} + \lambda^{n-2} B_{n-2} + \cdots + \lambda B_1 + B_0$ where $B_{n-1}, B_{n-2}, \ldots, B_1, B_0$ are n-square matrices with entries in F and free from λ.

Now, we are ready to prove our celebrated result.

Theorem 7.5.1 (Cayley–Hamilton theorem). *Every square matrix satisfies its own characteristic polynomial. That is, if A is a square matrix and $P_A(\lambda)$ is its characteristic polynomial then $P_A(A) = 0$.*

Proof. Let A be an n-square matrix. Then $P_A(\lambda) = (-1)^n(\lambda^n + a_{n-1}\lambda^{n-1} + a_{n-2}\lambda^{n-2} + \cdots + a_1\lambda + a_0)$.

Also, we have that

$$(A - \lambda I_n)\, \mathrm{adj}(A - \lambda I_n) = |A - \lambda I_n| I_n, \quad \text{or}$$
$$(A - \lambda I_n)\, \mathrm{adj}(A - \lambda I_n) = P_A(\lambda) I_n.$$

From the earlier discussion, we have that

$$\mathrm{adj}(A - \lambda I_n) = B_{n-1}\lambda^{n-1} + B_{n-2}\lambda^{n-2} + \cdots + B_1\lambda + B_0,$$

where $B_{n-1}, B_{n-2}, \ldots, B_1, B_0$ are n-square matrices independent from λ.
Then from above equation,

$$(A - \lambda I_n)(B_{n-1}\lambda^{n-1} + B_{n-2}\lambda^{n-2} + \cdots + B_1\lambda + B_0)$$
$$= (-1)^n(\lambda^n + a_{n-1}\lambda^{n-1} + a_{n-2}\lambda^{n-2} + \cdots + a_1\lambda + a_0)I_n,$$

or

$$-B_{n-1}\lambda^n + (AB_{n-1} - B_{n-2})\lambda^{n-1} + (AB_{n-2} - B_{n-3})\lambda^{n-2} + \cdots + (AB_1 - B_0)\lambda + AB_0$$
$$= (-1)^n(\lambda^n + a_{n-1}\lambda^{n-1} + a_{n-2}\lambda^{n-2} + \cdots + a_1\lambda + a_0)I_n.$$

Now, equating the coefficients of corresponding powers of λ, we get

$$(-1)^n I_n = -B_{n-1},$$
$$(-1)^n a_{n-1} I_n = AB_{n-1} - B_{n-2},$$
$$(-1)^n a_{n-2} I_n = AB_{n-2} - B_{n-3},$$
$$\vdots$$
$$(-1)^n a_1 I_n = AB_1 - B_0,$$
$$(-1)^n a_0 I_n = AB_0.$$

Multiplying above equations from first to n-th by $A^n, A^{n-1}, \ldots, A$ and I_n, respectively, we get

$$(-1)^n A^n = -A^n B_{n-1},$$
$$(-1)^n a_{n-1} A^{n-1} = A^n B_{n-1} - A^{n-1} B_{n-2},$$
$$(-1)^n a_{n-2} A^{n-2} = A^{n-1} B_{n-2} - A^{n-2} B_{n-3},$$
$$\vdots$$
$$(-1)^n a_1 A = A^2 B_1 - AB_0,$$
$$(-1)^n a_0 I_n = AB_0.$$

Adding the above equations, we get

$$(-1)^n \{A^n + a_{n-1}A^{n-1} + a_{n-2}A^{n-2} + \cdots + a_1 A + a_0 I_n\} = 0, \quad \text{or} \quad P_A(A) = 0$$

That is, A is a root of its characteristic polynomial $P_A(\lambda)$. ▢

The Cayley–Hamilton theorem is applicable to find the inverse of an invertible matrix.

Example 7.5.2. Let

$$A = \begin{bmatrix} 2 & 0 & 1 \\ 1 & 3 & 1 \\ 1 & 2 & 2 \end{bmatrix}.$$

Then the characteristic polynomial of A is $P_A(\lambda) = -\lambda^3 + 7\lambda^2 - 13\lambda + 7$.
From the Cayley–Hamilton theorem, we have

$$P_A(A) = 0 \Rightarrow -A^3 + 7A^2 - 13A + 7I_3 = 0$$

$$\Rightarrow \frac{1}{7}(A^3 - 7A^2 + 13A) = I_3$$

$$\Rightarrow A\frac{1}{7}(A^2 - 7A + 13I_3) = I_3$$

$$\Rightarrow A^{-1} = \frac{1}{7}(A^2 - 7A + 13I_3)$$

$$= \frac{1}{7}A^2 - A + \frac{13}{7}I_3$$

$$= \frac{1}{7}\begin{bmatrix} 5 & 2 & 4 \\ 6 & 11 & 6 \\ 6 & 10 & 7 \end{bmatrix} - \begin{bmatrix} 2 & 0 & 1 \\ 1 & 3 & 1 \\ 1 & 2 & 2 \end{bmatrix} + \frac{13}{7}\begin{bmatrix} 1 & 0 & 0 \\ 0 & 1 & 0 \\ 0 & 0 & 1 \end{bmatrix}$$

$$= \begin{bmatrix} \frac{4}{7} & \frac{2}{7} & \frac{-3}{7} \\ \frac{-1}{7} & \frac{3}{7} & \frac{-1}{7} \\ \frac{-1}{7} & \frac{-4}{7} & \frac{6}{7} \end{bmatrix}.$$

Note that if the characteristic polynomial of a matrix A does not contain the constant term, then A is not invertible. For example, suppose $P_A(\lambda) = -\lambda^3 + a_1\lambda^2 + a_2\lambda$ is the characteristic polynomial of any matrix A. Then $P_A(\lambda) = 0 \Rightarrow -\lambda^3 + a_1\lambda^2 + a_2\lambda = 0 \Rightarrow \lambda(-\lambda^2 + a_1\lambda + a_2) = 0$. That is, $\lambda = 0$ is an eigenvalue of A. Hence, A is noninvertible.

From the Cayley–Hamilton theorem, it is clear that for any given n-square matrix A there exists an n degree polynomial $P(x)$ such that $P(A) = 0$. But the question is that, "does there exist any n-square matrix A and a polynomial $P'(x)$ such that $\deg(P'(x)) < n$ and $P'(A) = 0$." The answer of this question is positive. Hence, we discuss about the existence and uniqueness of such a monic polynomial $m(x)$ of degree less than or equal to n such that $m(A) = 0$. Recall that a polynomial $P(x) = a_nx^n + a_{n-1}x^{n-1} + \cdots + a_1x + a_0$ is called monic if $a_n = 1$.

Definition 7.5.3. Let A be an n-square matrix with entries in a field F. Then the least degree monic polynomial $m_A(x)$ satisfying $m_A(A) = 0$ is called the minimal polynomial of A.

Theorem 7.5.4. *There exists a unique minimal polynomial of any n-square matrix A.*

Proof. Consider the set:

$$\Lambda = \{\deg(P(x)) : P(x) \text{ is a nonzero polynomial satisfying } P(A) = 0\}.$$

From the Cayley–Hamilton theorem, if $P_A(x)$ is the characteristic polynomial of A, then $P_A(A) = 0$. Hence, Λ is a nonempty subset of $\mathbb{N}$. By the well-ordering principle, there exist a smallest natural number k in Λ. Then without loss of any generality, we can assume a monic polynomial $m(x)$ such that $\deg(m(x)) = k$ and $m(A) = 0$. Since all the polynomials $P(x)$ of degree k such that $P(A) = 0$ will be of the form $P(x) = a_k m(x)$ where $a_k \neq 0 \in F$, $m(x)$ is unique, and hence minimal polynomial of A. □

Theorem 7.5.5. *Let A be an n-square matrix and let P(x) be a polynomial such that $P(A) = 0$. Then the minimal polynomial $m_A(x)$ divides $P(x)$.*

Proof. By division algorithm, there exist polynomials $q(x)$ and $r(x)$ such that $P(x) = m_A(x)q(x) + r(x)$, with $r(x) = 0$ or $\deg(r(x)) < \deg(m_A(x))$. Since $P(A) = 0$ and $m_A(A) = 0$, from above equation we have $0 = P(A) = m_A(A)q(A) + r(A) \Rightarrow r(A) = 0$. But $m_A(x)$ is the least degree monic polynomial satisfying $m_A(A) = 0$, $r(x)$ cannot be nonzero polynomial with $r(A) = 0$ and $\deg(r(x)) < \deg(m_A(x))$. Hence, $r(x) = 0$. It follows that $P(x) = m_A(x)q(x)$. Thus, $m_A(x)$ divides $P(x)$. □

Corollary 7.5.6. *If $P_A(x)$ and $m_A(x)$ are characteristic and minimal polynomials of a matrix A, respectively, then:*
(i) *$m_A(x)$ divides $P_A(x)$ and*
(ii) *every eigenvalue of A is a root of $m_A(x)$.*

Proof. (i) Since $P_A(x)$ is a polynomial satisfying $P_A(A) = 0$, from the above theorem we have that $m_A(x)$ divides $P_A(x)$. That is, every zero of $m_{(x)}$ is a zero of $P_A(x)$, also.
(ii) Let λ be an eigenvalue of A and X be its associated eigenvector. Also, we have that if $P(x)$ is any polynomial then the eigenvalue of matrix $P(A)$ is $P(\lambda)$, that is, $P(A)X = P(\lambda)X$. Hence, for the minimal polynomial $m_A(x)$ we have $m_A(\lambda)X = m_A(A)X = 0X = 0$. Since X is an eigenvector, $X \neq 0$, hence $m_A(\lambda) = 0$. This proves that λ is a root of minimal polynomial $m_A(x)$. That is, every zero of $P_A(x)$ is also a zero of $m_A(x)$. □

Thus, from the above corollary it is clear that the characteristic polynomial and the minimal polynomial of a matrix have the same zeros.

Theorem 7.5.7. *Similar matrices have the same minimal polynomial.*

Proof. Let A and B be similar matrices. Then there exist an invertible matrix P such that $B = P^{-1}AP$.

Let $f(x) = x^k + a_{k-1}x^{k-1} + a_{k-2}x^{k-2} + \cdots + a_1 x + a_0$, be any polynomial. Then

$$f(B) = f(P^{-1}AP) = (P^{-1}AP)^k + a_{k-1}(P^{-1}AP)^{k-1} + a_{k-2}(P^{-1}AP)^{k-2} + \cdots + a_1(P^{-1}AP) + a_0 I_n.$$

Since $(P^{-1}AP)^n = (P^{-1}AP)(P^{-1}AP)\cdots(P^{-1}AP) = (P^{-1}A^n P)$, we have

$$\begin{aligned} f(B) &= P^{-1}A^k P + a_{k-1}P^{-1}A^{k-1}P + a_{k-2}P^{-1}A^{k-2}P + \cdots + a_1 P^{-1}AP + a_0 I_n \\ &= P^{-1}(A^k + a_{k-1}A^{k-1} + a_{k-2}A^{k-2} + \cdots + a_1 A + a_0 I_n)P \\ &= P^{-1}f(A)P. \end{aligned}$$

Thus, $f(A) = 0$ if and only if $f(B) = 0$.

Hence, A and B have the same minimal polynomial. $\qquad\qquad\square$

Example 7.5.8. Consider the matrix

$$A = \begin{bmatrix} 1 & 0 & -1 \\ 1 & 2 & 1 \\ 2 & 2 & 3 \end{bmatrix}.$$

The characteristic polynomial of A is $P_A(\lambda) = |A - \lambda I_3| = -\lambda^3 + 6\lambda^2 - 11\lambda + 6 = -(\lambda - 1)(\lambda - 2)(\lambda - 3)$. By Corollary 7.5.6, we have that minimal polynomial contains all the eigenvalues of the matrix and divides the characteristic polynomial. Hence, the minimal polynomial is either $-(\lambda - 1)(\lambda - 2)(\lambda - 3)$ or $(\lambda - 1)(\lambda - 2)(\lambda - 3)$. Since the minimal polynomial is the monic polynomial, $m_A(\lambda) = (\lambda - 1)(\lambda - 2)(\lambda - 3)$.

Example 7.5.9. Let

$$A = \begin{bmatrix} 3 & 2 & -1 \\ 3 & 8 & -3 \\ 3 & 6 & -1 \end{bmatrix}.$$

Then $P_A(\lambda) = -\lambda^3 + 10\lambda^2 - 28\lambda + 24 = -(\lambda - 2)^2(\lambda - 6)$.

The minimal polynomial of A will be either $(\lambda - 2)(\lambda - 6)$ or $(\lambda - 2)^2(\lambda - 6)$. Now,

$$(A - 2I_3)(A - 6I_3) = \begin{bmatrix} 1 & 2 & -1 \\ 3 & 6 & -3 \\ 3 & 6 & -3 \end{bmatrix} \begin{bmatrix} -3 & 2 & -1 \\ 3 & 2 & -3 \\ 3 & 6 & -7 \end{bmatrix} = \begin{bmatrix} 0 & 0 & 0 \\ 0 & 0 & 0 \\ 0 & 0 & 0 \end{bmatrix}.$$

Hence, the minimal polynomial of A is $(\lambda - 2)(\lambda - 6)$.

ℹ️ Exercises

(7.5.1)   Verify the Cayley–Hamilton theorem for the following matrices:

$$A = \begin{bmatrix} 1 & 2 & 3 \\ 0 & -1 & 2 \\ 0 & 0 & -2 \end{bmatrix}, \quad B = \begin{bmatrix} 3 & 1 & 4 \\ 2 & 3 & 1 \\ 1 & 2 & 5 \end{bmatrix}.$$

(7.5.2)   Find the inverse of the following matrices using Cayley–Hamilton theorem:

$$A = \begin{bmatrix} 1 & 1 & 2 \\ 0 & 2 & 2 \\ -1 & 1 & 3 \end{bmatrix}, \quad B = \begin{bmatrix} 1 & 1 & 4 \\ 0 & -2 & 6 \\ 0 & 0 & 3 \end{bmatrix},$$

$$C = \begin{bmatrix} 3 & 2 & 0 \\ 3 & 3 & 1 \\ 4 & 4 & 1 \end{bmatrix}, \quad D = \begin{bmatrix} 1 & -1 & 0 \\ 2 & 3 & -4 \\ 2 & 3 & -3 \end{bmatrix}.$$

(7.5.3)   Find the minimal polynomial of the following matrices:

$$A = \begin{bmatrix} 2 & 1 & 1 \\ 1 & 2 & 1 \\ 0 & 0 & 1 \end{bmatrix}, \quad B = \begin{bmatrix} 1 & 2 & 2 \\ 2 & 1 & -2 \\ 2 & -2 & 1 \end{bmatrix},$$

$$C = \begin{bmatrix} 1 & 1 & 2 & 1 \\ 0 & 1 & 0 & -1 \\ 0 & 0 & 1 & 2 \\ 0 & 0 & 0 & 1 \end{bmatrix}, \quad D = \begin{bmatrix} 0 & 1 & 1 & 1 \\ 0 & 0 & 1 & 1 \\ 0 & 0 & 0 & 1 \\ 0 & 0 & 0 & 0 \end{bmatrix}.$$

(7.5.4)   Let $f(x)$ be the minimal polynomial of the matrix

$$A = \begin{bmatrix} 2 & 1 & 2 & 1 \\ 1 & 1 & 5 & -1 \\ 0 & 1 & -1 & -5 \\ -2 & 1 & -3 & -1 \end{bmatrix}.$$

Then find the rank and the determinant of the matrix $f(A)$.

(7.5.5)   Let $f(x)$ be the characteristic polynomial of the matrix

$$A = \begin{bmatrix} 2 & 3 & 4 & 5 \\ 6 & 3 & -2 & -1 \\ 0 & 1 & -1 & -5 \\ 2 & -1 & 3 & 1 \end{bmatrix}.$$

Then find the rank and the determinant of the matrix $f(A)$.

8 Bilinear and quadratic forms

In this chapter, we explore bilinear and quadratic forms. Bilinear forms are linear transformations that are linear in more than one variable, extending our understanding of linear phenomena. Although the field F is initially general, we will focus specifically on $F = \mathbb{R}$ or $\mathbb{C}$.

8.1 Bilinear forms

Definition 8.1.1. Let V be a vector space over a field F. A function $f : V \times V \to F$ is called a bilinear form on V if it satisfies the following conditions:

(i) $f(au + bv, w) = af(u, w) + bf(v, w)$;

(ii) $f(u, av + bw) = af(u, v) + bf(u, w)$

for all $a, b \in F$ and $u, v, w \in V$.

In the first part of the definition, fix w and denote $f(u, w) = f_w(u)$ for all $u \in V$. Then we have

$$f_w(au + bv) = f(au + bv, w) = af(u, w) + bf(v, w) = af_w(u) + bf_w(v).$$

This demonstrates that $f_w : V \to F$ is a linear functional. Similarly, in the second part, fix u and define $f_u(v) = f(u, v)$ for all $v \in V$. Then f_u is also a linear functional. Therefore, the bilinear form f on V is linear in each variable when the other variable is fixed, which is why it is termed bilinear.

It is evident that if f and g are bilinear forms on V, then $(f + g)$ and kf, defined by $(f + g)(u, v) = f(u, v) + g(u, v)$ and $(kf)(u, v) = kf(u, v)$, where $k \in F$ are also bilinear forms. Thus, $B_l(V)$, the set of bilinear forms on V, forms a vector space over F.

Example 8.1.2.

(a) Let V be a vector space over a field F. Then the map $f : V \times V \to F$ defined by $f(u, v) = 0$, called a zero map, is a bilinear form on V.

(b) Let V be a vector space over the field of real numbers $\mathbb{R}$. Then from the definition of inner product on V the map $f : V \times V \to \mathbb{R}$ defined by $f(u, v) = \langle u, v \rangle$ is a bilinear form.

In particular, $f : \mathbb{R}^n \times \mathbb{R}^n \to \mathbb{R}$ defined by

$$f(x, y) = \langle x, y \rangle = x_1 y_1 + x_2 y_2 + \cdots + x_n y_n,$$

where $x = (x_1, x_2, \ldots, x_n), y = (y_1, y_2, \ldots, y_n) \in \mathbb{R}^n$ is a bilinear form on $\mathbb{R}^n$.

Thus, an inner product on a real vector space is a bilinear form.

If V is a complex vector space, then the map $f : V \times V \to F$ defined by $f(u, v) = \langle u, v \rangle$ is not a bilinear form.

https://doi.org/10.1515/9783111516035-008

This is because $f(u, av + bw) = \langle u, av + bw \rangle = \bar{a}\langle u, v \rangle + \bar{b}\langle u, w \rangle = \bar{a}f(u, v) + \bar{b}f(u, w) \neq$
$af(u, v) + bf(u, w)$.

Example 8.1.3. Let V be a vector space over a field F and let f_1 and f_2 be linear functionals on V. Then the map $f : V \times V \to F$ defined by $f(u, v) = f_1(u)f_2(v)$ is a bilinear form
on V.

Let $u, v, w \in V$ and let $a, b \in F$. Then

$$f(au + bv, w) = f_1(au + bv)f_2(w) = (af_1(u) + bf_1(v))f_2(w) = af_1(u)f_2(w) + bf_1(v)f_2(w)$$
$$= af(u, w) + bf(v, w).$$

Similarly, $f(u, (av + bw)) = f_1(u)f_2(av + bw)) = f_1(u)(af_2(v) + bf_2(w)) = af_1(u)f_2(v) +$
$bf_1(u)f_2(w) = af(u, v) + bf(u, w)$.

Example 8.1.4. Let A be an n-square matrix with entries in a field F. Then the map $f :$
$F^n \times F^n \to F$ defined by $f(X, Y) = X^t A Y$, where X and Y are column vectors in F^n, is a
bilinear form on F^n.

Let $X, Y, Z \in F^n$ and $a, b \in F$. Then

$$f(aX + bY, Z) = (aX + bY)^t AZ$$
$$= (aX^t + bY^t)AZ$$
$$= aX^t AZ + bY^t AZ$$
$$= af(X, Z) + bf(Y, Z).$$

Similarly,

$$f(X, aY + bZ) = X^t A(aY + bZ)$$
$$= X^t AaY + X^t AbZ$$
$$= aX^t AY + bX^t AZ$$
$$= af(X, Y) + bf(X, Z).$$

In particular, if

$$A = \begin{bmatrix} -1 & 0 \\ 1 & 2 \end{bmatrix} \quad \text{and} \quad F = \mathbb{R}.$$

Then the bilinear form $f : \mathbb{R}^2 \times \mathbb{R}^2 \to \mathbb{R}$ is $f(X, Y) = X^t AY$.
If

$$X = \begin{bmatrix} x_1 \\ x_2 \end{bmatrix} \quad \text{and} \quad Y = \begin{bmatrix} y_1 \\ y_2 \end{bmatrix},$$

then

$$f(X, Y) = X^t A Y = \begin{bmatrix} x_1 & x_2 \end{bmatrix} \begin{bmatrix} -1 & 0 \\ 1 & 2 \end{bmatrix} \begin{bmatrix} y_1 \\ y_2 \end{bmatrix}$$

$$= \begin{bmatrix} -x_1 + x_2 & 2x_2 \end{bmatrix} \begin{bmatrix} y_1 \\ y_2 \end{bmatrix}$$

$$= -x_1 y_1 + x_2 y_1 + 2x_2 y_2.$$

In the next example, we explore all possible bilinear forms on a two-dimensional vector space.

Example 8.1.5. Let V be a 2-dimensional vector space over the field F and let $B = \{v_1, v_2\}$ be an ordered basis of V.

Suppose $f : V \times V \to F$ is a bilinear form.

Then for $u, w \in V$, $u = k_1 v_1 + k_2 v_2$ and $w = l_1 v_1 + l_2 v_2$ and so

$$\begin{aligned}
f(u, w) &= f(k_1 v_1 + k_2 v_2, w) \\
&= k_1 f(v_1, w) + k_2 f(v_2, w) \\
&= k_1 f(v_1, l_1 v_1 + l_2 v_2) + k_2 f(v_2, l_1 v_1 + l_2 v_2) \\
&= k_1 (l_1 f(v_1, v_1) + l_2 f(v_1, v_2)) + k_2 (l_1 f(v_2, v_1) + l_2 f(v_2, v_2)) \\
&= k_1 l_1 f(v_1, v_1) + k_1 l_2 f(v_1, v_2) + k_2 l_1 f(v_2, v_1) + k_2 l_2 f(v_2, v_2) \\
&= \begin{bmatrix} k_1 & k_2 \end{bmatrix} \begin{bmatrix} f(v_1, v_1) & f(v_1, v_2) \\ f(v_2, v_1) & f(v_2, v_2) \end{bmatrix} \begin{bmatrix} l_1 \\ l_2 \end{bmatrix},
\end{aligned}$$

where

$$\begin{bmatrix} k_1 \\ k_2 \end{bmatrix} \quad \text{and} \quad \begin{bmatrix} l_1 \\ l_2 \end{bmatrix}$$

are coordinate vectors of u and w in column form denoted by $[u]_B$ and $[w]_B$, respectively.

If $[f]_{B,B}$ denotes the matrix

$$\begin{bmatrix} f(v_1, v_1) & f(v_1, v_2) \\ f(v_2, v_1) & f(v_2, v_2) \end{bmatrix},$$

then $f(u, w) = [u]_B^t [f]_{B,B} [w]_B$.

Thus, f is completely determined by the matrix $[f]_{B,B}$ whose ij-th elements are $f(v_i, v_j)$.

Building on the previous discussion, we define the matrix associated with a bilinear form and extend the result to an n-dimensional vector space.

Definition 8.1.6. Let f be a bilinear form on an n-dimensional vector space V over a field F and let $B = \{v_1, v_2, \ldots, v_n\}$ be an ordered basis of V. Then the matrix of f with respect to the basis B is defined as

$$[f]_{B,B} = \begin{bmatrix} f(v_1, v_1) & f(v_1, v_2) & \cdots & f(v_1, v_n) \\ \vdots & \vdots & \ddots & \vdots \\ f(v_n, v_1) & f(v_n, v_2) & \cdots & f(v_n, v_n) \end{bmatrix}.$$

Similar to matrices associated with linear transformations, we can describe the relationship between the matrices of bilinear forms and coordinate vectors.

Theorem 8.1.7. *Let V be an n-dimensional vector space over a field F and let $B = \{v_1, v_2, \ldots, v_n\}$ be an ordered basis of V. Then:*
1. *Every bilinear form f on V is completely determined by its matrix $[f]_{B,B}$.*
2. *For every n-square matrix A with entries in F, there exists a bilinear form f on V such that $[f]_{B,B} = A$.*

Proof. 1.   Let $x = \sum_{i=1}^{n} x_i v_i$ and $y = \sum_{i=1}^{n} y_i v_i$ be any two vectors in V, where $x_i, y_i \in F$ for all i. Then

$$f(x, y) = f\left(\sum_{i=1}^{n} x_i v_i, y \right)$$

$$= \sum_{i=1}^{n} x_i f(v_i, y)$$

$$= \sum_{i=1}^{n} x_i f\left(v_i, \sum_{j=1}^{n} y_j v_j \right)$$

$$= \sum_{i=1}^{n} x_i \sum_{j=1}^{n} y_j f(v_i, v_j)$$

$$= \sum_{i=1}^{n} \sum_{j=1}^{n} x_i y_j f(v_i, v_j)$$

$$= \begin{bmatrix} x_1 & x_2 & \cdots & x_n \end{bmatrix} \begin{bmatrix} f(v_1, v_1) & f(v_1, v_2) & \cdots & f(v_1, v_n) \\ \vdots & \vdots & \ddots & \vdots \\ f(v_n, v_1) & f(v_n, v_2) & \cdots & f(v_n, v_n) \end{bmatrix} \begin{bmatrix} y_1 \\ y_2 \\ \vdots \\ y_n \end{bmatrix}$$

$$= [x]_B^t [f]_{B,B} [y]_B,$$

where

$$[x]_B = \begin{bmatrix} x_1 \\ x_2 \\ \vdots \\ x_n \end{bmatrix} \quad \text{and} \quad [y]_B = \begin{bmatrix} y_1 \\ y_2 \\ \vdots \\ y_n \end{bmatrix}$$

are the coordinate vectors of x and y in column form. Thus, f is completely determined by the matrix $[f]_{B,B}$.

2. Given $A \in M_{n,n}(F)$, define the map $f : V \times V \to F$ by

$$f(x,y) = [x]_B^t A [y]_B.$$

Since $[x]_B, [y]_B \in F^n$, from Example 8.1.4 it is clear that f is a bilinear form. For basis vectors v_i and v_j, have

$$v_i = 0v_1 + 0v_2 + \cdots + 0v_{i-1} + 1v_i + 0v_{i+1} + \cdots + 0v_n$$

and

$$v_j = 0v_1 + 0v_2 + \cdots + 0v_{j-1} + 1v_j + 0v_{j+1} + \cdots + 0v_n.$$

Then

$$[v_i]_B = \begin{bmatrix} 0 \\ \vdots \\ 0 \\ 1 \\ 0 \\ \vdots \\ 0 \end{bmatrix} \quad \text{and} \quad [v_j]_B = \begin{bmatrix} 0 \\ \vdots \\ 0 \\ 1 \\ 0 \\ \vdots \\ 0 \end{bmatrix},$$

and so

$$f(v_i, v_j) = [v_i]_B^t A [v_j]_B$$

$$= \begin{bmatrix} 0 & \cdots & 0 & 1 & 0 & \cdots & 0 \end{bmatrix} \begin{bmatrix} a_{11} & a_{12} & \cdots & a_{1n} \\ a_{21} & a_{22} & \cdots & a_{2n} \\ \vdots & & & \\ a_{n1} & a_{n2} & \cdots & a_{nn} \end{bmatrix} \begin{bmatrix} 0 \\ \vdots \\ 0 \\ 1 \\ 0 \\ \vdots \\ 0 \end{bmatrix}$$

$$= a_{ij}.$$

Thus, the ij-th element of the matrix A is equal to $f(v_i, v_j)$, which is the ij-th element of the matrix $[f]_{B,B}$. Therefore, $A = [f]_{B,B}$. $\square$

Corollary 8.1.8. *Let $B = \{v_1, v_2, \ldots, v_n\}$ be an ordered basis of an n-dimensional vector space V over the field F. Then the map $T : B_l(V) \to M_{n,n}(F)$, defined by $T(f) = [f]_{B,B}$, is a vector space isomorphism.*

Proof. Let $f_1, f_2 \in B_l(V)$ and $a_1, a_2 \in F$. The map T is a linear transformation if and only if

$$T(a_1 f_1 + a_2 f_2) = a_1 T(f_1) + a_2 T(f_2),$$

or

$$[a_1 f_1 + a_2 f_2]_{B,B} = a_1 [f_1]_{B,B} + a_2 [f_2]_{B,B}.$$

Now,

$$(a_1 f_1 + a_2 f_2)(v_i, v_j) = a_1 f_1(v_i, v_j) + a_2 f_2(v_i, v_j) \quad \forall i, j$$

shows that the ij-th entry of the matrix $[a_1 f_1 + a_2 f_2]_{B,B}$ is equal to a_1 times the ij-th entry of the matrix $[f_1]_{B,B} + a_2$ times the ij-th entry of the matrix $[f_2]_{B,B}$. Thus,

$$[a_1 f_1 + a_2 f_2]_{B,B} = a_1 [f_1]_{B,B} + a_2 [f_2]_{B,B}.$$

Hence, $T(a_1 f_1 + a_2 f_2) = a_1 T(f_1) + a_2 T(f_2)$. From the above theorem, it is clear that T is bijective. Therefore, T is an isomorphism. ☐

The following theorem describes the effect of a change of basis on the matrix representation of a bilinear form.

Theorem 8.1.9. *Let f be a bilinear form on a finite-dimensional vector space V over the field F and let $B = \{v_1, v_2, \ldots, v_n\}$, $B' = \{v_1', v_2', \ldots, v_n'\}$ be two ordered bases of V. Then $[f]_{B',B'} = [I]_{B',B}^t [f]_{B,B} [I]_{B',B}$, where $[I]_{B'B}$ is the matrix of identity linear transformation on V.*

Proof. Let $[I]_{B',B} = [q_{ij}]$. Then $v_j' = \sum_{i=1}^{n} q_{ij} v_i$ for all $j = 1, 2, \ldots, n$.
Suppose $[f]_{B,B} = [a_{ij}]$ and $[f]_{B',B'} = [a_{ij}']$, then $a_{ij} = f(v_i, v_j)$ and $a_{ij}' = f(v_i', v_j')$.
Hence,

$$a_{ij}' = f(v_i', v_j') = f\left(\sum_{k=1}^{n} q_{ki} v_k, \sum_{l=1}^{n} q_{lj} v_l\right)$$

$$= \sum_{k=1}^{n} \sum_{l=1}^{n} q_{ki} q_{lj} f(v_k, v_l)$$

$$= \sum_{k=1}^{n} \sum_{l=1}^{n} q_{ki} q_{lj} a_{kl}$$

$$= \sum_{k=1}^{n} q_{ki} \left(\sum_{l=1}^{n} a_{kl} q_{lj}\right).$$

From the property of matrix multiplication, this shows that a'_{ij} is the ij-th entry of the matrix $[I]^t_{B',B}[f]_{B,B}[I]_{B'B}$. Hence, $[f]_{B',B'} = [I]^t_{B',B}[f]_{B,B}[I]_{B'B}$. □

(Compare this with the similar result in the case of linear transformation.)

Example 8.1.10. Let

$$A = \begin{bmatrix} 1 & 2 & 3 \\ -1 & 0 & 2 \\ 4 & 5 & -2 \end{bmatrix}.$$

Then A gives rise to the bilinear form $f(X, Y) = X^t A Y$ on $\mathbb{R}^3$.
Let

$$X = \begin{bmatrix} x_1 \\ x_2 \\ x_3 \end{bmatrix} \quad \text{and} \quad Y = \begin{bmatrix} y_1 \\ y_2 \\ y_3 \end{bmatrix}.$$

Then

$$f(X, Y) = \begin{bmatrix} x_1 & x_2 & x_3 \end{bmatrix} \begin{bmatrix} 1 & 2 & 3 \\ -1 & 0 & 2 \\ 4 & 5 & -2 \end{bmatrix} \begin{bmatrix} y_1 \\ y_2 \\ y_3 \end{bmatrix}$$

$$= x_1 y_1 + 2 x_1 y_2 + 3 x_1 y_3 - x_2 y_1 + 2 x_2 y_3 + 4 x_3 y_1 + 5 x_3 y_2 - 2 x_3 y_3,$$

is the bilinear form on $\mathbb{R}^3$.
Note that, if

$$A = \begin{bmatrix} a_{11} & a_{12} & \cdots & a_{1n} \\ a_{21} & a_{22} & \cdots & a_{2n} \\ & \vdots & & \\ a_{n1} & a_{n2} & \cdots & a_{nn} \end{bmatrix}$$

is any matrix in $M_{n,n}(F)$, then the associated bilinear form $f(X, Y) = X^t A Y$ on F^n is defined as

$$f(X, Y) = a_{11} x_1 y_1 + a_{12} x_1 y_2 + \cdots + a_{1n} x_1 y_n$$
$$+ a_{21} x_2 y_1 + a_{22} x_2 y_2 + \cdots a_{2n} x_2 y_n$$
$$+$$
$$\vdots$$
$$+ a_{n1} x_n y_1 + a_{n2} x_n y_2 + \cdots + a_{nn} x_n y_n,$$

where $X^t = [x_1\, x_2 \cdots x_n]$ and $Y^t = [y_1\, y_2 \cdots y_n]$.

Thus, for a given matrix $A = [a_{ij}] \in M_{n,n}(F)$, the bilinear form f on F^n associated with A can be directly expressed using the entries a_{ij} as coefficients of $x_i y_j$. Specifically, $f(X, Y) = \sum_i \sum_j a_{ij} x_i y_j$.

Example 8.1.11. Let $f(x, y) = x_1 y_1 + 3x_1 y_3 - x_2 y_1 + 2x_2 y_2 + 5x_3 y_1 + 4x_3 y_2$ be a bilinear form on $\mathbb{R}^3$.

Let $B_1 = \{e_1 = (1, 0, 0), e_2 = (0, 1, 0), e_3(0, 0, 1)\}$ be the standard ordered basis of $\mathbb{R}^3$. Then from Definition 8.1.6,

$$a_{11} = f(e_1, e_1) = 1 = \text{ coefficient of } x_1 y_1,$$
$$a_{12} = f(e_1, e_2) = 0 = \text{ coefficient of } x_1 y_2,$$
$$\vdots$$
$$a_{33} = f(e_3, e_3) = 0 = \text{ coefficient of } x_3 y_3,$$

and we have

$$[f]_{B_1, B_1} = \begin{bmatrix} 1 & 0 & 3 \\ -1 & 2 & 0 \\ 5 & 4 & 0 \end{bmatrix}.$$

Thus, the matrix of a bilinear form f on $\mathbb{R}^n$ related to the standard ordered basis of $\mathbb{R}^n$ is $A = [a_{ij}]$, where $a_{ij} = $ coefficient of term $x_i y_j$ of $f(x, y)$.

Let $B_2 = \{v_1 = (1, 0, 0), v_2 = (1, -1, 0), v_3 = (1, 1, -1)\}$ be another ordered basis of $\mathbb{R}^3$. Then we find the matrix $[f]_{B_2, B_2}$ as follows:

$$a_{11} = f(v_1, v_1) = f((1, 0, 0), (1, 0, 0)) = 1,$$
$$a_{12} = f(v_1, v_2) = f((1, 0, 0), (1, -1, 0)) = 1,$$
$$a_{13} = f(v_1, v_3) = f((1, 0, 0), (1, 1, -1)) = -2,$$
$$a_{21} = f(v_2, v_1) = f((1, -1, 0), (1, 0, 0)) = 2,$$
$$a_{22} = f(v_2, v_2) = f((1, -1, 0), (1, -1, 0)) = 4,$$
$$a_{23} = f(v_2, v_3) = f((1, -1, 0), (1, 1, -1)) = -3,$$
$$a_{31} = f(v_3, v_1) = f((1, 1, -1), (1, 0, 0)) = -5,$$
$$a_{32} = f(v_3, v_2) = f((1, 1, -1), (1, -1, 0)) = -3,$$
$$a_{33} = f(v_3, v_3) = f((1, 1, -1), (1, 1, -1)) = -10,$$

hence

$$[f]_{B_2, B_2} = \begin{bmatrix} 1 & 1 & -2 \\ 2 & 4 & -3 \\ -5 & -3 & -10 \end{bmatrix}.$$

Now the transition matrix $[I]_{B_2,B_1}$ is given as

$$I(1,0,0) = (1,0,0) = 1(1,0,0) + 0(0,1,0) + 0(0,0,1),$$
$$I(1,-1,0) = (1,-1,0) = 1(1,0,0) + (-1)(0,1,0) + 0(0,0,1),$$
$$I(1,1,-1) = (1,1,-1) = 1(1,0,0) + 1(0,1,0) + (-1)(0,0,1),$$

$$[I]_{B_2,B_1} = \begin{bmatrix} 1 & 1 & 1 \\ 0 & -1 & 1 \\ 0 & 0 & -1 \end{bmatrix}.$$

Also, the transpose matrix is

$$[I]_{B_2,B_1}^t = \begin{bmatrix} 1 & 0 & 0 \\ 1 & -1 & 0 \\ 1 & 1 & -1 \end{bmatrix}.$$

Then

$$\begin{bmatrix} 1 & 0 & 0 \\ 1 & -1 & 0 \\ 1 & 1 & -1 \end{bmatrix} \begin{bmatrix} 1 & 0 & 3 \\ -1 & 2 & 0 \\ 5 & 4 & 0 \end{bmatrix} \begin{bmatrix} 1 & 1 & 1 \\ 0 & -1 & 1 \\ 0 & 0 & -1 \end{bmatrix}$$
$$= \begin{bmatrix} 1 & 1 & -2 \\ 2 & 4 & -3 \\ -5 & -3 & -10 \end{bmatrix}.$$

This shows that $[I]_{B_2,B_1}^t [f]_{B_1,B_1} [I]_{B_2,B_1} = [f]_{B_2,B_2}$.

Definition 8.1.12. Let A and B be n-square matrices. Then B is said to be congruent to A if there is a nonsingular matrix P such that $B = P^t A P$.

From Theorem 8.1.9, it is clear that the matrices of a bilinear form related to a different choice of bases are congruent. The relation of being congruent is an equivalence relation on $M_{n,n}(F)$.

The following example shows that any matrix congruent to a symmetric matrix is also symmetric.

Example 8.1.13. Let B be a matrix congruent to a symmetric matrix A. Then for some nonsingular matrix P, $B = P^t A P$.
Now, $B^t = (P^t A P)^t = P^t A^t (P^t)^t = P^t A P = B$.
This shows that B is a symmetric matrix.

We have that, if A and B are matrices of a bilinear form f related to a different choice of bases, then A and B are congruent to each other and so there is a nonsingular matrix P, such that $B = P^t A B$. Since P is nonsingular, $\text{rank}(B) = \text{rank}(A)$. Hence, the rank of a bilinear form is defined as follows.

Definition 8.1.14. Let f be a bilinear form on a finite-dimensional vector space V. The rank of f is defined as the rank of the matrix of f with respect to any basis of V.

If $\text{rank}(f) < \dim(V)$, then f is considered degenerate. Conversely, f is considered nondegenerate if $\text{rank}(f) = \dim(V)$.

Based on specific properties of bilinear forms, the following types of bilinear forms are defined.

Definition 8.1.15. Let f be a bilinear form on a vector space V over F. Then f is called:
(i)   symmetric if $f(v, w) = f(w, v) \ \forall v, w \in V$;
(ii) skew-symmetric if $f(v, w) = -f(w, v) \ \forall v, w \in v$; and
(iii) alternating if $f(v, v) = 0 \ \forall v \in V$.

Theorem 8.1.16. *Let f be a bilinear form on a vector space V over a field F of characteristic $\neq 2$. Then f is skew-symmetric if and only if f is alternating.*

Proof. Suppose f is skew-symmetric. Then

$$f(u, v) = -f(v, u) \quad \forall u, v \in V.$$

Taking $u = v$, $f(v, v) = -f(v, v)$, $\forall v \in V$. This implies

$$f(v, v) + f(v, v) = 0 \Rightarrow f(v, v)(1 + 1) = 0.$$

Since characteristic of F is $\neq 2$, $1 + 1 \neq 0$, and hence $f(v, v) = 0 \ \forall v \in V$.
This proves that f is an alternating bilinear form.
Conversely, suppose that f is an alternating bilinear form.
Then $f(v, v) = 0 \ \forall v \in V$.
Since

$$u + v \in V \ \forall u, v \in V, \quad f(u + v, u + v) = 0$$
$$\Rightarrow f(u, u) + f(u, v) + f(v, u) + f(v, v) = 0.$$

From $f(u, u) = f(v, v) = 0 \ \forall u, v \in V$, the above equation implies $f(u, v) + f(v, u) = 0$.
Hence, $f(u, v) = -f(v, u)$, $\forall u, v \in V$.
Thus, f is an skew-symmetric bilinear form. $\qquad\qquad\square$

ⓘ Exercises

(8.1.1)   Let $x = (x_1, x_2)$ and $y = (y_1, y_2)$. Then determine which of the following are bilinear forms on $\mathbb{R}^2$:
(i) $f_1(x, y) = x_1 y_1 + 2x_2 y_2$,
(ii) $f_2(x, y) = 2x_1 + y_2$,
(iii) $f_3(x, y) = x_1 x_2 + 2y_1 y_2$,
(iv) $f_4(x, y) = 4x_1 y_1 - x_1 y_2 - x_2 y_1 + 2x_2 y_2$,
(v) $f_5(x, y) = 1$.

(8.1.2) Show that the function F defined as $F(f,g) = \int_a^b f(x)g(x)dx, f, g \in C(\mathbb{R}^{[a,b]})$ is a bilinear form on the real vector space $C(\mathbb{R}^{[a,b]})$ of all continuous real valued functions on the closed interval $[a,b]$.

(8.1.3) Show that the function f defined as $f(A,B) = \text{tr}(B^tA)$ is a bilinear form on the vector space $M_{m,n}(\mathbb{R})$ of all $m \times n$ matrices over $\mathbb{R}$, where $A, B \in M_{m,n}(\mathbb{R})$.

(8.1.4) Let $f(x,y) = x_1y_1 - 2x_1y_2 - 3x_2y_1 + 4x_2y_2 + 5x_3y_1 + 2x_3y_3$ be a bilinear form on $\mathbb{R}^3$. Then find the matrix A such that $f(x,y) = x^t Ay$, where $x^t = [x_1\ x_2\ x_3]$ and $y^t = [y_1\ y_2\ y_3]$.

(8.1.5) Show that the function f defined as $f(p(x),q(x)) = \int_0^1 p(x)q(x)dx$ is a bilinear form on the polynomial space $P_2(x)$ over $\mathbb{R}$. Also, find the matrix $[f]_{B,B}$ of f, where $B = \{1, x, x^2\}$ is the basis of $P_2(x)$.

(8.1.6) Let $f(x,y) = 2x_1y_1 - x_1y_2 + 2x_1y_3 - 3x_2y_1 + 2x_2y_2 + 3x_3y_1 + 2x_3y_2$ be a bilinear form on $\mathbb{R}^3$, where $x = (x_1,x_2,x_3)$ and $y = (y_1,y_2,y_3)$. Then find the matrices $[f]_{B_1,B_1}$ and $[f]_{B_2,B_2}$ of f with respect to the bases

$$B_1 = \{e_1 = (1,0,0), e_2 = (0,1,0), e_3(0,0,1)\} \quad \text{and}$$
$$B_2 = \{v_1 = (1,0,0), v_2 = (1,-1,0), v_3 = (1,1,-1)\}.$$

Also, show that both the matrices are congruent, i. e., $[f]_{B_2,B_2} = [I]^t_{B_2,B_1}[f]_{B_1,B_1}[I]_{B_2,B_1}$.

(8.1.7) Let $f(x,y) = x_1y_1 - 2x_1y_3 - x_2y_1 + 2x_2y_2 + 3x_3y_1 + 3x_3y_3$ be a bilinear form on $\mathbb{R}^3$, where $x = (x_1,x_2,x_3)$ and $y = (y_1,y_2,y_3)$. Then find the matrices $[f]_{B_1,B_1}$ and $[f]_{B_2,B_2}$ of f with respect to the bases $B_1 = \{(1,0,0), (1,-1,0), (1,1,-1)\}$ and $B_2 = \{(1,0,-2), (-1,0,3), (0,1,-3)\}$. Also, show that both the matrices are congruent, i. e.,

$$[f]_{B_2,B_2} = [I]^t_{B_2,B_1}[f]_{B_1,B_1}[I]_{B_2,B_1}.$$

8.2 Symmetric bilinear forms and quadratic forms

In this section, we discuss the key concepts of symmetric bilinear forms and quadratic forms, along with their representation through symmetric matrices. Recall Definition 8.1.15 of a symmetric bilinear form f, which states: "A bilinear form on a vector space V over a field F is symmetric if $f(v,w) = f(w,v)\ \forall v, w \in V$." We will now prove some significant results and examine quadratic forms, which are closely related to symmetric bilinear forms.

Theorem 8.2.1. *A bilinear form f is symmetric if and only if its matrix representation $[f]_{B,B}$ associated to any basis B is symmetric.*

Proof. Let f be a symmetric bilinear form of a vector space V over a field F and let $B = \{v_1, v_2, \ldots, v_n\}$ be a basis of V.

Then ij-th entry of the matrix $[f]_{B,B} = f(v_i, v_j)$.

Since f is symmetric, $f(v_i, v_j) = f(v_j, v_i) = ji$-th entry of $[f]_{B,B}$.

Hence, matrix $[f]_{B,B}$ is symmetric.

Conversely, suppose that $[f]_{B,B}$ is a symmetric matrix.

Then from Theorem 8.1.7, $f(u,v) = [u]^t_B[f]_{B,B}[v]_B$, where $u, v \in V$ and $[u]_B$, $[v]_B$ are coordinate vectors in column form.

Since $[u]_B^t[f]_{B,B}[v]_B \in F$,

$$[u]_B^t[f]_{B,B}[v]_B = ([u]_B^t[f]_{B,B}[v]_B)^t = [v]_B^t[f]_{B,B}^t[u]_B = [v]_B^t[f]_{B,B}[u]_B = f(v,u).$$

Hence, f is a symmetric bilinear form.

Note that if the matrix $[f]_{B,B}$ of bilinear form f associated to a basis B is diagonal, then it is a symmetric matrix, and hence from the above theorem f is a symmetric bilinear form. □

In the following theorem, we shall show that the converse of this result is also true in case of field F with characteristic $\neq 2$.

Theorem 8.2.2. *Let f be a symmetric bilinear form on a finite-dimensional vector space V over a field F with the characteristic not equal to 2. Then there exists an ordered basis B of V such that the matrix $[f]_{B,B}$ of f is diagonal.*

Proof. We prove the theorem by induction on $n = \dim(V)$.

If $n = 0$ or 1, then the matrix of f is of order 1×1, and hence diagonal.

Assume the result is true for all symmetric bilinear forms on vector spaces of dimension less than n.

Let f be a symmetric bilinear form on a vector space V of dimension n.

If f is the zero bilinear form, then the matrix of f is the zero matrix for any basis of V, and hence diagonal.

Now, suppose f is not identically zero. Then we claim that there exists a nonzero vector $v \in V$ such that $f(v,v) \neq 0$.

Assume, to the contrary, that $f(v,v) = 0$ for all $v \in V$.

Then $f(u+v, u+v) = 0$ for all $u, v \in V$.

That is, $f(u,u) + f(u,v) + f(v,u) + f(v,v) = 0$.

Since f is symmetric, $f(u,v) = f(v,u)$.

Thus, $f(u,u) + f(v,v) + f(u,v)(1+1) = 0$. Since $f(u,u) = f(v,v) = 0$, we have $(1 + 1)f(u,v) = 0$.

Given that the characteristic of F is not 2, $1 + 1 \neq 0$, hence $f(u,v) = 0$ for all $u, v \in V$. This implies that f is identically zero, which is a contradiction.

Therefore, $f(v,v) \neq 0$ for some nonzero $v \in V$.

Let $0 \neq v_1 \in V$ such that $f(v_1, v_1) \neq 0$. Then $W_1 = \{kv_1 : k \in F\}$ is the subspace of V generated by v_1. Hence, $\dim(W_1) = 1$. Let $W_2 = \{x \in V : f(v_1, x) = 0\}$. It is straightforward to show that W_2 is a subspace of V.

Now, we claim that $V = W_1 \oplus W_2$.

Suppose $kv_1 \in W_2$. Then $f(v_1, kv_1) = 0 \Rightarrow kf(v_1, v_1) = 0$. Since $f(v_1, v_1) \neq 0$, it follows that $k = 0$. Hence, $kv_1 = 0$, and so $W_1 \cap W_2 = \{0\}$.

Next, let $v \in V$. If we take $y = v - \frac{f(v_1,v)}{f(v_1,v_1)}v_1$, then

$$f(v_1, y) = f\left(v_1, v - \frac{f(v_1,v)}{f(v_1,v_1)}v_1\right) = f(v_1,v) - \frac{f(v_1,v)}{f(v_1,v_1)}f(v_1,v_1) = f(v_1,v) - f(v_1,v) = 0.$$

This shows that $y \in W_2$ and $v = \frac{f(v_1,v)}{f(v_1,v_1)} v_1 + y \in W_1 + W_2$ for all v. Thus, $V = W_1 \oplus W_2$.

Since the restriction of f to W_2 is also a symmetric bilinear form on W_2 and $\dim(W_2) = n - 1$, it follows that there is an ordered basis $\{v_2, v_3, \ldots, v_n\}$ with respect to which f restricted to W_2 has a diagonal representation. Thus, $f(v_i, v_j) = 0$ for all $i \neq j$, $i \geq 2$ and $j \geq 2$. Also, since $v_j \in W_2, f(v_1, v_j) = 0$ for all $j \geq 2$. Thus, f has a diagonal representation with respect to the ordered basis $\{v_1, v_2, \ldots, v_n\}$. Hence, by mathematical induction, the result is proved. □

Corollary 8.2.3. *Let A be a symmetric matrix with entries in a field F of characteristic not equal to 2. Then there exists a nonsingular matrix P such that $P^t AP$ is a diagonal matrix.*

Proof. Since A is a symmetric matrix, there exists a symmetric bilinear form f on a vector space V over F and a basis of V such that $[f]_{B,B} = A$.

From the previous theorem, there exists a basis B' of V such that the matrix $[f]_{B',B'}$ is diagonal. This means A is congruent to a diagonal matrix. Hence, there exists a nonsingular matrix P such that $P^t AP$ is diagonal. □

We can apply the following algorithm to find an invertible matrix P such that $P^t AP = D$, where A is a given symmetric matrix and D is a diagonal matrix.

Let A be an n-square symmetric matrix with entries in a field F of characteristic not equal to 2.

Write $A = I_n A I_n$.

It is known that performing an elementary row operation on a matrix A is the same as multiplying A from the left by the corresponding elementary matrix E. Similarly, applying a column operation on A is equivalent to multiplying A from the right by the corresponding elementary matrix E'.

Note that if E is the elementary matrix corresponding to the elementary row operation $R_i \rightarrow R_i + kR_j$, then E^t is the elementary matrix corresponding to the column operation $C_i \rightarrow C_i + kC_j$.

Multiplying the elementary row matrix E_1 in the equation $A = I_n A I_n$ from the left and E_1^t from the right, we get

$$E_1 A E_1^t = E_1 A E_1^t.$$

Repeating the above process with elementary row matrices $E_2, E_3, \ldots, E_r$ so that the left-hand side becomes a diagonal matrix, we get

$$D = E_r \ldots E_2 E_1 A E_1^t E_2^t \cdots E_r^t.$$

Let $E_1^t E_2^t \cdots E_r^t = P$. Then $P^t = (E_1^t E_2^t \cdots E_r^t)^t = E_r E_{r-1} \cdots E_2 E_1$, and hence $D = P^t AP$. Also, P, being the product of elementary matrices, is nonsingular.

Let us illustrate the above algorithm with a couple of examples.

Example 8.2.4. Consider the symmetric matrix

$$A = \begin{bmatrix} 1 & 2 \\ 2 & 0 \end{bmatrix}.$$

Then $A = I_2 A I_2$,

$$\begin{bmatrix} 1 & 2 \\ 2 & 0 \end{bmatrix} = \begin{bmatrix} 1 & 0 \\ 0 & 1 \end{bmatrix} A \begin{bmatrix} 1 & 0 \\ 0 & 1 \end{bmatrix}.$$

Now, we apply a sequence of elementary row and column operations on the above equation such that the left-hand side converts into diagonal form. First, we apply the elementary row operation $R_2 \rightarrow R_2 - 2R_1$ on the equation. Since an elementary row operation is equivalent to the premultiplication of an elementary matrix, the operation will be applied on A on the left-hand side of the equation and on the left-side identity matrix only on the right-hand side of the equation.

Similarly, an elementary column operation will be applied on A and on the right-hand side of the equation on the right-side identity matrix only.

Then we get

$$\begin{bmatrix} 1 & 2 \\ 0 & -4 \end{bmatrix} = \begin{bmatrix} 1 & 0 \\ -2 & 1 \end{bmatrix} A \begin{bmatrix} 1 & 0 \\ 0 & 1 \end{bmatrix}.$$

Now, we apply the corresponding column operation $C_2 \rightarrow C_2 - 2C_1$. Then we get

$$\begin{bmatrix} 1 & 0 \\ 0 & -4 \end{bmatrix} = \begin{bmatrix} 1 & 0 \\ -2 & 1 \end{bmatrix} A \begin{bmatrix} 1 & -2 \\ 0 & 1 \end{bmatrix}.$$

Hence,

$$P^t A P = \begin{bmatrix} 1 & 0 \\ 0 & -4 \end{bmatrix}, \quad \text{where } P = \begin{bmatrix} 1 & -2 \\ 0 & 1 \end{bmatrix}.$$

Observing that the matrix P is obtained by elementary column operations on the right-side identity matrix I_2 of A in the equation $A = I_2 A I_2$, we write the block matrix

$$[A \quad : \quad I_2] = \begin{bmatrix} 1 & 2 & : & 1 & 0 \\ 2 & 0 & : & 0 & 1 \end{bmatrix}.$$

Now, we apply the elementary row operation on A only and then the corresponding elementary column operation on A and I_2 both.

Applying $R_2 \rightarrow R_2 - 2R_1$, we get

$$\begin{bmatrix} 1 & 2 & : & 1 & 0 \\ 0 & -4 & : & 0 & 1 \end{bmatrix}.$$

Now we apply the elementary column operation $C_2 \to C_2 - 2C_1$ on both matrices of the above block matrix, then we have

$$\begin{bmatrix} 1 & 0 & : & 1 & -2 \\ 0 & -4 & : & 0 & 1 \end{bmatrix}.$$

Then

$$P = \begin{bmatrix} 1 & -2 \\ 0 & 1 \end{bmatrix} \quad \text{and} \quad D = \begin{bmatrix} 1 & 0 \\ 0 & -4 \end{bmatrix}.$$

Example 8.2.5. Let

$$A = \begin{bmatrix} 0 & 1 & 2 \\ 1 & 3 & 4 \\ 2 & 4 & 5 \end{bmatrix}$$

be the symmetric matrix.

Consider the block matrix

$$[A \ : \ I_3] = \begin{bmatrix} 0 & 1 & 2 & : & 1 & 0 & 0 \\ 1 & 3 & 4 & : & 0 & 1 & 0 \\ 2 & 4 & 5 & : & 0 & 0 & 1 \end{bmatrix}.$$

Applying the row operation $R_1 \leftrightarrow R_2$ (note that the row operation will be applied on A only)

$$\begin{bmatrix} 1 & 3 & 4 & : & 1 & 0 & 0 \\ 0 & 1 & 2 & : & 0 & 1 & 0 \\ 2 & 4 & 5 & : & 0 & 0 & 1 \end{bmatrix}.$$

Applying the corresponding column operation $C_1 \leftrightarrow C_2$ (note that the column operation will be applied on both the part of above matrix),

$$\begin{bmatrix} 3 & 1 & 4 & : & 0 & 1 & 0 \\ 1 & 0 & 2 & : & 1 & 0 & 0 \\ 4 & 2 & 5 & : & 0 & 0 & 1 \end{bmatrix}.$$

Applying $R_2 \to R_2 - \frac{1}{3}R_1$ and $R_3 \to R_3 - \frac{4}{3}R_1$,

$$\begin{bmatrix} 3 & 1 & 4 & : & 0 & 1 & 0 \\ 0 & -\frac{1}{3} & \frac{2}{3} & : & 1 & 0 & 0 \\ 0 & \frac{2}{3} & -\frac{1}{3} & : & 0 & 0 & 1 \end{bmatrix}.$$

Applying corresponding column operations $C_2 \to C_2 - \frac{1}{3}C_1$ and $C_3 \to C_3 - \frac{2}{3}C_1$,

$$
\begin{bmatrix}
3 & 0 & 0 & : & 0 & 1 & 0 \\
0 & -\frac{1}{3} & \frac{2}{3} & : & 1 & -\frac{1}{3} & -\frac{4}{3} \\
0 & \frac{2}{3} & -\frac{1}{3} & : & 0 & 0 & 1
\end{bmatrix}.
$$

Applying $R_3 \to R_3 + 2R_2$,

$$
\begin{bmatrix}
3 & 0 & 0 & : & 0 & 1 & 0 \\
0 & -\frac{1}{3} & \frac{2}{3} & : & 1 & -\frac{1}{3} & -\frac{4}{3} \\
0 & 0 & 1 & : & 0 & 0 & 1
\end{bmatrix}.
$$

Applying corresponding column operation $C_3 \to C_3 + 2C_2$,

$$
\begin{bmatrix}
3 & 0 & 0 & : & 0 & 1 & 2 \\
0 & -\frac{1}{3} & 0 & : & 1 & -\frac{1}{3} & -2 \\
0 & 0 & 1 & : & 0 & 0 & 1
\end{bmatrix}.
$$

Then

$$
P = \begin{bmatrix}
0 & 1 & 2 \\
1 & -\frac{1}{3} & -2 \\
0 & 0 & 1
\end{bmatrix}
\quad \text{and} \quad
D = \begin{bmatrix}
3 & 0 & 0 \\
0 & -\frac{1}{3} & 0 \\
0 & 0 & 1
\end{bmatrix}
$$

such that $P^t A P = D$.

Theorem 8.2.6. *Let f be a bilinear form on a vector space V over a field F of characteristic not equal to two. Then f can be expressed uniquely as $f = f_s + f_{ss}$, where f_s and f_{ss} are symmetric and skew-symmetric bilinear forms on V, respectively.*

Proof. Define

$$
f_s(x,y) = \frac{1}{2}(f(x,y) + f(y,x)) \quad \text{and}
$$

$$
f_{ss}(x,y) = \frac{1}{2}(f(x,y) - f(y,x)).
$$

Since $1 + 1 \neq 0$, $\frac{1}{2}$ is well-defined. Also, it can be easily proved that f_s and f_{ss} are bilinear forms on V.

Further, $f_s(x,y) = f_s(y,x)$ implies that f_s is a symmetric bilinear form and from $f_{ss}(x,x) = \frac{1}{2}(f(x,x) - f(x,x)) = 0$ we have that f_{ss} is a skew- symmetric bilinear form.

Hence, $f_s(x,y) + f_{ss}(x,y) = f(x,y) \; \forall x,y \in V$.

Thus, $f = f_s + f_{ss}$.

That is, f is expressed as a sum of a symmetric and a skew-symmetric bilinear forms.

To prove that the representation of f is unique, let us suppose that $f = f'_s + f'_{ss}$, where f'_s is symmetric and f'_{ss} is skew-symmetric.

Then

$$f(x,y) + f(y,x) = f'_s(x,y) + f'_{ss}(x,y) + f'_s(y,x) + f'_{ss}(y,x)$$
$$= (f'_s(x,y) + f'_s(y,x)) + (f'_{ss}(x,y) + f'_{ss}(y,x))$$
$$= (f'_s(x,y) + f'_s(x,y)) + (f'_{ss}(x,y) - f'_{ss}(x,y))$$
$$= 2f'_s(x,y),$$

and hence $f'_s(x,y) = \frac{1}{2}(f(x,y) + f(y,x)) = f_s(x,y)$. Thus, $f'_s = f_s$.
  Also, from $f_s + f_{ss} = f'_s + f'_{ss}$, we have that $f_{ss} = f'_{ss}$.
  Hence, both the representations of f are the same. □

Theorem 8.2.7. *Let f be a symmetric bilinear form on a finite-dimensional vector space V. There exists an ordered basis B of V such that the matrix representation of f is a diagonal matrix of the form*

$$[f]_{B,B} = \begin{bmatrix} I_r & & \\ & -I_s & \\ & & 0 \end{bmatrix}.$$

Furthermore, the integers r and s are invariant regardless of the chosen basis.

Proof. By Theorem 8.2.2, there is a basis $\{u_1, u_2, \ldots, u_n\}$ for V such that the matrix representation of f is diagonal. Suppose there are r positive and s negative entries on the diagonal of this matrix. By rearranging the basis, we can assume $f(u_i, u_i) > 0$ for $1 \le i \le r$ and $f(u_j, u_j) \le 0$ for $r + 1 \le j \le r + s$.
  Let $v_i = \frac{u_i}{\sqrt{|f(u_i, u_i)|}}$ for $1 \le i \le r + s$. Then

$$f(v_i, v_i) = f\left(\frac{u_i}{\sqrt{|f(u_i, u_i)|}}, \frac{u_i}{\sqrt{|f(u_i, u_i)|}}\right) = \frac{f(u_i, u_i)}{|f(u_i, u_i)|} = \begin{cases} 1, & \text{if } 1 \le i \le r, \\ -1, & \text{if } r + 1 \le i \le r + s. \end{cases}$$

Let $B = \{v_1, v_2, \ldots, v_n\}$. Then the matrix of f with respect to B is

$$[f]_{B,B} = \begin{bmatrix} I_r & & \\ & -I_s & \\ & & 0 \end{bmatrix}.$$

Suppose $B' = \{v'_1, v'_2, \ldots, v'_n\}$ is another basis of V such that the matrix of f is

$$[f]_{B',B'} = \begin{bmatrix} I_{r'} & & \\ & -I_{s'} & \\ & & 0 \end{bmatrix}.$$

We need to show that $r = r'$ and $s = s'$. Since $[f]_{B,B}$ and $[f]_{B',B'}$ are congruent matrices, it follows that

$$\text{rank}([f]_{B,B}) = \text{rank}([f]_{B',B'}) \implies r + s = r' + s'.$$

Let W be the span of $\{v_1, v_2, \ldots, v_r\}$ and W' be the span of $\{v'_{r'+1}, v'_{r'+2}, \ldots, v'_n\}$. If $x = \sum_{i=1}^{r} a_i v_i$ is a nonzero vector in W, then

$$f(x,x) = f\left(\sum_{i=1}^{r} a_i v_i, \sum_{j=1}^{r} a_j v_j\right) = \sum_{i=1}^{r}\sum_{j=1}^{r} a_i a_j f(v_i, v_j) = \sum_{i=1}^{r} a_i^2 f(v_i, v_i).$$

Since $a_i \neq 0$ for all i and $f(v_i, v_i) = 1$, we have $f(x,x) > 0$. Similarly, $f(x,x) \leq 0$ for any nonzero $x \in W'$.

Therefore, $W \cap W' = \{0\}$. Since $\dim(W) = r$ and $\dim(W') = n - r'$,

$$\dim(W + W') = \dim(W) + \dim(W') - \dim(W \cap W') = r + (n - r') - 0.$$

Because $W + W' \subseteq V$, we have $\dim(W + W') \leq \dim(V) = n$. Thus,

$$r + (n - r') \leq n \implies r \leq r'.$$

By switching the roles of W and W', we obtain $r' \leq r$. Hence, $r = r'$.
    Given $r + s = r' + s'$, it follows that $s = s'$. □

Corollary 8.2.8. *Let A be an n-square real symmetric matrix. Then A is congruent to a unique diagonal matrix of the form $\text{diag}(I_r, -I_s, 0)$.*

Definition 8.2.9. Let V be a vector space over the field F. A map $q : V \to F$ is called a quadratic form if there exists a bilinear form f on V such that $q(x) = f(x,x)\ \forall x \in V$. The quadratic form q is called real if $F = \mathbb{R}$ and complex if $F = \mathbb{C}$.

Theorem 8.2.10. *Let V be a vector space over a field F with characteristic different from 2. Every quadratic form q on V determines a unique symmetric bilinear form f_s on V, obtainable from q by the identity $f_s(x,y) = \frac{1}{2}(q(x+y) - q(x) - q(y))$, and called the polar form of f_s.*

Proof. Let f be the bilinear form on V such that $q(x) = f(x,x)$ for all $x \in V$. By Theorem 8.2.2, f can be uniquely expressed as $f = f_s + f_{ss}$, where $f_s(x,y) = \frac{1}{2}(f(x,y) + f(y,x))$ is a symmetric bilinear form and $f_{ss}(x,y) = \frac{1}{2}(f(x,y) - f(y,x))$ is a skew-symmetric bilinear form. Then

$$q(x) = f(x,x) = f_s(x,x) + f_{ss}(x,x).$$

Since $f_{ss}(x,x) = 0$, we have $q(x) = f_s(x,x)$. Thus, q determines the unique symmetric bilinear form f_s.
    Now,

$$q(x+y) = f_s(x+y, x+y) = f_s(x,x) + f_s(x,y) + f_s(y,x) + f_s(y,y) = q(x) + 2f_s(x,y) + q(y).$$

Hence,

$$f_s(x,y) = \frac{1}{2}(q(x+y) - q(x) - q(y)).$$

Thus, the quadratic form $q(x)$ on a vector space V over a field F of characteristic different from 2 is defined as $q(x) = f(x,x)$ for all $x \in V$, where f is a symmetric bilinear form. □

Theorem 8.2.11. *Let V be an n-dimensional vector space over a field F with characteristic not equal to 2. Let $B = \{v_1, v_2, \ldots, v_n\}$ be an ordered basis of V. If q is a quadratic form with its associated symmetric bilinear form f, then $q(x) = [x]_B^t [f]_{B,B} [x]_B$, where $[x]_B$ is the coordinate vector of x.*

Proof. Given the basis $B = \{v_1, v_2, \ldots, v_n\}$ of V, any $x \in V$ can be expressed as $x = \sum_{i=1}^{n} x_i v_i$. Then

$$q(x) = f(x,x) = f\left(\sum_{i=1}^{n} x_i v_i, \sum_{j=1}^{n} x_j v_j\right) = \sum_{i=1}^{n}\sum_{j=1}^{n} x_i x_j f(v_i, v_j).$$

Let $[f]_{B,B} = [a_{ij}]$. Then $f(v_i, v_j) = a_{ij}$ for all $1 \leq i, j \leq n$. Therefore,

$$q(x) = \sum_{i=1}^{n}\sum_{j=1}^{n} x_i x_j a_{ij} = \begin{bmatrix} x_1 & x_2 & \cdots & x_n \end{bmatrix} \begin{bmatrix} a_{11} & a_{12} & \cdots & a_{1n} \\ a_{21} & a_{22} & \cdots & a_{2n} \\ \vdots & \vdots & \ddots & \vdots \\ a_{n1} & a_{n2} & \cdots & a_{nn} \end{bmatrix} \begin{bmatrix} x_1 \\ x_2 \\ \vdots \\ x_n \end{bmatrix}.$$

Thus, $q(x) = [x]_B^t [f]_{B,B} [x]_B$.

The symmetric matrix $[f]_{B,B}$ of f relative to the ordered basis B is called the matrix of the quadratic form q. □

Remark. From the above theorem, we have

$$q(x) = \sum_{i=1}^{n}\sum_{j=1}^{n} f(v_i, v_j) x_i x_j = \sum_{i} f(v_i, v_i) x_i^2 + \sum_{i<j} 2f(v_i, v_j) x_i x_j.$$

Given this formal expression, a quadratic form q can be defined as a polynomial in the variables $x_1, x_2, \ldots, x_n$, where the degree of every term is 2. Specifically, if $q(x) = \sum_i^n a_i x_i^2 + \sum_{i<j} b_{ij} x_i x_j$ is a quadratic form on the vector space V over a field F with the characteristic not equal to 2, then the associated symmetric matrix of the quadratic form q is $A = [a_{ij}]$, where $a_{ii} = a_i$ (the coefficient of x_i^2) and $a_{ij} = a_{ji} = \frac{1}{2}b_{ij}$ (half of the coefficient of $x_i x_j$ for $i \neq j$).

From Theorem 8.2.2, we know that every matrix representing a quadratic form can be brought to a normal form if $1+1 \neq 0$. Therefore, if the matrix representation $A = [a_{ij}]$ of q is diagonal, then q has a diagonal representation given by $q(x) = \sum_i^n a_{ii} x_i^2$.

Based on Theorem 8.2.7, we have the following corresponding result for quadratic forms.

Corollary 8.2.12. *Let q be a quadratic form on an n-dimensional real vector space V. Then q can be uniquely expressed as*

$$q(x_1, x_2, \ldots, x_{r+s}) = x_1^2 + x_2^2 + \cdots + x_r^2 - x_{r+1}^2 - \cdots - x_{r+s}^2,$$

where r + s is the rank of q.

Proof. Let f be the unique symmetric bilinear form associated with the quadratic form q on V. According to Theorem 8.2.7, there exists an ordered basis B of V such that the matrix representation of f is diagonal and has the form

$$[f]_{B,B} = \begin{bmatrix} I_r & & \\ & -I_s & \\ & & 0 \end{bmatrix}.$$

From Theorem 8.2.11, $q(x)$ can be expressed as $q(x) = [x]_B^t [f]_{B,B} [x]_B$, where

$$[x]_B = \begin{bmatrix} x_1 \\ x_2 \\ \vdots \\ x_n \end{bmatrix}$$

is the coordinate vector of x. Hence,

$$q(x) = \begin{bmatrix} x_1 & x_2 & \cdots & x_n \end{bmatrix} \begin{bmatrix} I_r & & \\ & -I_s & \\ & & 0 \end{bmatrix} \begin{bmatrix} x_1 \\ x_2 \\ \vdots \\ x_n \end{bmatrix} = x_1^2 + x_2^2 + \cdots + x_r^2 - x_{r+1}^2 - \cdots - x_{r+s}^2.$$

Since every diagonal matrix representation of a symmetric bilinear form f (or quadratic form q) with respect to different ordered bases has the same number r of positive entries and the same number s of negative entries, $r + s$ and $r - s$ are called the rank and signature of f (or q), respectively. □

Example 8.2.13. Let $q(x_1, x_2, x_3) = 2x_1x_2 + 3x_2^2 + 4x_1x_3 + 8x_2x_3 + 5x_3^2$ be a quadratic form on the vector space $\mathbb{R}^3$ over $\mathbb{R}$.

In this example, we determine the following: (a) the symmetric matrix A of q relative to the standard basis of $\mathbb{R}^3$, (b) the normal or diagonal form of q, (c) rank and the signature of q.

(a) To find the matrix $A = [a_{ij}]$, we use the method described in the remark. Hence,

$$a_{11} = \text{the coefficient of } x_1^2 = 0,$$

$$a_{12} = a_{21} = \frac{1}{2} \times \text{the coefficient of } x_1 x_2 = 1,$$

$$a_{13} = a_{31} = \frac{1}{2} \times \text{the coefficient of } x_1 x_3 = 2,$$

$$a_{22} = \text{the coefficient of } x_2^2 = 3,$$

$$a_{23} = a_{32} = \frac{1}{2} \times \text{the coefficient of } x_2 x_3 = 4,$$

$$a_{33} = \text{the coefficient of } x_2^2 = 5.$$

Thus,

$$A = \begin{bmatrix} 0 & 1 & 2 \\ 1 & 3 & 4 \\ 2 & 4 & 5 \end{bmatrix}$$

is the matrix of quadratic form q.

(b) Now, to find the diagonal matrix D congruent to matrix A, we consider the block matrix

$$\begin{bmatrix} A & : & I_3 \end{bmatrix} = \begin{bmatrix} 0 & 1 & 2 & : & 1 & 0 & 0 \\ 1 & 3 & 4 & : & 0 & 1 & 0 \\ 2 & 4 & 5 & : & 0 & 0 & 1 \end{bmatrix}.$$

Since the matrix A is same as given in Example 8.2.5, the diagonal matrix

$$D = \begin{bmatrix} 3 & 0 & 0 \\ 0 & -\frac{1}{3} & 0 \\ 0 & 0 & 1 \end{bmatrix}.$$

The normal form of q is

$$q(x_1, x_2, x_3) = \begin{bmatrix} x_1 & x_2 & x_3 \end{bmatrix} D \begin{bmatrix} x_1 \\ x_2 \\ x_3 \end{bmatrix}$$

$$= \begin{bmatrix} x_1 & x_2 & x_3 \end{bmatrix} \begin{bmatrix} 3 & 0 & 0 \\ 0 & -\frac{1}{3} & 0 \\ 0 & 0 & 1 \end{bmatrix} \begin{bmatrix} x_1 \\ x_2 \\ x_3 \end{bmatrix}$$

$$= 3x_1^2 - \frac{1}{3}x_2^2 + x_3^2.$$

(c) Since the number of positive entries r in the diagonal matrix D is 2 and the number of negative entries s in the diagonal matrix D is 1, the rank of the quadratic form $q = 2 + 1 = 3$ and the signature of $q = r - s = 2 - 1 = 1$.

Exercises

(8.2.1) Find the symmetric matrix A associated to each of the following quadratic forms q on $\mathbb{R}^3$ relative to the standard basis:

(i) $q(x_1, x_2, x_3) = x_1^2 - 4x_1x_2 + 2x_2^2 - 8x_1x_3 + 10x_2x_3 - 3x_3^2$,

(ii) $q(x_1, x_2, x_3) = 3x_1x_2 - x_2^2 - 10x_1x_3 + 5x_3^2$,

(iii) $q(x_1, x_2, x_3) = x_1^2 - 6x_1x_3 + 7x_2^2$,

(iv) $q(x_1, x_2, x_3) = x_1^2 - 2x_3^2 - 2x_1x_2 + 10x_2x_3$.

(8.2.2) For the following symmetric matrices A_1, A_2, A_3, A_4 find nonsingular matrices P_1, P_2, P_3, P_4 such that the matrices $P_1^t A_1 P_1, P_2^t A_2 P_2, P_3^t A_3 P_3, P_4^t A_4 P_4$ are diagonal.

$$A_1 = \begin{bmatrix} 2 & 3 \\ 3 & -1 \end{bmatrix}, \quad A_2 = \begin{bmatrix} 1 & -3 & 2 \\ -3 & 2 & 4 \\ 2 & 4 & 0 \end{bmatrix},$$

$$A_3 = \begin{bmatrix} 2 & -2 & 1 \\ -2 & 4 & 0 \\ 1 & 0 & 3 \end{bmatrix}, \quad A_4 = \begin{bmatrix} 1 & 2 & 0 & 3 \\ 2 & -2 & 4 & 0 \\ 0 & 4 & -1 & 1 \\ 3 & 0 & 1 & 3 \end{bmatrix}.$$

(8.2.3) Find the normal (or diagonal) form of the following quadratic forms over $\mathbb{R}^3$. Also, find their rank and signature:

(i) $q(x_1, x_2, x_3) = x_1^2 + x_2^2 + x_3^2 + 2x_1x_2 - 4x_2x_3 + 3x_1x_3$,

(ii) $q(x_1, x_2, x_3) = 2x_1^2 - 6x_1x_2 + x_2^2 - 12x_1x_3 + 14x_2x_3 + 3x_3^2$,

(iii) $q(x_1, x_2, x_3) = x_1^2 - 4x_1x_2 + 6x_2^2$,

(iv) $q(x_1, x_2, x_3) = x_1x_2 - 2x_1x_3 - 4x_2x_3$.

Index

https://doi.org/10.1515/9783111516035-009

www.ingramcontent.com/pod-product-compliance
Lightning Source LLC
Chambersburg PA
CBHW061353210326
41598CB00035B/5972

* 9 7 8 3 1 1 1 5 1 5 7 0 0 *